Compendium of Tomato Diseases and Pests

SECOND EDITION

Edited by

Jeffrey B. Jones
Plant Pathology Department, University of Florida, Gainesville

Thomas A. Zitter
Department of Plant Pathology and Plant-Microbe Biology, Cornell University, Ithaca, New York

Timur M. Momol
Plant Pathology Department, University of Florida, Gainesville

Sally A. Miller
Department of Plant Pathology, The Ohio State University, Wooster

APS PRESS
The American Phytopathological Society

Front cover photograph: Cora Mueller, Shutterstock.
Back cover photographs: Top left, top right, and bottom right: Courtesy T. A. Zitter; Bottom left: Courtesy J. P. Jones.

Reference in this publication to a trademark, proprietary product, or company name by personnel of the U.S. Department of Agriculture or anyone else is intended for explicit description only and does not imply approval or recommendation to the exclusion of others that may be suitable.

Library of Congress Control Number: 2013935899
International Standard Book Numbers:
Print: 978-0-89054-424-2
Online: 978-0-89054-434-1
Mobi: 978-0-89054-425-9
ePub: 978-0-89054-557-7

First edition published 1991
Second edition published 2014

Printed in the United States of America on acid-free paper

The American Phytopathological Society
3285 Northwood Circle, Suite 100
St. Paul, Minnesota 55121, U.S.A.

Preface

Compendium of Tomato Diseases and Pests, Second Edition, is intended as a general reference and diagnostic guide for a diverse group of readers. This guide will be useful not only to plant pathologists and other agricultural scientists and professionals for whom disease diagnosis is an integral part of their work but also to backyard gardeners who want to grow healthy tomato plants that bear ample fruit.

This new edition is organized similarly to the first edition but has been significantly updated and expanded:

- The Introduction provides background information about tomato botany and culture, seed production and quality assurance, and container production of transplants. Among the new topics addressed are the change in nomenclature, in which the genus *Lycopersicon* was classified as *Solanum* section *Lycopersicon,* and the sequencing of the tomato genome.
- Part I outlines the infectious diseases that affect tomato—including those caused by fungi and oomycetes, bacteria, phytoplasmas, viruses, and viroids—along with postharvest diseases and disorders and diseases caused by nematodes. Each disease section has been updated from the first edition, and sections have been added to address approximately 20 new diseases and disorders.
- Part II covers arthropod pests: namely, mites, insects, and "worm" pests. The coverage of pests has been expanded significantly in this edition and includes the addition of 23 color photos that illustrate pests and the damage they cause.
- Part III examines noninfectious diseases, disorders, and damage: namely, physiological disorders, nutritional disorders, herbicide damage, and genetic diseases. This part now includes 40 color photos to assist readers in recognizing the damage caused by these various diseases and disorders, for which the signs and symptoms are often similar. The discussion of herbicides, in particular, has been updated to reflect the full range of available products.
- Part IV, which discusses diseases of unconfirmed etiology, has been reorganized to adopt the format used elsewhere in the book.

Throughout *Compendium of Tomato Diseases and Pests,* diseases and disorders are discussed in sections that follow a consistent structure: a statement of the importance and distribution of the disease or disorder, identification of its common signs and symptoms, description of its causal organism or agent, explanation of its cycle and epidemiology, and suggestions for management. In most cases, this discussion is followed by a list of selected references, which readers can use to supplement their reading of this book. The information presented in this compendium is intended to be introductory, so readers are encouraged to consult these resources and other pertinent information.

In sections that provide guidelines for management, discussions of specific chemical treatments (e.g., fungicides, bactericides), resistant cultivars, and other timely subjects have been written in a general manner intentionally to reduce the likelihood of presenting outdated information. Again, readers are encouraged to review recent literature on disease management and to consult extension specialists and other professionals in the field of disease management to obtain the most current information about these topics.

The text in this new edition is enhanced by nearly 250 high-quality color photographs that capture typical signs and symptoms of important tomato diseases and pests occurring throughout the world—in the greenhouse, in the field, and in transit to the market. This presentation of text and illustrations will support individuals with both broad and limited experience in making accurate diagnoses of tomato diseases and disorders.

Acknowledgments

Compendium of Tomato Diseases and Pests, Second Edition, has been prepared by a group of international experts. The contributing authors are acknowledged both on pages v and vi and with their respective sections of text throughout the book.

The editors also acknowledge those individuals and organizations that contributed additional information or photographs or reviewed an earlier version of this book:

J. Arboleya, Instituto Nacional de Investigacíon Agropecuaria (INIA) Las Brujas Research Station, Rincon del Colorado, Uruguay
AVRDC—The World Vegetable Center, Shanhua, Taiwan
D. Blancard, French National Institute for Agricultural Research (INRA), Villenave D'Ornon, France
M. Bledsoe, Village Farms International, Heathrow, Florida, U.S.A.
H. Bouzar, Sakata Seed America, Salinas, California, U.S.A.
J. K. Clark, University of California Statewide IPM Program, University of California–Davis, Davis, California, U.S.A.
H. Dankers, North Florida Research and Education Center (NFREC), University of Florida, Quincy, Florida, U.S.A.
D. Fulbright, Department of Plant Pathology, Michigan State University, East Lansing, Michigan, U.S.A.
J. P. Gilreath, PhytoServices, Greenville, South Carolina, U.S.A.
R. Hammond, U.S. Department of Agriculture, Agricultural Research Service (ARS), Beltsville Agricultural Research Center (BARC), Beltsville, Maryland, U.S.A.
T. K. Hartz, Department of Plant Sciences, University of California–Davis, Davis, California, U.S.A.
D. B. Langston, Department of Plant Pathology, University of Georgia, Tifton, Georgia, U.S.A.
H. Laterrot, Caumont Sur Durance, France
R. X. Latin, Department of Botany and Plant Pathology, Purdue University, West Lafayette, Indiana, U.S.A.
D. P. Maxwell, Professor Emeritus, Department of Plant Pathology, University of Wisconsin–Madison, Madison, Wisconsin, U.S.A.
M. T. McGrath, Long Island Horticultural Research and Extension Center (LIHREC), Cornell University, Riverhead, New York, U.S.A.

S. Rideout, Virginia Tech–Eastern Shore AREC, Painter, Virginia
C. Sandlin, Syngenta, Davis, California, U.S.A.
K. Tifft, Greenhouse Vegetable Consultants, Tucson, Arizona, U.S.A.
B. Watt, UMaine Cooperative Extension, University of Maine, Orono, Maine, U.S.A.
G. C. Wisler, U.S. Department of Agriculture, Agricultural Research Service (ARS), Beltsville, Maryland, U.S.A.

J. B. Jones
T. A. Zitter
T. M. Momol
S. A. Miller

Contributors

Yesim Aysan
Department of Plant Protection
Faculty of Agriculture
Cukurova University
Adana, Turkey

Faith E. Bartz
Department of Plant Pathology
North Carolina State University
Raleigh, North Carolina, U.S.A.

Jerry A. Bartz
Plant Pathology Department
University of Florida
Gainesville, Florida, U.S.A.

Ozgur Batuman
Department of Plant Pathology
University of California–Davis
Davis, California, U.S.A.

Hasan Bolkan
Hb Agricultural Consulting
Davis, California, U.S.A.

Judith K. Brown
School of Plant Sciences
The University of Arizona
Tucson, Arizona, U.S.A.

Dan O. Chellemi
U.S. Department of Agriculture
Agricultural Research Service (ARS)
U.S. Horticultural Research Laboratory
Fort Pierce, Florida, U.S.A.

Roger T. Chetelat
C. M. Rick Tomato Genetics Resource Center
Department of Plant Sciences
University of California–Davis
Davis, California, U.S.A.

James Correll
Department of Plant Pathology
University of Arkansas
Fayetteville, Arkansas, U.S.A.

Marc A. Cubeta
Department of Plant Pathology
North Carolina State University
Raleigh, North Carolina, U.S.A.

R. Michael Davis
Department of Plant Pathology
University of California–Davis
Davis, California, U.S.A.

Katherine F. Dobinson
Agriculture and Agri-Food Canada
University of Western Ontario
London, Ontario, Canada

Ronald D. French-Monar
Department of Plant Pathology and Microbiology
Texas A&M AgriLife Extension Service
Amarillo, Texas, U.S.A.

Randolph G. Gardner
Department of Horticultural Science
North Carolina State University
Mountain Horticultural Crops Research and Extension Center
Mills River, North Carolina, U.S.A.

Robert L. Gilbertson
Department of Plant Pathology
University of California–Davis
Davis, California, U.S.A.

Phyllis R. Gilreath
Horticultural Consultant
Greenville, South Carolina, U.S.A.

Daniel Gilrein
College of Agriculture and Life Sciences
Long Island Horticultural Research and Extension Center
Cornell University
Riverhead, New York, U.S.A.

Mark L. Gleason
Department of Plant Pathology
Iowa State University
Ames, Iowa, U.S.A.

Helen M. Griffiths
Department of Plant Pathology and Plant-Microbe Biology
Cornell University
Ithaca, New York, U.S.A.

George J. Hochmuth
Soil and Water Science Department
University of Florida
Gainesville, Florida, U.S.A.

David M. Ingram
Central Mississippi Research and Extension Center
Mississippi State University
Raymond, Mississippi, U.S.A.

Kelly L. Ivors
Department of Plant Pathology
North Carolina State University
Mountain Horticultural Crops Research and Extension Center
Mills River, North Carolina, U.S.A.

Pingsheng Ji
Plant Pathology Department
Coastal Plain Experiment Station
University of Georgia
Tifton, Georgia, U.S.A.

Jeffrey B. Jones
Plant Pathology Department
University of Florida
Gainesville, Florida, U.S.A.

Steven T. Koike
Cooperative Extension Monterey County
University of California Agricultural and Natural Resources (UCANR)
Salinas, California, U.S.A.

Y.-W. Kuo
Department of Plant Pathology
University of California–Davis
Davis, California, U.S.A.

Kai Shu Ling
U.S. Department of Agriculture
Agricultural Research Service (ARS)
U.S. Vegetable Laboratory
Charleston, South Carolina, U.S.A.

Sally A. Miller
Department of Plant Pathology
The Ohio State University
Wooster, Ohio, U.S.A.

Timur M. Momol
Plant Pathology Department
University of Florida
Gainesville, Florida, U.S.A.

Francisco J. Morales
International Center for Tropical Agriculture (CIAT) (Scientist Emeritus)
Palmira, Colombia

John F. Murphy
Department of Entomology and Plant Pathology
Auburn University
Auburn, Alabama, U.S.A.

Joseph W. Noling
Citrus Research and Education Center (CREC)
University of Florida
Lake Alfred, Florida, U.S.A.

Aleksa Obradovic
Plant Pathology Department
Faculty of Agriculture
University of Belgrade
Belgrade, Serbia

Stephen M. Olson
North Florida Research and Education Center (NFREC)
University of Florida
Quincy, Florida, U.S.A.

Margarita Palmieri
Department of Plant Protection
University of Valle de Guatemala
Guatemala City, Republic of Guatemala

Albert O. Paulus
Department of Plant Pathology and Microbiology (Professor Emeritus)
University of California–Riverside
Riverside, California, U.S.A.

Kenneth L. Pernezny
Everglades Research and Education Center (EREC)
University of Florida
Belle Glade, Florida, U.S.A.

Jane E. Polston
Plant Pathology Department
University of Florida
Gainesville, Florida, U.S.A.

Pamela D. Roberts
Plant Pathology Department
Southwest Florida Research and Education Center (SWFREC)
University of Florida
Immokalee, Florida, U.S.A.

Fikrettin Sahin
Department of Genetics and Bioengineering
Faculty of Engineering and Architecture
Yeditepe University
Istanbul, Turkey

Bielinski M. Santos
Gulf Coast Research and Education Center (GCREC)
University of Florida
Wimauma, Florida, U.S.A.

Hikmet H. Saygili
Department of Plant Protection
Faculty of Agriculture
Ege University
Izmir, Turkey

Robert L. Schlub
Western Pacific Tropical Research Center
University of Guam, UOG Station
Mangilao, Guam

David J. Schuster
Entomology and Nematology Department (Professor Emeritus)
Gulf Coast Research and Education Center (GCREC)
University of Florida
Bradenton, Florida, U.S.A.

John W. Scott
Gulf Coast Research and Education Center (GCREC)
University of Florida
Wimauma, Florida, U.S.A.

Larry J. Smith
Northwest Research and Outreach Center
University of Minnesota–Crookston
Crookston, Minnesota, U.S.A.

David Stead
Central Science Laboratory
Ministry of Agriculture, Fisheries, and Food
Sand Hutton, York, U.K.

Walter R. Stevenson
Department of Plant Pathology
University of Wisconsin–Madison
Madison, Wisconsin, U.S.A.

Gary E. Vallad
Gulf Coast Research and Education Center (GCREC)
University of Florida
Wimauma, Florida, U.S.A.

Jon C. Watterson
Watterson Agricultural Consultants
Davis, California, U.S.A.

Simon A. Weller
Detection Department
Defence Science and Technology Laboratory
Porton Down, Salisbury, U.K.

William M. Wintermantel
U.S. Department of Agriculture
Agricultural Research Service (ARS)
Salinas, California, U.S.A.

Frank Zalom
Entomology Department
University of California–Davis
Davis, California, U.S.A.

Thomas A. Zitter
Department of Plant Pathology and Plant-Microbe Biology
Cornell University
Ithaca, New York, U.S.A.

Contents

Introduction

Tomato Diseases, Pests, and Disorders

In many parts of the world, plant diseases often become the limiting factor in tomato production when disease-resistant cultivars are not used. There are nearly 200 known diseases of tomato, and they have diverse causes and etiologies. Management of these diseases requires incorporating host resistance, exclusion, eradication, and protection into a program of integrated pest management (IPM):

- **Resistance** involves the development and use of cultivars that prevent or impede the activity of the pathogen and exploit the genetically controlled differences in susceptibility to disease within *Solanum lycopersicum* L. and other *Solanum* species.
- **Exclusion** involves preventing the pathogen from coming into contact with the host or reducing contact to a minimum. Doing so involves inspection and quarantines, certification of planting materials, use of disease-free seedlings and seed, and production in areas where the pathogen is minimized or excluded.
- **Eradication** involves the elimination of the pathogen after it has been established in the area in which the host is growing. This involves removal of alternative and wild overseasoning hosts; destruction of overseasoning debris by rotation, sterilization, or sanitation; and removal of the pathogen from seed and other plant parts.
- **Protection** involves the use of cultural practices, the manipulation of greenhouse environments, the regulation of soil moisture, adjustment of soil reaction and fertility, the control of insect vectors, the manipulation of planting time, and the use of protective chemicals.

Infectious Diseases

Infectious diseases are those caused by bacteria and phytoplasmas, fungi and oomycetes, viruses and viroids, nematodes, and parasitic phanerogams. For disease to develop and result in damage to the host plant, a susceptible part of a susceptible host, a virulent pathogen, and the proper environment must be present.

Fungi

Fungi are organisms with somatic structures that are generally filamentous and branched. They have cell walls and nuclei and typically reproduce both sexually and asexually. More than 100,000 species of fungi have been described, and 20,000 of them are pathogenic to plants or animals. Traditional classification of fungi is based on spore production and morphology, providing many challenges and differences of opinion.

Fungal spores are easily disseminated by wind, rain, and splashing water and on people, machinery, animals, and plants. Fungi overseason in living and dead plants (including fruit and seed), in the soil, and in or on mites and insects.

Bacteria and Phytoplasmas

Bacteria are microscopic organisms that consist of individual cells ranging from 0.2 to 10.0 μm in size. One characteristic that separates these organisms from higher organisms is that the genetic material of bacteria is not separated from the cytoplasm by a unit membrane system.

In the first edition of this compendium, five main genera of true bacteria accounted for most of the bacterial species that incite plant diseases. That list has expanded to include nine genera: *Ralstonia, Pseudomonas, Erwinia, Pectobacterium, Agrobacterium, Clavibacter, Acidovorax, Pantoea,* and *Xanthomonas.* These bacteria consist of cells, each of which is surrounded by a cell wall, and may be referred to as "walled prokaryotes."

Bacteria survive on crop residue, seed, and volunteer plants (epiphytes or in diseased tissue) and in soil and water. They may be spread by mechanical means, by insects, and by seed. Water from both rain and irrigation is also an effective means for dissemination of bacteria.

Phytoplasmas (formerly referred to as mycoplasma-like organisms (MLOs)) are unculturable, wall-less bacteria that are obligate parasites of plant phloem tissue. They cause diseases of many important crops and induce a wide variety of symptoms, ranging from mild yellowing to death. Phytoplasmas are transmitted from plant to plant by the feeding activities of certain insects (mostly leafhoppers). They have been identified by host specificity and symptomatology, insect vector, and, more recently, by DNA analysis.

Viruses and Viroids

More than 600 plant diseases are thought to be caused by viruses; however, two-thirds of these diseases have not been conclusively proven to have viral causal agents. Viroids, phytoplasmas, and spiroplasmas have been shown to cause some of these diseases.

Viruses are tiny entities of nucleic acid (either RNA or DNA), each surrounded by a protein coat. Viroids consist solely of small, circular, single-stranded RNA, which is unencapsidated and encodes no known proteins.

Viruses are differentiated and identified by host specificity, mode of transmission, purification properties, electron microscopy, electrophoresis, serology, and complementary DNA (cDNA) technology. Viruses are transmitted by insects, mites, fungi, nematodes, mechanical means, grafting, and dodder and occasionally by seed.

Nematodes

Nematodes are semihyaline eelworms, microscopic in size and aquatic by nature, that inhabit fresh and salt water, soils, and animals of the world. More than 15,000 species of nematodes have been identified, nearly 2,500 of which are parasitic on plants. Parasitic nematodes can be classified as endoparasites, which invade plant tissues and spend part of their life cycle isolated from the soil, and ectoparasites, which may introduce the anterior portion of the body into the root but generally spend their entire life cycle in the soil.

Temperature, moisture, aeration, substrate composition, food supply, and competition for feeding sites greatly affect the communities that occupy specific environmental niches. Nematodes can be disseminated by any agent that moves infested soil (e.g., wind, flood water, animals, humans, tools, equipment) and by passive movement in water, soil, and infected plants.

Insects and Mites

More than 1 million species of insects and mites have been described, and most are beneficial or not considered pests. However, phytophagous species attack the roots, stems, petioles, leaves, and reproductive parts of plants.

Insects and mites both have chitin exoskeletons and jointed appendages. They occupy numerous diverse habitats and feed on plants, animals, and decaying organic matter.

Mites have two body portions: cephalothorax and abdomen. They generally go through four life stages: egg, larva, nymph, and adult. The nymphal stage may consist of two or more instars. The larval and nymphal stages resemble the adult.

Insects have three body portions: head, thorax, and abdomen. Insects generally have six legs, although the legs may be absent in the immature life stages of some species or supplemented with prolegs in others. Metamorphosis may be holometabolous or hemimetabolous:

- Holometabolous insects go through four distinct life stages: egg, larva, pupa, and adult. None of the immature life stages resembles the adult. Mouthparts of larval insects differ from those of adults, and larvae and adults feed on different hosts or host parts. Pupae do not feed.
- Hemimetabolous insects go through three life stages: egg, nymph, and adult. Nymphs generally resemble adults and usually possess the same type of mouthparts and feed on the same host as the adults. Some may go through a nonfeeding resting stage.

Insects and mites can be disseminated passively by wind, soil, infested plant parts, plant containers, and machinery. Both insects and mites may disperse actively by walking, and insects may also disperse by flying.

Noninfectious Diseases, Disorders, and Damage

Noninfectious diseases, disorders, and damage are caused by extremes in light, heat, soil moisture, and soil reaction and by nutritional imbalances, unfavorable oxygen relations, atmospheric pollutants, herbicides and other pesticides, and lightning. Genetic diseases (e.g., pox and fleck) and allelopathic diseases (e.g., walnut wilt and frenching) are also examples of noninfectious diseases. Symptoms caused by noninfectious diseases often mimic those caused by infectious diseases and add to the difficulty of making an accurate diagnosis.

Botany and Culture

Importance of the Crop

The tomato—a fruit generally treated as a vegetable—is a popular food and an important source of vitamins and antioxidants in the American diet. Tomatoes are rich in the carotenoids lycopene and β-carotene (provitamin A), which are reported to have anticancer properties. Because they are so widely consumed in the United States, tomatoes account for approximately 10% of Americans' total dietary intake of vitamin C and are the leading source of phenolics, another category of antioxidants.

Tomato is widely cultivated as an annual crop in many regions of the world; China and the United States are the leading producers (Table 1). In the United States, production of tomatoes in 2007 occupied approximately 170,700 ha and yielded more than 13 million metric tons of produce at harvest, worth $2.1 billion and second only to potatoes among vegetables in total farm gate value. Florida and California are the principal fresh-market tomato-growing states. California is also the leading source of processing tomatoes, accounting for more than 95% of the U.S. supply and a large share of the world supply.

Cytotaxonomic Background

The cultivated tomato, *Solanum lycopersicum* L., and its wild relatives are a group of related species (Table 2) that were formerly recognized as the genus *Lycopersicon* and are now classified as *Solanum* section *Lycopersicon*. Some species epithets under the *Lycopersicon* system were already in use in *Solanum*, so new names were created—for example, *S. neorickii* D. M. Spooner, G. J. Anderson, & R. K. Jansen for *L. parviflorum* C. M. Rick and *S. habrochaites* S. Knapp & D. M. Spooner for *L. hirsutum* Humb. & Bonpl. In addition, four taxa previously recognized as subspecies of *L. cheesmanii* Riley or *L. peruvianum* C. H. Muller are now recognized as distinct species in the *Solanum* nomenclature: *S. galapagense* S. C. Darwin & Peralta, *S. corneliomulleri* J. F. Macbr., *S. arcanum* Peralta, and *S. huaylasense* Peralta. (For the sake of consistency with the prior literature, both the *Solanum* names and their *Lycopersicon* equivalents are listed in Table 2.)

TABLE 1. Top-10 Tomato-Producing Countries (2007)[a]

	Production (in 1,000 t)	Area (in 1,000 ha)	Yield (in t/ha)
China	33,597	1,454.9	23.1
United States	14,185	170.7	83.1
California, processing	10,961	119.8	91.5
California, fresh market	503	15.0	33.6
Florida, fresh market	604	14.0	43.1
India	10,055	596.0	16.9
Turkey	9,945	270.0	36.8
Egypt	8,639	537.2	16.1
Italy	6,530	125.3	52.1
Iran	5,000	140.0	35.7
Brazil	3,431	58.4	58.8
Spain	3,664	54.1	67.7
Mexico	3,150	116.7	27.0
World total	133,260	5,122.0	26.0

[a] Data from http://usda.mannlib.cornell.edu/MannUsda/viewDocumentInfo.do?documentID=1210

TABLE 2. Overview of Geographic Distribution, Reproductive Biology, and Principal Disease Resistance of Wild Tomatoes

Solanum Species[a] (*Lycopersicon* equivalent)	Distribution	Elevation[b]	Mating System[c]	Crosses to Tomato[d]	Principal Disease Resistance(s)[e,f]
S. lycopersicum '*cerasiforme*' (*L. esculentum* var. *cerasiforme*)	Andean region; adventive in tropics worldwide	20–2,400	Selfing (SC)	Very easy	Fungal/Oomycete: anthracnose, late blight, Phytophthora root rot, powdery mildew, Septoria leaf spot, Verticillium wilt Bacterial: bacterial wilt, tomato pith necrosis, bacterial spot
S. pimpinellifolium (*L. pimpinellifolium*)	Lowland Ecuador and coastal Peru	5–1,500	Facultative or selfing (SC)	Very easy	Fungal/Oomycete: Fusarium wilt, gray leaf spot, late blight, leaf mold, Septoria leaf spot, target spot, Verticillium wilt Bacterial: bacterial speck Nematode: potato cyst nematode
S. cheesmaniae (*L. cheesmanii*)	Volcanic slopes of Galapagos Islands	10–1,200	Selfing (SC)	Very easy	Fungal/Oomycete: black mold Bacterial: bacterial spot, tomato pith necrosis
S. galapagense (*L. cheesmanii* f. *minor*)	Rocky coast of Galapagos Islands	5–650	Selfing (SC)	Very easy	Fungal/Oomycete: gray mold
S. chmielewskii (*L. chmielewskii*)	Inter-Andean valleys of southern Peru	1,900–3,100	Facultative (SC)	Easy (UI)	Fungal/Oomycete: powdery mildew Nematode: false root-knot nematode
S. neorickii (*L. parviflorum*)	Inter-Andean valleys of central Peru to central Ecuador	600–2,600	Selfing (SC)	Easy (UI)	Fungal/Oomycete: Powdery mildew, gray mold
S. arcanum, *S. peruvianum*, *S. corneliomulleri*, *S. huaylasense* (all *L. peruvianum*)	Coastal, inland, and inter-Andean valleys from northern Peru to northern Chile	0–3,000	Mostly out-crossing (SI), some SC	Difficult (UI, EC)	Fungal/Oomycete: corky root, Fusarium wilt, Septoria leaf spot Bacterial: bacterial canker, bacterial speck, bacterial spot Nematode: root-knot nematodes Viral: CMV, ToMV, TSWV, TYLCV
S. chilense (*L. chilense*)	Coastal and western slopes of Andes from southern Peru to northern Chile	50–3,500	Outcrossing (SI)	Difficult (UI, EC)	Fungal/Oomycete: powdery mildew, corky root Bacterial: bacterial canker, bacterial speck Viral: BCTV, CMV, ToMV, TSWV, TYLCV
S. habrochaites (*L. hirsutum*)	Western slopes of Andes from central Peru to southern Ecuador	40–3,300	Outcrossing (SI), facultative (SC)	Easy (UI)	Fungal/Oomycete: Cercospora leaf mold, Didymella stem rot, early blight, late blight, powdery mildew, gray mold Bacterial: bacterial canker, bacterial spot Viral: AMV, CMV, PVY, TYLCV
S. pennellii (*L. pennellii*)	Dry, rocky slopes of coastal Peru	50–1,900	Mostly out-crossing (SI)	Easy (UI)	Fungal/Oomycete: Fusarium wilt, powdery mildew Bacterial: bacterial spot
S. lycopersicoides	High elevations in southern Peru and northern Chile (very limited)	1,500–3,600	Outcrossing (SI)	Very difficult (UI, EC)	Fungal/Oomycete: gray mold, Phytophthora root rot Bacterial: bacterial spot, bacterial canker Nematode: root-knot nematodes Viral: CMV, PepMV, TICV, ToMV, TYLCV
S. sitiens	Atacama Desert of northern Chile (very limited)	2,500–3,000	Outcrossing (SI)	Very difficult (UI, EC)	Largely untested
S. juglandifolium	Temperate rain forests of Colombia and Ecuador	1,300–3,000	Outcrossing (SI)	No	Fungal/Oomycete: root diseases
S. ochranthum	Temperate rain forests from southern Peru to central Colombia	1,900–3,500	Outcrossing (SI)	No	Fungal/Oomycete: late blight, root diseases

[a] *Solanum* species names and their equivalents under *Lycopersicon* are from Peralta et al. (2008). The species are listed in approximate order of increasing phylogenetic distance from cultivated tomato. For simplicity, the species included in the *Lycopersicon peruvianum* complex are listed together.
[b] Meters above sea level.
[c] SI = self-incompatible; SC = self-compatible.
[d] UI = unilaterally incompatible with *S. lycopersicum*; EC = embryo or ovule culture usually required to obtain hybrids with tomato.
[e] See Rick and Chetelat (1995) for references to most disease-resistance traits.
[f] AMV = *Alfalfa mosaic virus;* BCTV = *Beet curly top virus;* CMV = *Cucumber mosaic virus;* PepMV = *Pepino mosaic virus;* PVY = *Potato virus Y;* TICV = *Tomato infectious chlorosis virus;* ToMV = *Tomato mosaic virus;* TSWV = *Tomato spotted wilt virus;* TYLCV = *Tomato yellow leaf curl virus.*

The nearest relatives of tomato outside the *Lycopersicon* clade are four tomato-like nightshades, which show affinities to tomatoes on the basis of morphology, crossing relationships, and molecular systematics. They represent two pairs of sister taxa: (1) *S. lycopersicoides* Dunal and *S. sitiens* I. M. Johnst. and (2) *S. juglandifolium* Dunal and *S. ochranthum* Dunal. The species in each pair are closely related to one another but very different from the other group. Within the much larger *Solanum* genus, which contains approximately 1,500 species, tomato is closely related to wild and domesticated species of potato (*Solanum tuberosum* L.).

The tomato species are short-lived perennial herbs native to the western Andean region and the Galapagos Islands of South America. They are almost exclusively diploid, with the same basic chromosome number ($2n = 2\times = 24$), although a few instances of spontaneous autotetraploidy have been recorded. The chromosomes of *Solanum* section *Lycopersicon* species are similar in morphology at the pachytene stage of meiosis and synapse normally in F_1 interspecific hybrids. The 12 chromosomes are metacentric or submetacentric, except for chromosome 2, which is acrocentric.

The tomato genome consists of approximately 1,300 centimorgans and 900 megabase pairs (MBp). Comparative linkage maps show that the order of genes along chromosomes is strongly conserved among the tomato species; however, short chromosomal inversions have been detected by linkage analysis or bacterial artificial chromosome (BAC) fluorescence in situ hybridization (FISH) (BAC-FISH). Coding genes are concentrated in the euchromatin regions, which are located in the distal portion of each chromosome arm. A large fraction of heterochromatin, containing approximately 75% of the nuclear DNA, is concentrated around the centromeres and telomeres. The pericentromeric regions have a relatively low density of coding genes and are recombinationally inactive.

The *S. lycopersicum* genome has been sequenced and provides a high-quality reference genome, integrated with genetic maps and other data. A genome sequence for *S. pimpinellifolium* L. is also available, and the other tomato species will be sequenced in the near future.

Culture

Although tomato can be grown under a wide range of environmental conditions, it requires a warm season and prefers well-drained soil and full sunlight. A daily maximum temperature of approximately 25–35°C is optimal for growth and reproduction. Tomato is sensitive to frost at any stage, and prolonged exposure to chilling temperatures (0–10°C) adversely affects seed germination, vegetative growth, and fruit ripening and quality. When provided with sufficient moisture, mature plants can tolerate temperatures in excess of 38°C, but fruit set is reduced in most cultivars.

Under cultivation, tomato is grown as an annual; however, it is naturally a perennial plant. The normal growth habit is originally indeterminate, but many modern cultivars are determinate, including those bred for mechanical harvest. While the cultivated tomato is relatively insensitive to photoperiod, certain wild species require short days to induce flowering, particularly under greenhouse conditions. Once flowering begins, shoots are composed of a series of sympodia, each made up of two to three leaves and an inflorescence. Tomato leaves are pinnately compound, consisting of two or more pairs of laterals and one terminal leaflet. Plants are almost always grown from seed, either sown directly or transplanted at the seedling stage. Grafting onto disease-resistant rootstock is a relatively common practice in greenhouse production systems.

Mating systems vary widely among tomato species. Whereas the cultivated tomato and some wild species are primarily self-pollinated (autogamous), other species are facultative or obligate outcrossers (allogamous) (Table 2). Self-fertilization in the allogamous species is prevented by gametophytic self-incompatibility. Tomato bears perfect flowers with a basic five-parted structure, although many cultivars have more highly subdivided (fasciated) flowers with extra carpels, anthers, petals, and sepals. Petals and sepals are arranged in whorls of fused segments, forming the corolla and calyx, respectively. Anthers are fused in a cone, typically with a sterile tip, and pollen is released via inner longitudinal slits. Wild tomatoes are naturally cross-pollinated by native bees that "buzz" the flowers to collect pollen. In the facultative or allogamous species, the stigma surface is usually exposed (exserted) beyond the end of the anther cone, which promotes outcrossing. In the autogamous species, the stigma surface is generally level with the tip of the anther cone, and in modern cultivars, it is recessed (inserted), which promotes selfing and fruit set. Tomato flowers are relatively easy to emasculate and cross pollinate, and each pollinated flower yields large quantities of seed (approximately 25–50 per fruit) of controlled parentage. The inflorescences of most cultivars are simple cymes (unbranched), whereas some of the wild species produce dichotomously branched, compound inflorescences.

Fruit growth begins shortly after fertilization and involves an initial phase of cell division followed by cell expansion. Tomato fruit are berries composed of an outer wall (pericarp) surrounding the locules and made up of seeds and placental tissues separated by inner walls (septa). Fruit of the wild species are generally small (less than 2 cm in diameter) and bilocular, whereas fruit of cultivated tomatoes are larger and often multilocular.

Tomato fruit are composed mainly of water (90–95%), as well as sugars, acids, and various other inorganic and organic constituents. The flavor of tomato is determined in large part by the balance of soluble sugars (mainly glucose and fructose) and organic acids (citric and malic), as well as a complex mix of volatile aromatic compounds.

Fruit color is conferred by the composition and distribution of carotenoids, chlorophyll, and anthocyanins. During ripening, chloroplasts undergo a transformation into chromoplasts, during which large amounts of carotenoids—mainly lycopene (responsible for the red color) and β-carotene—are accumulated. The extent of red pigmentation in tomato fruit is used to categorize the ripening stages for commercial grading: immature, mature green, breaker, turning, pink, light red, and red. The wild species fall into two categories with respect to fruit color: those that produce red or orange fruit (e.g., *S. pimpinellifolium, S. cheesmaniae* (L. Riley) Fosberg, and *S. galapagense*) and those that produce green or yellow green fruit with variable green or purple stripes or shading (all the other species).

Fruit maturation and ripening generally lasts 6–10 weeks after pollination in most cultivars and substantially longer in the wild relatives. At maturity, cells in the abscission layer of the pedicel articulation (joint) separate, leading to detachment of the fruit from the plant.

History of Domestication

Although archeological evidence concerning the domestication of tomato is scarce, linguistic, historical, and other evidence points to Mexico as the likely region of original cultivation and domestication. The common name for tomato is a derivative of *tomatl,* which, in the Nahuatl language of the Aztec, was applied to the tomatillo and other species of *Physalis* (*tomatl, xaltomatl,* and *tepetomatl*), as well as to tomato (*xitomatl*).

The immediate progenitor of cultivated tomato was probably the feral or wild cherry tomato—a member of the same species

(*S. lycopersicum*) that differs primarily in fruit size. Originally native to the Andean region of South America, the cherry tomato spread into Central America and is now found in tropical and subtropical America and throughout the tropics of the Old World.

From Mesoamerica, tomatoes were brought to Europe in the sixteenth century. Early botanical references, such as woodcuts and descriptions in European herbals from the 1500s, depict plants with large, highly subdivided fruit, indicating that selection for increased fruit size was accomplished before the Spanish conquest.

Genetic Resources and Modern Breeding

Genetic variation within *S. lycopersicum* has been reduced to a great extent from that present in its wild progenitors. The processes of migration out of the native area followed by domestication, early utilization, and migration to Europe and North America would have tended to reduce genetic variation. Until the 1920s, genetic improvement of tomato depended on selection of variants arising from spontaneous mutation, accidental outcrossing, or reassortment of existing variation. Since that time, breeders have also utilized variation in the wild species and primitive forms of *S. lycopersicum* to an increasing extent. Introgression of genes from exotic sources is thought to have contributed to the steady increase in average yields (per hectare) attained in commercial tomato fields in recent decades. Since the 1930s, the average yield of processing tomatoes in the United States has increased more than eightfold.

The cultivated tomato can be hybridized with all other species of *Solanum* sect. *Lycopersicon*, as well as *S. lycopersicoides* and *S. sitiens*, with varying degrees of difficulty. The red- or orange-fruited species are fully compatible with *S. lycopersicum*, and interspecific hybrids are viable and fertile. In contrast, crosses to the green-fruited species, in general, succeed only when cultivated tomato is used as the female parent; in the reciprocal cross, pollen tubes of *S. lycopersicum* are inhibited in styles of the wild species (unilateral incompatibility or incongruity), which prevents fertilization. Crosses between *S. lycopersicum* and *S. peruvianum* L., *S. corneliomulleri*, *S. huaylasense*, *S. arcanum*, or *S. chilense* (Dunal) Reiche typically result in zygotic failure during early embryo development, but hybrids can be rescued by embryo or ovule culture. Of the tomato-like nightshades, *S. lycopersicoides* and *S. sitiens* have been crossed to *S. lycopersicum*, but *S. ochranthum* and *S. juglandifolium* appear to be sexually incompatible with cultivated tomato.

In few other crops have wild relatives been exploited for genetic improvement to as great an extent as in tomato. Wild species and other exotics provided the first major genes for disease resistance bred into new cultivars, which, before the 1940s, were virtually devoid of such variation. Resistances to at least 44 major diseases has been identified in wild relatives, and at least 20 of these have been bred into horticultural tomatoes. Tomatoes could not be grown commercially in many areas without these disease-resistant species. Other traits—including insect resistance and abiotic stress tolerances, as well as fruit-quality parameters, such as the concentration and composition of sugars, acids, and pigments—have also been improved by utilizing exotics. Use of the wild species has been facilitated by the development of near-isogenic lines, introgression lines, and other prebred lines containing more alien chromosome segments in a cultivated tomato background.

The principal germplasm repositories for tomatoes in the United States are the U.S. Department of Agriculture (USDA) Plant Genetic Resources Unit (PGRU) at Geneva, New York, and the C. M. Rick Tomato Genetics Resource Center (TGRC) at Davis, California. The PGRU maintains approximately 5,800 tomato accessions, of which approximately 90% are stocks of *S. lycopersicum* (primarily open-pollinated cultivars); the remainder are wild species. The TGRC maintains more than 3,500 accessions of wild species, monogenic mutants, and miscellaneous genetic stocks in roughly equal proportions. Information on available accessions and forms for submitting seed requests are accessible via the USDA Germplasm Resources Information Network (GRIN) database[1] and the TGRC website.[2] Other tomato genebanks are maintained by the Centre for Genetic Resources (CGN)[3] in the Netherlands; the Institute of Plant Genetics and Crop Plant Research (IPK)[4] in Germany; the Conservación y Mejora de la Agrodiversidad Valenciana (COMAV)[5] in Spain; and the World Vegetable Center (AVRDC)[6] in Taiwan. The Solanaceae Genomics Network (SGN),[7] hosted by Cornell University, maintains a database of genetic information, including sequences, markers, and mutants, pertaining to tomato and other solanaceous crops.

Selected References

Arumuganathan, K., and Earle, E. D. 1991. Nuclear DNA content of some important plant species. Plant Mol. Biol. Rep. 9:208-218.

Gerrior, S., and Bente, L. 2002. Nutrient content of the U.S. food supply, 1909-1999: A summary report. Home Economics Report 55. U.S. Department of Agriculture, Center for Nutrition Policy and Promotion.

Jenkins, J. A. 1948. The origin of the cultivated tomato. Econ. Bot. 2:379-392.

McCue, G. A. 1952. The history of the use of the tomato: An annotated bibliography. Ann. Mo. Bot. Gard. 39:289-348.

Peralta, I. E., Spooner, D. M., and Knapp, S. 2009. Taxonomy of Wild Tomatoes and Their Relatives (*Solanum* sect. *Lycopersicoides*, sect. *Juglandifolia*, sect. *Lycopersicon*; Solanaceae). Monogr. Syst. Bot. 84:1-186.

Peterson, D. G., Price, H. J., Johnston, J. S., and Stack, S. M. 1996. DNA content of heterochromatin and euchromatin in tomato (*Lycopersicon esculentum*) pachytene chromosomes. Genome 39:77-82.

Rick, C. M. 1988. Tomato-like nightshades: Affinities, autecology, and breeders' opportunities. Econ. Bot. 42:145-154.

Rick, C. M. 1995. Tomato, *Lycopersicon esculentum* (Solanaceae). Pages 452-457 in: Evolution of Crop Plants. 2nd ed. J. Smartt and N. W. Simmonds, eds. Longman House, Essex, U.K.

Rick, C. M., and Chetelat, R. T. 1995. Utilization of related wild species for tomato improvement. Acta Hortic. 412:21-38.

Spooner, D. A., Peralta, I. E., and Knapp, S. 2005. Comparison of AFLPs with other markers for phylogenetic inference in wild tomatoes [*Solanum* L. section *Lycopersicon* (Mill.) Wettst.]. Taxon 54:43-61.

Szinay, D., Wijnker, E., van den Berg, R., Visser, R. G. F., de Jong, H., and Bai, Y. 2012. Chromosome evolution in *Solanum* traced by cross-species BAC-FISH. New Phytol. 195:688-698.

Tomato Genome Consortium. 2012. The tomato genome sequence provides insights into fleshy fruit evolution. Nature 485:635-641.

Vinson, J. A., Hao, Y., Su, X., and Zubik, L. 1998. Phenol antioxidant quantity and quality in foods: Vegetables. J. Agric. Food Chem. 46:3630-3634.

Zamir, D. 2001. Improving plant breeding with exotic genetic libraries. Nat. Rev. Genet. 2:983-989.

(Prepared by R. T. Chetelat)

[1] GRIN: www.ars-grin.gov
[2] TGRC: http://tgrc.ucdavis.edu
[3] CGN: www.genebank.nl
[4] IPK: http://gbis.ipk-gatersleben.de/gbis_i
[5] COMAV: www.comav.upv.es
[6] AVRDC: http://avrdc.org
[7] SGN: http://solgenomics.net

Seed Production and Quality Assurance

Seed Production

Tomato is predominantly a self-pollinating crop, with seed formation occurring via pollination of the pistil by pollen from the same flower. Until World War II, tomato varieties consisted of pure-line, open-pollinated selections. The first hybrids were developed for fresh-market production and home garden use in the 1950s, but it took several decades before the advantages of hybrid seed became widely recognized. Beginning in the early 1970s, hybrids gained acceptance by the tomato industry. These new hybrids—with increased vigor, yield, and quality—gradually replaced most open-pollinated tomato varieties.

A hybrid results from the deliberate crossing of two different parent varieties, usually inbreds. Hybrids can provide increased yield, stronger resistance to disease, greater adaptability, uniformity in maturity, and other fruit-quality traits (e.g., higher-soluble solids, better color).

The technology of producing hybrid tomato seed involves specialized skills and requires a thorough understanding of crop management and hybridization practices to ensure high quality yet reasonable seed costs. Production of hybrid varieties is labor intensive, and consequently, most hybrid seed is produced in regions where labor costs are low.

Fig. 1. An emasculated tomato flower, with the sepals, ovary, style, and stigma of the female plant. (Courtesy H. Bolkan)

Hybrid Seed Production

Hybrid tomato seed production requires the planting of inbred breeding lines that have originated from single plants chosen for their special traits. Like hybrid seeds of other vegetable crops, hybrid tomato seeds are produced by emasculating the flowers on female-designated plants (Fig. 1) followed by application to the stigma of pollen from male-designated plants (Fig. 2). The male parent is typically planted 7–10 days earlier than the female parent to ensure sufficient flowers are available for pollen collection before and during the flowering of the female.

In hybrid seed production, the purity of the parents is extremely important. Therefore, every effort is made to avoid seed mixture during planting and transplanting. In addition, any off-type plants are removed by trained personnel who walk through the male and female plant rows in the field. This process is repeated two to three times prior to pollen extraction and pollination.

For small-scale hybridization, pollen may be collected from the male plants manually using forceps or a mechanical hand-held vibrator (Fig. 3). For commercial hybrid seed production, a large amount of pollen is needed and thus extracted from dry flowers. The pollen-extraction process varies according to each hybrid seed producer.

The most common practice is to collect the staminate cones from male plants and then dry them in a desiccator overnight (Fig. 4). Pollen is extracted by agitating the staminate cones over a fine net (Fig. 5) or sieve and collecting the pollen in a container placed below the net (Fig. 6). Alternatively, flowers are collected from the male plant, placed on screens in a thin layer, and then placed in an oven at 30°C for 24 h or until completely dry. The fresh flowers can be dried under direct sunlight as long as the temperature does not exceed 30°C. After the flowers have dried, they are manually or mechanically crushed and passed through a sieve. For manual extraction after passing through the sieve, the crushed flowers are placed in a plastic bag. The bag is closed, leaving some air in it, and then shaken by hand. The shaking action causes the pollen and some small parts of the crushed flowers to stick to the inside surfaces of the plastic bag. Pollen is then collected by scraping the inside

Fig. 2. Transferring pollen collected from a male parent to the stigma of a female flower that was previously emasculated. (Courtesy H. Bolkan)

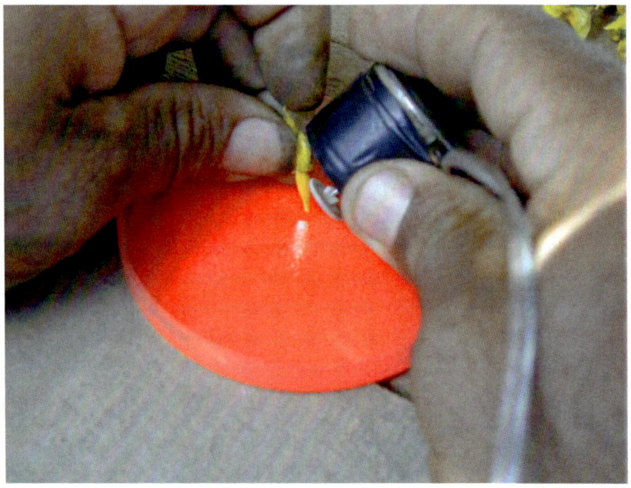

Fig. 3. A mechanical hand-held vibrator used for collecting pollen from male parent flowers. (Courtesy H. Bolkan)

of the bag into a petri dish or similar vessel. The collected pollen and crushed flower parts are then passed through a testing sieve. When most of the pollen passes through the sieve, a bright-yellow color appears in the container under the sieve. To maintain viability, the collected pollen is stored in a cool place in an airtight container with a desiccant. Pollen stored in a refrigerator maintains its viability up to 1 month.

Emasculation and Hybridization

Hybridization begins in the morning by selecting flowers from female plants for emasculation. Flowers must be in the late bud stage of development (i.e., are closed and have not turned yellow) (Fig. 7). The flower bud is opened with a pair of fine forceps or fingernails, and the staminate cone is removed (Fig. 8). Care must be taken to ensure that the receptive stigma is not damaged during the cone removal. Since the female plant will have many flowers at different stages of development, two sepals are removed from emasculated flowers (Fig. 9) to mark the crossed flower and to document that the developing fruit is a hybrid. Open flowers and flowers that are past the accepted stage of emasculation are removed.

Generally, transfer of the male pollen is accomplished by dabbing the pollen with a fine brush or finger and then dabbing the brush or finger on the stigma of the emasculated flower. An alternative way to transfer pollen to the stigma of the emasculated flower is to use a small plastic tube. The plastic tube containing the male pollen is worn like a ring by the pollinator, and

Fig. 4. Staminate cones collected from male-designated plants. (Courtesy H. Bolkan)

Fig. 5. Staminate cones placed over a fine nylon net for pollen extraction. (Courtesy H. Bolkan)

Fig. 6. Pollen extraction by agitating the staminate cones over a nylon net. (Courtesy H. Bolkan)

Fig. 7. A tomato flower at the correct stage for emasculation: closed and not yet yellow. (Courtesy H. Bouzar)

Fig. 8. The emasculation process, involving removal of the staminate cone. (Courtesy H. Bolkan)

pollination is accomplished by holding the emasculated flower and dipping the stigma into the plastic tube containing the pollen (Fig. 10). In either method, care must be taken not to break the pistil. The number of hybrid fruit that will be produced varies from three or four fruit per cluster and from 20 to 25 fruit per plant, depending on the vigor of the plant.

Harvesting

Fruit is harvested manually and transported to the seed-extraction area in crates or bags. Only fruit with marked sepals and red color are harvested. Fruit without marked sepals are discarded prior to harvest to prevent accidental mixture of fruit from open flowers with hybridized fruit. Similarly, fruit with symptoms of blossom-end rot are removed, because seeds from these fruit have poor quality. Fruit with pink color are sometimes harvested for seed extraction, but generally, seed from fully ripened fruit gives the highest seed quality.

Seed Extraction

Seed can be hand or machine extracted. Hand extraction consists of cutting the fruit in half and either squeezing out the seeds or removing them with a spoon (Fig. 11). Mechanical seed extraction involves placing the fruit into a crusher to separate the gelatinous seed matrix from the remaining fruit tissue and then pressing it through a sieve or screen (Fig. 12).

Separation of the gel from the seed is accomplished naturally by fermentation or accelerated with the use of chemicals. When natural fermentation is used, the seed is kept in the tomato extract until microorganisms degrade the gelatinous material surrounding it. Depending on air temperature, fermentation takes 24–48 h. Chemical extraction involves the use of hydrochloric acid. The acid is added to the fermenting extract, which allows rapid degradation of the gelatinous material surrounding the seed. When acid is used, the fermentation time is shortened to a few hours. Fermentation longer than 4 h may adversely affect seed quality.

Regardless of the method used, the seed is immediately washed to remove digested material and any remaining chemicals. Clean water is added to the container with the seed extract. Viable seeds sink, and digested material along with nonviable seed float. The digested material can be decanted, leaving good-quality seed in the bottom of the container. The washing process uses clean, fresh water and is repeated three or four times or until the seed is clean and free of unwanted tomato tissue.

Following washing, the seed is air dried to a moisture content of 6–8% on a fine nylon net (Fig. 13). During the drying process, the seed is hand stirred two or three times daily to separate clumps and to ensure uniform drying. As an alternative to air drying, seed can be placed in fine nylon-net bags and dried in an air tunnel or a seed dryer.

After the seed has dried to the desired moisture content, it is placed in cloth bags or plastic containers and delivered according to the specifications of the seed company. Each seed container is labeled with the name or code of the hybrid, the year produced, and, most importantly, the lot number, which identifies the production location and the grower. The seed is then stored in a cool, dry place. For a large quantity of seed, as is produced in commercial operations, a special room with

Fig. 9. A pollinated tomato flower with removed sepals, allowing easy recognition of hybridized fruit. (Courtesy H. Bolkan)

Fig. 10. Pollinating a previously emasculated flower by dipping the stigma into a container of pollen collected from a male parent. (Courtesy H. Bolkan)

Fig. 11. Hand harvesting of seed from cut tomato fruit. (Courtesy H. Bolkan)

Fig. 12. Mechanical harvesting of seed, in which fruit are passed through a crusher. Fruit and seed with gelatinous material are then passed through a screen. (Courtesy H. Bolkan)

Fig. 13. Drying seeds on a fine nylon mesh and exposing them to air or direct sunlight. (Courtesy H. Bouzar)

controlled humidity (less than 30%) and temperature (less than 20°C) is used. Under these conditions, tomato seed can be stored for at least 3–5 years without losing viability.

Open-Pollinated Seed Production

With the exception of emasculation, the steps involved in open-pollinated seed production are the same as those described for hybrid seed production.

Certification and Quality Assurance

In tomato seed production, maximum attention is paid to producing hybrid seed that is free from seedborne pathogens and that produces high levels of germination, vigor, and genetic purity. The strategies outlined in the following sections are representative of the seed-quality assurance measures practiced by most European and U.S. seed companies. Most private seed companies employ in-house seed technologists, pathologists, and agronomists, who conduct field inspections and laboratory assays and develop and implement seed assay techniques. Most seed certification programs for tomato hybrid seed production involve two interrelated components: field certification and laboratory assays.

Field Certification

Locations for hybrid seed production are selected based on the absence of seedborne pathogens in the region. Because little outcrossing occurs in tomatoes, minimal isolation of production fields is required.

Fields for hybrid seed production are inspected by trained personnel or by plant pathologists to identify self-fruit and seedborne diseases. Field certification is based on visual field inspections of the growing crop to ensure that no self-fruit are found on female plants and to detect the presence of diseases that have the potential to become seedborne. If self-fruit are found, the grower is asked to immediately reinspect each plant in the field and to remove any self-fruit.

Production fields with no evidence of disease symptoms of seedborne pathogens are given the highest field rating. Fields showing symptoms of diseases caused by pathogens that can become seedborne are "red flagged." Each production field is scouted at the critical periods of emasculation/pollination and at the start of fruit harvest.

Laboratory Assays

Following the final cleaning, seed sizing, and density gradient separation, each commercial production lot is tested in the laboratory for germination, vigor, and physical purity. A controlled environment is used to ensure reliability of the germination and vigor tests.

In some cases, seed lots may be evaluated for vigor under greenhouse or artificially imposed stress conditions. This allows the technologist to more accurately assess and categorize the quality of each seed lot. Genetic purity testing may be done in the lab using protein or DNA separation techniques for identification of hybrid seed as well as any inbreds. Alternatively, seed lots may be grown in the plant nursery or field (purity grow-out test) until typical characteristics of the hybrid can be distinguished.

(Prepared by H. Bolkan and J. Watterson)

Containerized Production of Tomato Transplants

Containerized tomato transplants are an alternative to field-produced transplants and direct seeding for crop establishment. They are generally used for fresh-market tomato establishment, but in some production areas, much of the processing crop is produced with containerized transplants. Using transplants provides several benefits, including reduction in seed costs, planting to stand, optimal spacing, and earliness of harvest.

The ideal tomato transplant is stocky, green, and pest free and has a well-developed root system. Once transplanted, the plant should tolerate environmental challenges and continue growth to optimum yield. The conditions and procedures for production of high-quality containerized tomato transplants are outlined in the following sections.

Plant-Growing Structures

Location

If possible, the transplant production facility should be located away from tomato production areas. Tomato fields are a frequent source of insects and diseases, which can easily find their way into the transplant production facility. In addition, the site should be level and well drained and have ready access to an abundant supply of high-quality water.

The plant-growing structure (PGS) should be located a sufficient distance from surrounding trees and buildings to prevent falling under shadows during the day. Other important considerations for the PGS location include (1) the directional orientation (i.e., an east-facing orientation supplies more overall light, but a north–south orientation supplies more uniform light), (2) the availability of labor, (3) the proximity to shipping routes, (4) the ease of access to utilities, (5) local zoning and tax laws, and (6) the availability of room for expansion.

Structure

Most tomato transplant PGSs are a modified Quonset-style, with metal trusses and metal or wood sidewall support structures covered with polyethylene film. This type of structure has a life expectancy of 3 years. The movable sidewalls (often called curtains) can be moved up (opened) to allow cross ventilation or down (closed) to conserve heat and to provide protection from wind and rain. This style of construction can be used when having a cooling system is too expensive for the production budget.

The roof of the PGS is generally double-layered polyethylene inflated with a small fan to reduce chafing on the trusses and to increase insulating value. Having an inflated roof is also important because maintaining constant tension on the polyethylene reduces the tearing action of the wind. Algae and particulate matter may build up on the outer surface of the polyethylene over time, but they can be removed by gently scrubbing the surface with a mild soap-and-water solution.

The front and rear walls of the PGS are generally constructed of a more permanent material, such as fiberglass or polycarbonates, for strength and persistence. End-wall doors should be large enough to allow the passage of small equipment.

Irrigation Equipment

Two types of irrigation are presently used in tomato transplant production: overhead and ebb and flow (or ebb and flood). Overhead irrigation can be accomplished simply by using a garden hose (usually used for touch-up watering). However, the majority of growers rely on automated watering systems for uniform application. These machines require electricity and come as ground or elevated-track models. Both models irrigate via a boom system, which can be equipped with an additional boom for pesticide application.

Ground irrigation units occupy approximately 0.8–1.1 m^2 (9–12 square feet) of walkway and are generally stored at one end of the greenhouse after completion of watering. Ground units are constructed in such a way that the grower must step through the unit when walking through the greenhouse. Elevated irrigation units provide all of the options of ground units except that during power outages, they are harder to operate manually. Exposed portions of elevated units should be padded to reduce the risk of head injuries to workers.

Ebb and flow irrigation is accomplished by building modified troughs or pools in which the growing trays are floated. (Expanded polystyrene (EPS) trays perform best.) The troughs can be easily constructed with wooden planks. A four-sided structure should be pressed into a floor filled or partially filled with sand. The floor of the trough should be sloped to facilitate drainage to one end for emptying. A liner of heavy-gauge plastic will serve to hold the water. Having sand beneath the liner will reduce the tendency for punctures. An ebb and flow system requires a high-velocity pump to facilitate filling and emptying and a reservoir into which the irrigation water can be pumped.

Plants that are given unrestricted access to water exhibit unruly and uncontrolled growth, which results in transplants that are difficult to plant and establish. Therefore, ebb and flow systems should allow for daily water removal and be equipped with some support structure on which the growing trays can come to rest and drain properly.

(For more information, see also the section Irrigation, later in the Introduction.)

Tray Support Systems

The structure that supports the growing container (a flat or tray) corresponds with the type of irrigation used. For overhead irrigation, growers use the T-rail system for bench construction. On both sides of the central walkway are rows of square posts, about 1 m (3 feet) tall, topped with strips of extruded aluminum in the shape of an inverted T. The flats of containers are slid between (and supported by) pairs of T-rails.

For ebb and flow irrigation and bottom-watering facilities, growers use different tray support systems. Guide wires that extend through the walls of the trough are often used. These wires are anchored to I-beams outside the trough and tightened via turnbuckles. A series of such wires are spaced appropriately to prevent the growing trays from falling into the trough.

Other support structures, such as PVC pipe racks and cinder blocks and hog wire, can also work well in small systems. When designing an in-rack system, care should be taken to avoid the possibility of puncturing the liner.

Heating

Heat buildup within the greenhouse from sunlight during the day may be enough to provide sufficient warmth on most nights. Even so, it is important to have a source of heat should heating become necessary.

Portable propane or kerosene heaters can be used, but they do not provide uniform heat distribution throughout the greenhouse. An overhead unit, connected to a polyethylene tube for balanced heat distribution, is well worth the investment. Growers should not install an undersized heating unit in an effort to save money. They should also make sure that the heater is properly vented or that adequate fresh air is available to prevent ethylene levels from building up to damaging levels.

The minimum temperature for providing adequate plant protection during a freeze emergency is, of course, crop dependent. However, tomatoes can tolerate a temperature as low as 5°C for 2–3 h.

Seed

Growers should purchase the best-quality seed available. Using high-quality seed ensures good germination, rapid emergence, and vigorous growth. Poor-quality seed can have low germination or vigor and can result in smaller plant size, lack of uniformity within the tray, reduced overall greenhouse productivity, and thus decreased revenue. Poor-quality seed also results in "skips" or excessive seeding to compensate for poor stand establishment, both of which diminish profitability.

To reduce the risk of contamination by disease-causing organisms, growers should purchase only high-quality seed that is certified as disease free or treated. The cost generally ranges from 3 to 6 cents per seed to as much as 20 cents per seed for some greenhouse, grape, and cherry hybrids.

Primed Seed

Priming (enhancing) good seed can improve seed quality and vigor. Primed seed is biologically altered through the addition of just enough water or other liquid (and sometimes plant growth regulators) to allow the seed to undergo the early stages of germination without the radicle emerging from the seed coat. In the primed state, seeds germinate more rapidly and emerge more uniformly over a greater range of temperature, soil moisture, and various stresses. This results in greater vigor, increased uniformity, and more rapid establishment. The added cost for priming ranges from a few cents to several dollars per 1,000 seeds.

Priming technology was developed to overcome germination and emergence problems in field plantings caused by adverse conditions (e.g., cold soil, excessive soil moisture, etc.). Using enhanced seed in the greenhouse can be cost effective, as well, by providing plants of uniform size and physiological age when shipped. The rapidness of emergence means that production programs (fertilizers, pesticides, etc.) can be quickly implemented. Moreover, better production scheduling can be achieved because of the reduction in wasted space from poor germination and the reduced labor needed for thinning overseeded orders or combining flats to increase plant percentage in the tray.

Pelleted and Coated Seed

Pelletization is a technology that uses a material such as clay to enlarge the seed and to make it spherical for ease in mechanical seeding. Pelletizing increases the speed of seeding and the accuracy of singulation (i.e., placing one seed per cell) within the flat.

Seeding

The rotary drum, vacuum-type seeder is the most commonly used device for placing seed into tray cells. Even so, at least four alternative seeding methods may be used:

1. sowing seed into a flat and moving (pricking out) seedlings to a growing tray at the cotyledon stage
2. direct seeding to a final stand, which requires the use of high-quality seed and precision in all phases of the operation
3. overseeding into flats, followed by thinning to a single plant per cell
4. seeding into small-celled plug trays, followed by transplanting into larger-celled growing trays 2–3 weeks later

Seed Germination

Of the many factors that influence germination, water and temperature are most significant. Extremes in either factor can result in poor germination. However, different seed lots sown in the same environment may not perform alike.

For best results, a controlled temperature room should be used for germination. Such a room is insulated to allow for heating or cooling according to the specific germination needs of the seed. Using a controlled room reduces the fluctuations in temperature and moisture that can occur in an open house. Room design should allow forklift access for moving trays on pallets, if large-scale production is anticipated.

Temperature in the germination chamber is maintained at 25°C for 3 days. After that, trays are placed in the growing house, which should be partially shaded in summer. Plant-growing house design, open side walls, ventilators, shading, and fans are all used to reduce growing-house temperatures. In cooler weather, night temperatures are maintained at 15°C for 1 week after seeding or transplanting, followed by 13°C thereafter.

High Temperature Effects

The high soil temperatures (greater than 35°C) that occur in summer or fall production can suppress germination via a process called thermoinhibition. Seed affected by thermoinhibition will readily germinate when the soil temperature drops below the inhibitory level.

Planting Depth

In most commercial facilities, flats are filled mechanically with soilless mix; "dibbled," or pressed to a depth of approximately 0.6 cm (¼ inch); and seeded. Once seeded, the planting container may be top-dressed with vermiculite. Adding vermiculite reduces moisture loss from evaporation and prevents large shifts in media temperature, thus providing a more stable germination environment. Variable planting depth can result in problems such as uneven stands and "leggy" plants.

Media

Because of the difficulty of achieving the correct composition (i.e., peat, perlite, vermiculite, fertilizer, etc.) of a self-prepared medium, new and small growers are encouraged to use a commercial preblended, soilless mix. Using a medium in which these ingredients are distributed unevenly may affect germination and subsequent seedling growth and result in size variability among the transplants.

Soilless mixes that contain bark or peat moss with twigs are composed of large elements, which permit gaps in the cell (incomplete filling). Given this composition, such mixes are not appropriate for small-cell tomato transplant production. Also, sand should not be included in the medium because of the weight it adds. Commercial vegetable mixes can be obtained from local agricultural product supply houses.

Soilless mixes vary widely in composition. Many include a micronutrient "charge," and all contain a surfactant to improve their wetting capability. Wetting agents degrade rapidly (in 3–6 months), which means growers should order only enough mix to complete a given planting cycle. Most companies that produce soilless mixes will change their standard mix to conform to grower specifications. For example, some growers may want to have granular phosphorus, lime, or specific fertilizers added.

Some mixes allow growing "drier" or "wetter" according to growers' individual needs. Quantity and frequency of water application will change the physical (pore space) and chemical (pH) properties of the mix, which may or may not have a negative effect on plant growth. Strategies for crop management will likely need to change if another mix is used.

Containers

Tray or Flat Composition

Transplant production trays are made from expanded polystyrene (EPS) and various plastics. Containers made from materials other than EPS and plastic (e.g., clay, wood, peat, etc.) are not recommended for large-scale commercial production because of the high cost involved in shipping them, essentially. Growers should select trays based on the compositional strength required for their needs.

In many areas, transplants grown for local tomato production are shipped in the trays, and the trays are returned for reuse after the seedlings have been transplanted into the field. The quality of reused trays should be monitored for several reasons. One reason is that age cracks in older EPS trays allow fertilizer and disease organisms to penetrate and collect. These "pockets" of fertilizer and disease may lead to uneven growth or even plant death. Root damage may also occur when roots that penetrate the EPS in search of fertilizer are later "pulled." Vacuum-molded plastic inserts have recently become available to alleviate the age-crack phenomenon.

Drainage

The hole in the bottom of the tray should be large enough to allow good drainage when watering from overhead and to provide good capillarity (i.e., upward movement of water through the spaces within the soilless mix) when using ebb and flow irrigation. The physical properties of soilless mixes are extremely important when considering drainage. Over the 4- to 6-week time frame of transplant production, some soilless mixes compact. Compaction can lead to waterlogging of the medium, which will reduce the supply of oxygen reaching the roots. Lack of oxygen can lead to physiological problems, such as root-hair loss and spindly growth. Preventing these conditions requires good drainage within the plug.

Tray Cells

Different shapes of tray cells are available, including the inverted pyramid, the truncated cone, and the bullet, as well as assorted beveled shapes. In terms of the number of cells, the trend in most commercial transplant houses is toward having more cells per tray or having smaller cells, thus increasing the level of production without having to build more houses. In addition to maximizing the number of cells per tray, tray designers are also striving to maximize plug volume.

Cell size is generally a matter of economics, and the effects of cell size on yield in the field have been shown to vary. Despite these findings, larger cells are recommended for longer-cycle transplants (more than 5 weeks to fruit). When large-celled trays are used during a short growing season, root growth may be insufficient to completely fill the cell, and damage may occur when plugs are "pulled," as the soil falls away and exposes the roots.

Cleaning

Transplant trays should be cleaned and sterilized before reuse to avoid the buildup of soilborne pathogens. Adequate cleaning can be achieved by washing trays with soap and water, followed by a chemical dip in either a quaternary ammonium (2–5%) or a sodium hypochlorite (1–2%) solution for at least 20 min. Trays should then be rinsed in clean water and allowed to dry thoroughly before reuse. All traces of the disinfestant must be removed to avoid potential seedling damage later.

Steam sterilization at 70–82°C for 40–60 min is the most effective method of eliminating tray-borne pathogens, but purchasing sterilization equipment can be costly. In addition, steam sterilization causes the EPS bead structure to expand and separate, accelerating age cracks and resulting in fissures. The young roots that penetrate these fissures are often torn off during plug removal. The excessive temperatures and run times involved in the steaming process can even melt EPS trays. Cleaning EPS trays with steam sterilization will result in a tray life of four or five uses, whereas nonsteam cleaning methods will allow approximately 20 uses.

Tray Inserts

Tray deterioration caused by steam sterilization has prompted the design of vacuum-molded plastic tray inserts. These inserts fit inside a standard EPS tray, cost about $1.00 each, and have a life expectancy of approximately 2 years. Inserts can be steam sterilized in EPS trays without being damaged, provided that sterilization temperatures are not excessive.

Scheduling

Accurate schedules must be developed to ensure the orderly movement of plants through the transplant operation. Transplant producers usually specify a narrow window of delivery for plants, so each grower must develop a workable schedule to fit the production time frame. Here is a sample tomato plant-growing schedule for Florida growers:

Seeding Practice	Summer	Winter
Direct	4 weeks	7 weeks
Direct with thinning	4½ weeks	7½ weeks
Transplant	5 weeks	8 weeks

Time to the finished product can vary depending on seed quality, growing temperatures, and cell size of growing trays.

Culture and Management

Sanitation in and around the transplant house is essential to the production of high-quality transplants. Anything that may come in contact with growing trays should be treated with chlorine or bromine compound or another disinfestant labeled for this use.

Weeds in and around the transplant house should be mechanically or chemically controlled. (Very few herbicides are labeled for use inside houses, and if chemicals are used around houses, care should be taken to prevent drift into the houses.) Many weeds can harbor insects and other disease-causing agents that can cause problems in houses.

Employees that come in contact with any materials used in growing transplants should not be allowed to smoke in the house or during work hours. All employees should disinfect their hands with an approved disinfectant before working with plants, and they should disinfest their shoes before entering the house.

Irrigation

The volume and frequency of watering are influenced by production season and cell size. Knowing when to water is generally determined by experience. Water requirements change as plants grow and under varying weather conditions.

Seed should be kept moist but not soaking wet, as saturated plugs generally have less oxygen. (Oxygen is very important during germination.) Established transplants should generally be watered only when the surface of the medium is dry to the touch. Excessive watering produces succulent plants with restricted root growth.

Initial watering after seeding and seed covering consists of spraying with a hollow-cone nozzle to settle the covering, followed by a thorough watering with a larger nozzle in a watering tunnel. Trays should not be watered again until they are placed in transplant houses after spending 3 days in a seed-germination chamber. Water stress should be avoided during the early seedling development stage. During this growth stage, watering frequency increases from two or three times a day to five or six times. During the finishing stage, thorough watering is reduced to once per day and supplemented by light watering as needed.

Water Quality

Water quality is an often overlooked but extremely important factor in growing vegetable transplants. Water quality should be checked by a reputable lab before attempting transplant production and then at least yearly to ensure continued quality.

Water should have a pH of 6.0–7.0 and be low in organic particulate and dissolved salts. High levels of carbonates and bicarbonates in the water will result in rapid upward pH swings in the plug, which will eventually cause nutrient deficiencies (particularly iron and boron).

To reduce bacterial contamination, water may need to be chlorinated (1 ppm chlorine). Chlorination is difficult without a proper monitoring system, however, and over- and underchlorination can both be disastrous.

Overhead Irrigation

A standard technique in vegetable transplant production is to water until "run through" is achieved. This means to apply enough water so that each cell of the tray or flat has received sufficient water to allow some of it to drain out the bottom. This technique is intended to saturate each cell, which will provide for more uniform growth within the flat. (That is, growth differences caused by an inequality of water among neighboring cells will be reduced.)

Overhead irrigation should be done in the morning, if possible, to allow the foliage to dry before night. Prolonged periods of wet foliage encourage the development of disease.

Ebb and Flow Irrigation

With ebb and flow irrigation, the duration of tray contact with the water should be based on the wetting characteristics of the soilless mix. Standard practice is to allow water contact with the trays for 30–60 min and then hand weigh the trays to determine if they have become saturated. If more water is needed to reach saturation, the trays should continue to float. However, continuous flotation of vegetable transplants may result in loss of height control, as unrestricted access to water will result in rampant growth. Chlorination (1–2 ppm chlorine) is suggested for ebb and flow systems to reduce bacterial and fungal contamination in irrigation water.

Plant Height

Water management is the major technique used to control plant height, but too little or too much water will result in problems. Plastic trays tend to stay wetter than EPS trays, so controlling height in a plastic tray may require more attention. Under cloudy conditions, maintaining low soil moisture is preferred, as plants will tend to become "leggy" in low light conditions.

Root Pruning

In a house with an overhead irrigation system, when plants' roots get sufficiently long and grow out of the trays through the drainage holes, the roots prune naturally upon exposure to the air. Pruning occurs because air dehydrates or dries out the roots. Branch roots will then begin growing in the soilless mix of the plug.

In a house with a continuous float system, however, root growth goes unchecked. Sometimes, even if the water source is removed daily, the humidity of the air space beneath the tray may be high enough to support root growth. (To allow for good airflow to dissipate the humidity after drainage, the flotation troughs must be sufficiently deep (more than 20.3 cm).) The mat of roots that extends below the tray interferes with shipping and handling, and if it is mechanically removed, it will create numerous sites for the entry of disease-causing agents. Further damage may occur upon "pulling" the plants, resulting in greater dieback after transplanting. Given the potential for damage to the plants, use of a continuous float system is not advised for tomatoes.

Fertilization

The goal in tomato transplant production is to produce a sturdy, compact plant that, when transplanted, will establish and grow quickly and produce an optimum yield. To help achieve that goal, fertilization should begin only after the first true leaf has expanded from 1.3 to 1.9 cm in length.

Excessive fertilization, especially early in transplant growth, will result in "leggy" plants that may defoliate and establish slowly in the field. Minimal fertilization will result in stunted plants with retarded growth in the field. Proper fertilization directly affects field performance.

No fertilizer formula or prescribed application is available that covers all transplant crops or all environmental conditions. Growers generally develop their own blends and application schedules. Some growers prefer to have phosphorus, lime, starter nitrogen (N), and micronutrients provided in the soilless mix. Others prefer to dispense all nutrients through liquid fertilization with the intent of achieving greater control of plant growth. Good vegetable transplant crops can be produced using either method. The method selected should be determined by specific growing needs. For example, if leaching of fertilizer is necessary to slow excessive growth, then liquid fertilizers are preferable, as they are more easily washed out of a mix than granular forms.

Slow-release N sources (nitrate source is best) are not generally used for tomato transplant production, since plant height is difficult to control with constant N. A single application of a full-complement micronutrient package usually provides all of the elements required for seedling growth.

Fertilizers used for liquid applications should be 100% soluble in water. The proper rate of application depends on the nozzle size and trolley speed in an automatic watering system. Frequency of fertilizer application varies from grower to grower. Fertilizer added to the growing mix will be sufficient to carry the plants to the cotyledon expansion stage. Some systems deliver 250–400 ppm N once a week or 125 ppm N twice a week. However, 30 ppm N of a completely water-soluble fertilizer applied at every watering should meet most demands for crop nutrients. Generally, summer- and fall-grown transplants should receive more frequent but lower-rate feedings (i.e., 30 ppm), and winter- and spring-grown transplants should receive less frequent but higher-rate feedings. Fertilization is usually reduced during the finishing stage.

Fertilization, watering, temperature, and sunlight are all important factors in controlling plant height. The grower's ability to be flexible in scheduling fertilizer applications will usually be rewarded with plants of greater acceptability for commercial use.

Hardening

Most transplants will be sufficiently hardened for field setting if strict measures are applied to maintain plant height, particularly under Florida growing conditions. However, additional hardening can be accomplished by increasing the light intensity, reducing the irrigation inputs, or lowering the greenhouse temperatures 3–6°C. Experienced growers may manipulate or reduce fertilizer inputs to achieve hardening. This technique is not advised for novice growers because of the risk of losing quality. Overhardening can result in slow field establishment and reduced yield.

Growers should harden plants slowly by letting them acclimate to new conditions, beginning about 1 week before transplanting. Preparation of transplants for field planting is accomplished by reducing fertilizer applications to 50 ppm N for 2 weeks before the delivery date in winter and for 1 week before the delivery date in summer. Water application should be reduced concomitantly to one thorough watering per day, so that plants are slightly water stressed. In the winter, gradually lowering the greenhouse temperature will assist in the hardening process.

No legal chemical tool is available to growers for controlling plant height. Certain chemicals can be used to accomplish height control, but none is currently registered with the U.S. Environmental Protection Agency (EPA) for use on tomato.

Plant height can be controlled to a certain extent with the judicious use of water and fertilizer. Learning when to withhold water and fertilizer for height control takes experience, and making an error with either can greatly reduce the quality of plants.

Sanitation

Sanitation is paramount to successful pest management in a transplant production operation and should be practiced in every phase of production. Strict sanitary practices should be followed in the greenhouse and after crop removal from the house. Workers should not touch plants unless necessary. Alcohol spray bottles should be made available and used frequently to sterilize hands, if plants and flats must be handled. When filling flats, workers should use only clean tools and uncontaminated or sterilized materials. Growing areas should be sprayed with a bromine or chlorine solution (according to label specifications) after production to sanitize benches, walkways, and the like.

Transplant materials (e.g., containers, media, seed, trays, fertilizers, etc.) should be stored separately from pesticides and growth regulators, because vapors from these compounds can contaminate transplant items and be absorbed by EPS trays. Ventilation systems should be shut off and windows closed when field spraying occurs adjacent to the greenhouse (e.g., for weed control) to prevent herbicide drift and risk damaging the transplants.

Preventative applications of crop protection chemicals should be made on a regular basis to control plant diseases such as bacterial spot and speck, early and late blight, and target spot. Insecticides should be applied as needed to control whiteflies, leafminers, and worms. Selecting protective chemicals for use in plant-growing houses is challenging, because few are specifically registered for greenhouse use. Also, there is some ambiguity as to the difference between a *transplant house* and a *greenhouse* in determining which materials are and are not registered.

Shipping

Shipping factors—such as mechanical injury, environmental conditions, and length of storage—can affect plant vigor and establishment. Mechanical injury occurs most often when plants are pulled from trays and packed in boxes. Researchers have reported that tomato transplants shipped in trays yield more extra-large fruit than transplants pulled from trays and packed in boxes.

Densely packed transplants do not fare as well as loosely packed transplants in terms of both field survival and yield. Furthermore, plants held at 10°C have been shown to have greater field survival than plants held at 21°C or higher.

Exposure to ethylene can also damage vegetable transplants. Ethylene is a plant hormone that, when present at high levels, can cause leaf drop, root decay, senescence, and loss of chlorophyll. Sources of ethylene gas include the exhaust of propane-powered forklifts and heaters and senescing (overripe) fruits and plants.

Plants that will be "pulled" should be watered thoroughly the afternoon before the day of shipment. Plants that will be sold locally in returnable trays should be watered thoroughly the morning of the shipment day. A commercial antitranspirant is sometimes applied prior to shipping.

Tomato plants intended for distant locations are pulled and placed in well-ventilated, corrugated, waxed cartons. In summer, the plants are precooled to 10°C before being shipped. Truck shipments should maintain the same temperature and not mix in tomatoes with loads of ethylene-generating fruits. Air shipments should be made via direct flights whenever possible.

(Prepared by S. M. Olson)

Part I. Infectious Diseases

Diseases Caused by Fungi and Oomycetes

Alternaria Stem Canker

Alternaria stem canker is primarily a problem on tomato grown in the coastal regions of California. The disease first occurred in San Diego County in southern California in the early 1960s, and it was later found in other coastal counties. The disease is of limited distribution and is rarely found elsewhere in the state. It can still occur, however, if susceptible cultivars are used.

Symptoms

Alternaria stem canker is characterized by the formation of large, irregularly shaped, dark-brown to almost black cankers on stems near the soil line (Fig. 14). Stem cankers typically contain light and dark concentric zonation within their borders. Cankers continue to enlarge as the plant develops, resulting in girdled stems, stem death, and plant collapse. Vascular and pith tissue a few centimeters above and below the cankers may show brown streaking and become dry and cracked.

Toxins produced in the cankers move up through the plant and cause leaf necrosis. Affected leaves develop irregularly shaped, dark-brown to black areas of dead tissue, which are primarily interveinal (Fig. 15). Necrotic leaf areas may have yellow borders. Diseased plants may be stunted.

Fruit symptoms always begin on green, unripe fruit and appear as brown, circular to oval, sunken lesions, which also have concentric rings (Fig. 16). These lesions usually stop expanding when the fruit ripens. Symptomless fruit can develop these ring symptoms 3–5 days after harvest.

Alternaria stem canker on green fruit is distinct from the black mold caused by isolates of the *Alternaria* pathogen, which affects only ripe tomato fruit.

Fig. 14. Alternaria canker on a plant stem. (Courtesy J. C. Watterson)

Fig. 15. Alternaria stem canker on a tomato leaflet. (Courtesy J. C. Watterson)

Fig. 16. Alternaria stem canker on tomato fruit. (Courtesy J. C. Watterson)

Causal Organism

The pathogen that causes Alternaria stem canker is *Alternaria alternata* (Fr.:Fr.) Keissl. f. sp. *lycopersici* Keissl. The pathogen can be isolated on standard microbiological media. On potato dextrose agar, colonies are initially whitish and floccose. With age, colonies darken and become gray to almost black and show appressed, flattened growth. Conidiophores are light brown to olive and are septate. Conidia are light olive brown to dark brown and cylindrical and have three to five cross septa, longitudinal septa in the second and third cells, and a short, beaked apical cell on the terminal end. Conidia are produced in chains of three to five spores and measure 7–18 × 18–50 μm.

This fungus is pathogenic only to tomato and produces a host-selective toxin (HST), called AAL-toxin, that occurs as two analogs (TA and TB). This toxin is water soluble and moves systemically from stem cankers to leaves, causing interveinal necrotic lesions. The toxin is quite stable and not degraded by boiling (up to 20 min) or autoclaving (up to 10 min) at pH 3 or 7. Purified AAL-toxin alone can cause the symptoms, and toxin-deficient mutants of *A. alternata* f. sp. *lycopersici* do not cause symptoms in susceptible tomato cultivars. Tomato black mold and nonpathogenic isolates of *A. alternata* do not produce the AAL-toxin.

The Alternaria stem canker pathogen does not appear to be morphologically distinct from the *A. alternata* that causes black mold. It is interesting to note that different HST-producing isolates of *A. alternata* are pathogenic on and host specific to other crops, such as sorghum (*Sorghum* Moench), sunflower (*Helianthus* L.), wheat (*Triticum* L.), and perhaps cucurbits (family Cucurbitaceae).

Disease Cycle and Epidemiology

A. alternata f. sp. *lycopersici* can survive indefinitely in the soil, as it is an excellent soil-inhabiting saprophyte. Infection takes place when plants come in contact with infested residue or soil or when conidia are blown or splashed onto susceptible tomato plants. Cankers also occur around wounds, such as pruning cuts on pole tomato.

Spore germination and infection requires free moisture on the host tissue and occurs by direct penetration through the plant cuticle. Disease is more likely to occur if rain, heavy dew, or sprinkler irrigation is present. The optimum temperature for disease development is 25°C.

Management

Planting resistant cultivars essentially eliminates serious outbreaks of Alternaria stem canker. Resistance is achieved by a single gene (*Asc*) with two alleles, which confers complete protection against the *Alternaria* pathogen.

If susceptible cultivars are used, irrigating with overhead sprinklers should be avoided. Crop rotation is not an effective control strategy. Fumigants do not effectively eliminate the pathogen from the soil, and fungicides do not efficiently control the disease on susceptible cultivars.

Selected References

Akamatsu, H., Itoh, Y., Kodama, M., Otani, H., and Kohmoto, K. 1997. AAL-toxin-deficient mutants of *Alternaria alternata* tomato pathotype by restriction enzyme-mediated integration. Phytopathology 87:967-972.

Clouse, S. D., and Gilchrist, D. G. 1987. Interaction of the *asc* locus in F_8 paired lines of tomato with *Alternaria alternata* f. sp. *lycopersici* and AAL-toxin. Phytopathology 77:80-82.

Fuson, G. B., and Pratt, D. 1988. Effects of the host-selective toxins of *Alternaria alternata* f. sp. *lycopersici* on suspension-cultured tomato cells. Phytopathology 78:1641-1648.

Gilchrist, D. G., and Grogan, R. G. 1976. Production and nature of a host-specific toxin from *Alternaria alternata* f. sp. *lycopersici*. Phytopathology 66:165-171.

Grogan, R. G., Kimble, K. A., and Misaghi, I. 1975. A stem canker disease of tomato caused by *Alternaria alternata* f. sp. *lycopersici*. Phytopathology 65:880-886.

Morris, P. F., Connolly, M. S., and St. Clair, D. A. 2000. Genetic diversity of *Alternaria alternata* isolated from tomato in California assessed using RAPDs. Mycol. Res. 104:286-292.

Simmons, E. G. 1999. *Alternaria* themes and variations (236-243). Host-specific toxin producers. Mycotaxon 70:325-369.

Vakalounakis, D. J. 1990. *Alternaria alternata* f. sp. *cucurbitae*, the cause of a new leaf spot disease of melon (*Cucumis melo*). Ann. Appl. Biol. 117:507-513.

Wang, H., Li, J., Bostock, R. M., and Gilchrist, D. G. 1996. Apoptosis: A functional paradigm for programmed plant cell death induced by a host-selective phytotoxin and invoked during development. Plant Cell 8:375-391.

(Prepared by S. T. Koike, R. M. Davis, and A. O. Paulus)

Anthracnose

Anthracnose is primarily a disease of ripe and overripe tomato fruit. If left uncontrolled, it results in serious losses in fruit quality and yield. Anthracnose pathogens are widely distributed, and disease losses are common worldwide in areas with moist conditions, which favor disease development. Fruit on plants partially defoliated by leaf diseases are especially vulnerable to infection. Root infection (often called black dot root rot) sometimes causes serious losses, especially in greenhouses in Europe, Canada, and the United States. The anthracnose pathogens also are a component of the brown root rot complex in Great Britain.

Symptoms

Immature fruit can be infected with anthracnose, but symptoms do not appear until the fruit begin to ripen (Fig. 17). On ripe fruit, symptoms first appear as small, slightly depressed, circular lesions. A lesion may enlarge to 12 mm in diameter, become more sunken, and develop a concentric ring. The flesh beneath a lesion may be more lightly colored than the surrounding tissue and granular in texture. The center of a lesion is usually tan, and as the lesion matures, it becomes dotted with small, black specks (microsclerotia) formed by the pathogen. The surface of a mature lesion generally remains smooth and intact. During moist weather, masses of salmon-colored spores may be present on the lesion surface. Individual fruit lesions often coalesce to form blemishes that cover large areas. Infected fruit have a very short storage life and are often invaded by a wide variety of secondary organisms, including yeast and soft-rotting bacteria.

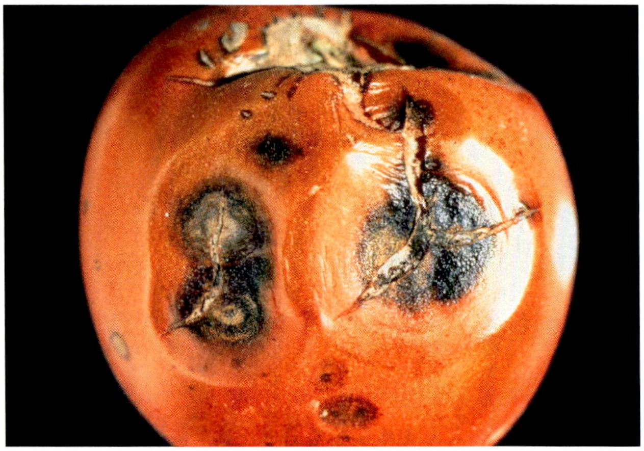

Fig. 17. Anthracnose on ripened fruit. (Courtesy W. R. Stevenson)

Infection may also occur on stems, leaves, and roots. Leaf infections are characterized by small, circular, brown lesions surrounded by yellow halos. Root symptoms become evident on plants about the time the fruit begin to ripen. Infected roots exhibit brown lesions, and as the roots begin to senesce, black microsclerotia form on their surfaces. Large roots are easily decorticated. Secondary roots are stunted, and many are nearly decomposed. Diseased plants can often be pulled easily from the soil because of their damaged root systems.

Causal Organisms

Anthracnose of tomato may be caused by several fungal species of the genus *Colletotrichum,* including *C. coccodes* (Wallr.) S. Hughes, *C. gloeosporioides* (Penz.) Penz. & Sacc., *C. dematium* (Pers.:Fr.) Grove, and others.

C. coccodes is most commonly associated with anthracnose symptoms on fruit and is the only species associated with black dot root rot. Fruiting bodies called acervuli are often present and have conspicuous black setae. Conidiophores are tapered to slightly clavate, occasionally branched, and 10–30 μm long. Conidia (phialospores) in the acervuli are pink en masse but hyaline individually. They are one celled, somewhat attenuated at the basal end and rounded at the apical end, slightly curved on occasion, and guttulate, measuring 3.0–7.5 × 17.5–22.0 μm. Numerous small, black microsclerotia (approximately 0.5 mm in diameter) form both in nature and in culture. A diversity of vegetative compatibility groups within this species may be related to pathogenic characteristics or specialization.

Disease Cycle and Epidemiology

C. coccodes is capable of attacking various crop and weed species (68 species in 19 families) and producing both acervuli and microsclerotia on infected tissue. *C. coccodes* survives from year to year as conidia and microsclerotia, often in association with plant debris, such as skins of infected tomato fruit and plant roots. The level of fruit infection is correlated with the amount and duration of rainfall.

Under wet and warm conditions, microsclerotia are both myceliogenic (i.e., germinating by producing hyphae) and sporogenous (i.e., germinating by producing conidia in acervuli on the surfaces of microsclerotia). In addition, secondary microsclerotia may be formed by the intertwining of hyphae produced by germinating primary microsclerotia. Primary microsclerotia are capable of several consecutive germinations during repeated cycles of wet and dry conditions. They survive for long periods in the soil under adverse weather conditions and produce large numbers of conidia during periods with moist conditions, which favor fruit infection.

Anthracnose fungi may infect fruit resting on the soil surface, and microsclerotia and conidia may also be splashed onto foliage and fruit. Once in contact with the fruit, conidia germinate and form appressoria, which become anchored to the fruit surface. The appressorium serves as a survival structure until an infection peg penetrates the fruit cuticle.

Anthracnose fungi also cause latent infections between the cuticle and epidermal tissues of immature tomato fruit. Latent infections are activated after exposure of the fruit to low temperatures or after fruit maturation and plant senescence. Once latent infections have been activated, symptoms develop rapidly. Ripe and overripe tomatoes are highly susceptible, and symptoms progress rapidly on them.

Root infection occurs primarily in greenhouses where inoculum levels are high and environmental conditions favor disease development. The continuous cropping of tomato provides an ideal scenario for buildup of the pathogen. Other conditions, such as soluble salt injury, may also predispose plants to root rot caused by this pathogen.

Conditions that favor plant infection include temperatures of 10–30°C (with an optimum of 20–24°C), together with free moisture. At temperatures of 16–28°C, appressoria are formed within as few as 6 h. At 15–30°C, the severity of anthracnose on fruit increases, as the duration of the dew period and inoculum density increase. The distribution of microsclerotia and conidia is favored by splashing rain and overhead irrigation.

Management

The fungi that cause anthracnose survive from crop to crop on infected plant refuse and in the soil. Lengthy crop rotations with nonsolanaceous crops are needed to significantly decrease levels of viable inoculum of *C. coccodes.* Overhead irrigation should be applied only as needed and never in excess. Weed control is very important, since the pathogen has a broad host range that includes many common weeds.

Fungicide sprays promote acceptable control if applied periodically from first fruit set to harvest. Fungicide treatment is especially important as the fruit begin to ripen. However, because latent infections can occur even on green fruit, spray coverage is needed for most of the growing period. Staking plants and mulching (e.g., with straw, plastic, etc.) helps to reduce losses from anthracnose. In home gardens, the destruction of infected fruit helps to prevent a buildup of inoculum in the soil.

Currently available tomato cultivars are mostly susceptible to fruit infection, but resistance seems to be controlled by multiple genes. Since the rate of lesion expansion appears related to fruit maturation, cultivars with a slow-ripening or nonripening gene appear most resistant.

The root rot phase of this disease may be controlled in the greenhouse by excluding infested soil, avoiding closed-recirculation irrigation/fertilization systems, pasteurizing the soil with steam or broad-spectrum fumigants, and preventing root injury, which may predispose roots to infection. For field control of root rot, broad-spectrum fumigants should be applied and root injury should be avoided.

Selected References

Barksdale, T. H. 1972. Resistance in tomato to six anthracnose fungi. Phytopathology 62:660-663.

Batson, W. E., and Roy, K. W. 1982. Species of *Colletotrichum* and *Glomerella* pathogenic to tomato fruit. Plant Dis. 66:1153-1155.

Blakeman, J. P., and Hornby, D. 1966. The persistence of *Colletotrichum coccodes* and *Mycosphaerella ligulicola* in soil with special reference to sclerotia and conidia. Trans. Br. Mycol. Soc. 49:227-240.

Fulling, B. A., Tigchelaar, E. C., and Latin, R. 1995. Integration of host resistance and weather-based fungicide scheduling for control of anthracnose of tomato fruit. Plant Dis. 79:228-233.

Fulton, J. P. 1948. Infection of tomato fruits by *Colletotrichum phomoides.* Phytopathology 38:235-246.

Kendrick, J. B., Jr., and Walker, J. C. 1948. Anthracnose of tomato. Phytopathology 38:247-260.

Last, F. T., and Ebben, M. H. 1966. The epidemiology of tomato brown root rot. Ann. Appl. Biol. 57:95-112.

Nitzan, N., Hazanovsky, M., and Tsror (Lahkim), L. 2002. Vegetative compatibility groups in *Colletotrichum coccodes,* the causal agent of black dot on potato. Phytopathology 92:827-832.

Raid, R. N., and Pennypacker, S. P. 1987. Weeds as hosts for *Colletotrichum coccodes.* Plant Dis. 71:643-646.

Sanogo, S., Pennypacker, S. P., Stevenson, R. E., and MacNab, A. A. 1997. Weather variables associated with infection of tomato fruit by *Colletotrichum coccodes.* Plant Dis. 81:753-756.

Sanogo, S., Stevenson, R. E., and Pennypacker, S. P. 2003. Appressorium formation and tomato fruit infection by *Colletotrichum coccodes.* Plant Dis. 87:336-340.

Schneider, R. W., Grogan, R. G., and Kimble, K. A. 1978. Colletotrichum root rot of greenhouse tomatoes in California. Plant Dis. Rep. 62:969-971.

Stommel, J. R., and Haynes, K. G. 1998. Inheritance of resistance to anthracnose caused by *Colletotrichum coccodes* in tomato. J. Am. Soc. Hortic. Sci. 123:832-836.

(Prepared by W. R. Stevenson and K. L. Pernezny)

Black Mold

Black mold occurs wherever tomatoes are grown and is primarily a disease of senescent fruit. Hence, the incidence of black mold is greatest when fruit are allowed to overripen on the vine. Processing tomatoes, which are often subject to this practice, are particularly susceptible. In addition to causing potential yield reductions of up to 15%, black mold can negatively impact fruit quality. Loads of fruit with infection levels beyond certain thresholds (determined by local industry standards) can be rejected, resulting in significant economic losses to growers.

Symptoms

Symptoms of black mold vary from small, superficial, brown flecks to large, sunken, black lesions that may develop on one-third or more of the fruit surface (Fig. 18). On ripe fruit, the lesions begin as water-soaked, firm, flattened areas. As the lesions age, the affected areas become sunken and tan or bleached. In warm, humid weather, the fungus produces a black, velvety layer of spores on the surfaces of lesions.

Causal Organisms

Black mold is caused primarily by the fungus *Alternaria alternata* (Fr.:Fr.) Keissl. and to a lesser extent by *Stemphylium botryosum* Wallr. Other fungi, such as *A. arborescens* E. G. Simmons, are occasionally associated with this disease.

Conidia produced by *A. alternata* are generally obclavate or obpyriform in shape, with a beak or apical cell that is shorter than the rest of the spore. The olive-green to black conidia usually possess both longitudinal and transverse septa and are often borne in chains. Conidia of *S. botryosum,* which are never borne in chains, are globose to broadly ellipsoidal and do not have a beaklike terminal cell. Both *A. alternata* and *S. botryosum* produce masses of dark spores on black mold lesions, creating similar signs and symptoms of the disease.

Disease Cycle and Epidemiology

The fungi that cause black mold are ubiquitous saprophytes of decaying plant debris and are often present on dead or senescent tomato leaves. Damage caused by these relatively nonaggressive fungi is generally limited to senescent or wounded fruit (e.g., sunscald, insect injury, blossom-end rot). The fungi may infect green fruit through wounds or cracks in the fruit cuticle, but infections remain latent until the fruit ripen.

Conidia of the causal fungi are dispersed by wind and splashing water. Because free water on the surface of the fruit is required for spore germination and infection, the disease is closely associated with heavy dews or periods of rainy weather. Increasing relative humidity and duration of fruit wetness are positively correlated with disease severity. At least 12 h of fruit wetness are needed for consistent development of disease, and maximum disease severity occurs at 100% relative humidity. The disease progresses most rapidly at temperatures of 24–28°C, although black mold can occur between 10 and 35°C.

Canopy coverage may be an important factor in disease development. Fruit exposed to the sun are also exposed to dew; hence, black mold and sunscald often occur simultaneously. Practices that encourage free water on the surface of fruit—such as using overhead irrigation and planting during times of the year when mature fruit are subject to rains—may promote disease incidence.

Fig. 18. Black mold, caused by *Alternaria alternata.* (Courtesy Jack Kelly Clark, University of California Statewide IPM Program)

Management

If mature fruit are not harvested in a timely manner, the use of fungicides may reduce disease incidence and severity. Fungicides should be applied before rain or dew promotes infection. For maximum effectiveness, applications should begin 5–6 weeks before harvest. Applications made later than 2 weeks before harvest will be of little value.

The susceptibility of fruit increases the longer they stay on the vines. In addition, delays in harvest increase the chance of fruit exposure to rain or dew. A dense leaf canopy protects fruit from dew deposits, so cultivars that have a good canopy may have a lower incidence of black mold. However, a dense canopy also retains high humidity, which may favor other fruit molds.

Selected References

Butler, E. E. 1959. Fungi and rots in California canning tomatoes. Plant Dis. Rep. 43:187-192.

Davis, R. M., Miyao, E. M., Mullen, R. J., Valencia, J., May, D. M., and Gwynne, B. J. 1997. Benefits of applications of chlorothalonil for the control of black mold of tomato. Plant Dis. 81:601-603.

Miyao, E. M., Hall, D. H., Somerville, P., and Blaker, N. 1986. Fungicidal control of tomato blackmold under rainy conditions. Calif. Agric. 40:7-8.

Pearson, R. C., and Hall, D. H. 1975. Factors affecting the occurrence and severity of blackmold of ripe tomato fruit caused by *Alternaria alternata*. Phytopathology 65:1352-1359.

(Prepared by R. M. Davis and A. O. Paulus)

Black Root Rot

Black root rot is an occasional disease of tomato that mostly affects transplants being grown in trays or young seedlings and transplants placed in production fields. Severe outbreaks of black root rot have occurred on tomato transplants produced in California; losses of 50–60% have been experienced, although such a high incidence is unusual. The disease apparently does not affect mature, established tomato plants.

Symptoms

The first noticeable symptoms of black root rot are infections of the root cortex that result in small (2–10 mm), discrete, dark-brown to black lesions on small feeder roots. These lesions later decay and the cortex breaks down, exposing the inner root stele. If root infections on young plants are extensive, tomato transplants and seedlings can show reduced root systems, poor growth, stunting, chlorosis, and foliar symptoms that resemble those of nutrient deficiencies and general stress.

Causal Organism

The causal agent of black root rot is *Thielaviopsis basicola* (Berk. & Broome) Ferraris (syn. *Chalara elegans* Nag Raj & W. B. Kendr.). This soilborne fungus has a broad host range and is best known as a pathogen of tobacco (*Nicotiana* L.), cotton

(*Gossypium* L.), and other crops. *T. basicola* is widespread in soils throughout the world.

On standard agar media, colonies of *T. basicola* are gray in color and often have a velvety texture. The fungus produces two kinds of conidia. Endoconidia are hyaline, thin walled, single celled, and cylindrical with truncate ends; they measure 3–6 × 7–17 μm. Slender phialides produce these spores endogenously, resulting in endoconidial chains that are pushed out from the tips of the conidiophores. Such chains are fragile and do not remain intact for long. *T. basicola* also produces aleurioconidia that are dark brown, thick walled, single celled, and subrectangular; they measure 5–8 × 10–16 μm. Aleurioconidia are formed in stacks or chains of five to eight spores, with the top spore in the stack having a dome-shaped apex. Aleurioconidia chains remain intact until spores reach maturity, at which point the chains break apart and the aleurioconidia become separated.

There is evidence that *T. basicola* is nutritionally a hemibiotroph with a predominant biotrophic phase. The fungus apparently has limited activity as a saprobe in the soil.

Disease Cycle and Epidemiology

T. basicola survives in soil in the form of aleurioconidia for extended periods (more than 1 year) if soil conditions are dry, but under moist soil conditions, survival is greatly reduced (7 months or less). Black root rot is typically a cool temperature disease, with optimum temperatures for infection and disease development ranging from 15 to 20°C.

In transplant facilities, *T. basicola* has been recovered from peat-based rooting media; such materials may therefore be the source of primary inoculum for the transplant phase of the disease. The adults and larvae of shore flies (*Scatella stagnalis* Fallen) can be internally infested with *T. basicola* aleurioconidia and vector the pathogen between plants in greenhouses.

Management

To manage black root rot in transplant facilities, clean peat moss that is not infested with the pathogen should be used. In addition, bench tops and transplant trays must be properly sanitized, including empty trays that are returned from a field planting (which may have been exposed to infested soil). The shore fly population must be kept under control, as well.

Because black root rot is a minor disease in production fields, control measures in the field are not currently warranted. Even so, useful management methods include soil fumigation, planting when soil temperatures are greater than 16°C, and planting tomatoes in fields that do not have a history of problems with *T. basicola*. Soil populations of *T. basicola* are not significantly reduced by the presence of nonhost crops, which means crop rotation might have limited benefits in controlling the disease.

Selected References

Hood, M. E., and Shew, H. D. 1997. Initial cellular interactions between *Thielaviopsis basicola* and tobacco root hairs. Phytopathology 87:228-235.

Hood, M. E., and Shew, H. D. 1997. Reassessment of the role of saprophytic activity in the ecology of *Thielaviopsis basicola*. Phytopathology 87:1214-1219.

Hsi, D. C. H. 1978. Effect of crop sequence, previous peanut blackhull severity, and time of sampling on soil populations of *Thielaviopsis basicola*. Phytopathology 68:1442-1445.

Luttrell, E. S. 1946. A pycnidial strain of *Macrophomina phaseoli*. Phytopathology 36:978-980.

Mathre, D. E., and Ravenscroft, A. V. 1966. Physiology of germination of chlamydospores and endoconidia of *Thielaviopsis basicola*. Phytopathology 56:337-342.

Stanghellini, M. E., Rasmussen, S. L., and Kim, D. H. 1999. Aerial transmission of *Thielaviopsis basicola*, a pathogen of corn-salad, by adult shore flies. Phytopathology 89:476-479.

Tabachnik, M., DeVay, J. E., Garber, R. H., and Wakeman, R. J. 1979. Influence of soil inoculum concentrations on host range and disease reactions caused by isolates of *Thielaviopsis basicola* and comparison of soil assay methods. Phytopathology 69:974-977.

(Prepared by S. T. Koike and K. L. Pernezny)

Buckeye Rot and Phytophthora Root Rot

Buckeye fruit rot is reported in all parts of the world with high relative humidity, abundant soil moisture, and warm weather—the conditions that favor fruit infection. In the United States, the disease is most common in the southeastern and south-central areas. Phytophthora root rot is a major disease of tomato in California.

Symptoms

The first fruit symptom of buckeye rot appears as a brownish spot, often at the point of contact between the fruit and soil. Either green or ripe fruit may be infected through the uninjured skin. As the spot enlarges, the surface of the lesion assumes a pattern of concentric rings of narrow, dark-brown bands and wide, light-brown bands (Fig. 19). The rings are similar to those in fruit lesions that occur in soil rot and late blight. Lesions of buckeye rot resemble those of late blight, except that buckeye rot lesions remain firm and smooth, whereas late blight lesions become rough and are slightly sunken at the margins. Buckeye rot lesions may cover half or more of the fruit. The margins of lesions are smooth but not sharply defined. Initially, infected fruit remain firm, but eventually, the decay progresses rapidly. Under moist conditions, a white, cottony fungal growth appears on the lesion. The pathogen does not affect the foliage.

Symptoms of Phytophthora root rot on plants with reduced vigor include water-soaked lesions, which gradually dry and turn dark brown. The root xylem may also turn brown above a lesion, and the discoloration may extend into the lower part of the stem. Severely infected roots may be girdled by lesions, and the decay may be extensive.

Causal Organisms

Buckeye rot of tomato fruit is caused by three species of *Phytophthora: P. nicotianae* var. *parasitica* (Dastur) G. M. Waterhouse, *P. capsici* Leonian, and *P. drechsleri* Tucker. Phytophthora root rot is attributed to the first two species. *Phytophthora* spp. have a broad host range and are reported on a wide range of crop and weed species.

Fig. 19. Buckeye rot, caused by *Phytophthora nicotianae* var. *parasitica*. (Courtesy W. R. Stevenson)

Disease Cycle and Epidemiology

Buckeye rot is most common during periods of prolonged warm, wet weather. The fungi are spread by surface water, spattering rain, and furrow irrigation. Excess soil moisture and a soil temperature of 18–30°C are essential for the development of widespread losses.

Rot is often observed in fruit in contact with soil. Infection also occurs in association with soil residue where mud containing inoculum has been splashed onto the fruit by rainfall or overhead irrigation.

Phytophthora root rot is most damaging to tomato planted in compacted soils or soils with poor drainage. Soil saturation for 5 h or longer is sufficient to promote infection by these soilborne fungi.

Management

Tomato should not be planted in fields with heavy, poorly drained soils. Practices that reduce or alleviate soil compaction may improve management, as may growing plants in raised beds. Low areas in fields are particularly prone to heavy losses from buckeye fruit rot and Phytophthora root rot.

Rotation with crops other than those belonging to the tomato family aids control, although *P. capsici* also attacks cucurbits (family Cucurbitaceae) and beans (*Phaseolus* spp.). Staking and mulching (e.g., with plastic, newspapers, straw, etc.) also help to reduce losses.

Solarization and soil fumigation help to reduce the soilborne density of inoculum. Fungicides applied to fruit and foliage for late blight control also assist in the control of buckeye fruit rot.

Selected References

Café-Filho, A. C., and Duniway, J. M. 1995. Dispersal of *Phytophthora capsici* and *P. parasitica* in furrow-irrigated rows of bell pepper, tomato and squash. Plant Pathol. 44:1025-1032.

Chellemi, D. O., Olson, S. M., and Mitchell, D. J. 1994. Effects of soil solarization and fumigation on survival of soilborne pathogens of tomato in northern Florida. Plant Dis. 78:1167-1172.

Fulton, J. P., and Fulton, N. D. 1951. Phytophthora root rot of tomatoes. Phytopathology 41:99-103.

Kreutzer, W. A., and Bryant, L. R. 1946. Certain aspects of the epiphytology and control of tomato fruit rot caused by *Phytophthora capsici* Leonian. Phytopathology 36:329-339.

Obreno, R. C., and Aragaki, M. 1965. Some factors influencing infection and disease development of *Phytophthora parasitica* Dast. on tomato. Plant Dis. Rep. 49:327-331.

Sherf, A. F., and MacNab, A. A. 1986. Vegetable Diseases and Their Control. John Wiley & Sons, New York.

Statewide Integrated Pest Management (IPM) Program. 1998. Integrated Pest Management for Tomatoes. 4th ed. Publ. 3274. Statewide IPM Program, Agriculture and Natural Resources, University of California.

(Prepared by W. R. Stevenson)

Cercospora Leaf Mold

Cercospora leaf mold (also called tomato black mold) can be a serious foliar disease of tomato in the warm, humid environment of the tropics. The disease is widespread in Asia, having been reported on tomato in China, India, Japan, Malaysia, the Philippines, Taiwan, Thailand, and Vietnam. It is found less frequently in Mexico, Nigeria, and the southern United States. Yield reductions of up to 34% have been documented for tomato cultivars susceptible to the disease.

Symptoms

Cercospora leaf mold begins with the development of small lesions in the interveinal leaf areas. The lesions are first noticed as light-green to pale-yellow patches without distinct margins (Fig. 20). As the lesions develop, they appear as light-green zones surrounding light-gray to black centers. More advanced lesions remain entire and appear as uncoalesced brown areas (Fig. 21).

Profuse black sporulation develops first on the lower leaf surfaces. At advanced stages of the disease, the leaves may twist and curl upward, revealing dense, black mold on the lower surfaces (Fig. 22). The upper leaf surfaces can also support sporulation under favorable conditions.

The fungus infects primarily young leaf tissue, although elongated lesions also may appear on leaf petioles and pedicels of flowers and fruit. Foliar blight of the plant can develop, with disease progressing from the older to the younger leaves. Infected leaves generally remain on the plant, giving it a sooty appearance. No symptoms are seen on fruit.

Causal Organism

The causal organism of Cercospora leaf mold is *Pseudocercospora fuligena* (Roldan) Deighton. The fruiting of the fungus is effuse, hypophyllous, and brown, with small, angular colonies of fasciculate, spreading stalks (two to seven or more) with few or no stromata. Conidiophores (3.5–5.0 × 25.0–70.0 µm)

Fig. 20. Early symptoms of Cercospora leaf mold. (Courtesy C. H. Blazquez)

Fig. 21. Cercospora leaf mold affecting the upper surfaces of tomato leaflets. (Courtesy The World Vegetable Center)

are pale to medium brown, septate, not branched, and slightly geniculate. Conidia (3.5–5.0 × 15.0–120.0 µm) are obclavate to cylindro-obclavate, straight to mildly curved, subhyaline, and indistinctly multiseptate and have round to long obconic bases and conical tips.

Disease Cycle and Epidemiology

The fungus that causes Cercospora leaf mold can survive on infected plants in abandoned tomato fields or on alternative host plants. In addition to tomato, black nightshade (*Solanum nigrum* L.) and other solanaceous crops—such as sweet pepper and chili pepper (*Capsicum annuum* L.) and eggplant (*S. melongena* L.)—can support infection and sporulation of the fungus.

The spores are disseminated by wind, splashing rain, machinery, and worker activity. The optimum temperature for development of the fungus is 28°C, and under warm conditions, infection can occur within 72 h. Free water or dew on the leaves is critical for penetration of the fungus, which occurs through the stomata. Following infection, lesions can be seen within 6 days, and the first black conidia emerge through the stomata on the undersides of leaves after 12 days.

Management

Volunteer tomatoes and abandoned tomato fields are frequently reservoirs for *P. fuligena*. All crop residue should be plowed under immediately after the final tomato crop harvest. Solanaceous weeds, such as black nightshade, should be eliminated from around and within the field to prevent buildup of the fungus, and tomato crops should be planted some distance from older crops of pepper and eggplant.

A number of fungicides have been found effective in controlling Cercospora leaf mold if applied on a regular schedule during extended periods of dew or wet foliage. Those products that are most resistant to the disease should be applied whenever possible.

Selected References

Blazquez, C. H., and Alfieri, S. A., Jr. 1974. Cercospora leaf mold of tomato. Phytopathology 64:443-445.

Mohanty, U. N., and Mohanty, N. N. 1955. Cercospora leaf mold of tomato. Sci. Cult. 21:269-270.

Roger, L. 1936. Quelques champignons exotiques nouveaux ou peu connus. II. (In French.) Bull. Trimest. Soc. Mycol. Fr. 52:80-84.

Roldan, E. F. 1938. New or noteworthy lower fungi of the Philippine Islands. II. Philipp. J. Sci. 66:1-7.

Wang, T., Hartman, G. L., Hsieh, W. H., and Black, L. L. 1995. Reactions of Solanaceous species to *Pseudocercospora fuligena*, the causal agent of tomato black leaf mold. Plant Dis. 79:661-665.

(Prepared by J. C. Watterson)

Fig. 22. Cercospora leaf mold affecting the lower surfaces of tomato leaflets. (Courtesy The World Vegetable Center)

Charcoal Rot

Charcoal rot has been reported as a disease of tomato at high temperatures (32°C or higher). The pathogen, *Macrophomina phaseolina* (Tassi) Goid. (syn. *M. phaseoli* (Maubl.) S. F. Ashby), is a polyphagous species in the tropics and warm, temperate areas. Several other crops, particularly legumes, appear to be more suitable hosts for *M. phaseolina* than tomato.

The pathogen can cause black, sunken cankers below the cotyledonary node at the time of emergence. Mature plants that are attacked will have stunted growth. They may also wilt, and their lower leaves will likely become chlorotic. When stems are cut open, the interior is gray to black because of the presence of small, black sclerotia (approximately 0.5 mm in diameter).

M. phaseolina produces pycnidia in diseased stems. Pycnidia (100–200 µm in diameter) are membranous to subcarbonaceous, erumpent, and globose and have inconspicuous truncate ostioles. Conidiophores (10–15 µm in length) are simple and rod shaped. Conidia (6–9 × 16–29 µm in size) are one celled, hyaline, thin walled, and elliptical to oval.

(Prepared by K. L. Pernezny)

Corky Root Rot

Corky root rot (also known as brown root rot) is a soilborne disease of tomato that has been reported in European greenhouses since the early 1960s. Yield losses of up to 75% have been noted in European greenhouses. The disease has subsequently been reported in greenhouses in the U.S. state of Massachusetts and in field-grown plants in the states of Florida and California.

Symptoms

The first indication of corky root rot is the appearance of unthrifty plants in localized areas of plantings. Affected plants may wilt. Other symptoms include interveinal and peripheral chlorosis and premature defoliation.

Root volume is also decreased. When roots of mature plants are examined, three distinct symptoms are seen. Small, feeder roots are completely rotted away. Small roots may also show smooth, brown lesions (hence, the name brown root rot). Older, larger roots exhibit the typical corky lesions, which are characterized by dry, furrowed bark (Fig. 23).

Causal Organism

Corky root rot is caused by *Pyrenochaeta lycopersici* R. Schneid. & Gerlach. For a long time, the pathogen was referred to as the gray sterile fungus (GSF), because it did not sporulate well in culture. On general purpose media, the fungus is slow growing and medium to dark gray. It competes poorly with saprophytic fungi, making routine isolations difficult.

Isolation and identification of *P. lycopersici* have been enhanced by the development of selective media and in vitro sporulation techniques. In particular, the CRM medium of Grove and Campbell allows for recovery of *P. lycopersici* colonies with little contamination problem. Suspect colonies can subsequently be transferred to double-strength V8 juice agar, incubated as described, and examined for fruiting structures.

The fungus produces setose pycnidia and one-celled, hyaline conidia (2.0 × 4.5 µm) on ramose conidiophores. Microsclerotia (63.5 × 448.0 µm) are formed in infected roots.

Disease Cycle and Epidemiology

P. lycopersici survives in soil as microsclerotia. In California, the fungus survives between cropping seasons as microsclerotia in buried tomato root tissue. Potential alternative hosts include pepper (*Capsicum annuum* L.), melon (*Cucumis* L.), eggplant (*Solanum melongena* L.), squash (*Cucurbita* L.), spinach (*Spinacia oleracea* L.), and safflower (*Carthamus tinctorius* L.).

Corky root rot is a cool weather disease. Tomatoes grown during the warm months can conceivably escape the disease. Disease development is optimal at 15–20°C and declines above 28°C (for strains from temperate regions).

P. lycopersici is probably moved short distances by cultivation and other farm practices that transport soil. Corky root rot has also been shown to be less severe in soils on organic farms than on conventional farms.

Management

The most practical management of corky root rot is based on the application of broad-spectrum sterilants to the soil before planting. Steam is often used in the greenhouse, and fumigants provide excellent control in field tomatoes, when fumigation is economically practical. The hardiness of the microsclerotia and the relatively wide host range likely preclude crop rotation as a practical management strategy in intensively farmed areas.

Commercial cultivars of both processing and fresh-market tomatoes are susceptible to corky root rot. Several plant introductions are known to have varying levels of resistance, and crosses to commercial lines have shown a polygenic resistance controlled by four to eight genes. High soil and tissue levels of nitrogen (especially nitrate N) are associated with increased corky root rot severity.

Selected References

Campbell, R. N., Schweers, V. H., and Hall, D. H. 1982. Corky root rot of tomato in California caused by *Pyrenochaeta lycopersici* and control by fumigation. Plant Dis. 66:657-661.

Ebben, M. H., and Williams, P. H. 1956. Brown root rot of tomatoes. I. The associated fungal flora. Ann. Appl. Biol. 44:425-435.

Grove, G. G., and Campbell, R. N. 1987. Host range and survival in soil of *Pyrenochaeta lycopersici*. Plant Dis. 71:806-809.

Hockey, A. G., and Jeves, T. M. 1984. Isolation and identification of *Pyrenochaeta lycopersici,* causal agent of tomato brown root rot. Trans. Br. Mycol. Soc. 82:151-152.

Last, F. T., and Ebben, M. H. 1966. The epidemiology of tomato brown root rot. Ann. Appl. Biol. 57:95-112.

Last, F. T., Ebben, M. H., Rothwell, J. B., and Jones, D. A. G. 1968. Effects of cultural treatments on the incidence of and damage done by tomato brown root rot. Ann. Appl. Biol. 62:55-75.

Manning, W. J., and Vardaro, P. M. 1974. Brown root rot of tomato in Massachusetts. Plant Dis. Rep. 58:483-486.

Preece, T. F. 1964. Observations on the corky root disease of tomatoes in England. Trans. Br. Mycol. Soc. 47:375-379.

Shishkoff, N., and Campbell, R. N. 1990. Survival of *Pyrenochaeta lycopersici* and the influence of temperature and cultivar resistance on the development of corky root of tomato. Plant Dis. 74:889-894.

Volin, R. B., and McMillan, R. T., Jr. 1977. Inheritance of resistance to *Pyrenochaeta lycopersici* in tomato. Euphytica 24:75-79.

Workneh, F., and van Bruggen, A. H. C. 1994. Suppression of corky root of tomatoes in soils from organic farms associated with soil microbial activity and nitrogen status of soil and tomato tissue. Phytopathology 84:688-694.

Fig. 23. Corky root rot lesions. (Courtesy R. M. Davis)

(Prepared by K. L. Pernezny and R. B. Volin)

Didymella Stem Rot

Didymella stem rot is a major disease of greenhouse tomatoes in western Europe, and it occurs occasionally on tomatoes grown outdoors, as well. This disease has also been reported in Canada, India, Morocco, New Zealand, and Russia, and it has been reported but not confirmed in the United States. Following artificial inoculation, a number of solanaceous plants can be infected with the fungus, including pepper (*Capsicum annuum* L.), eggplant (*Solanum melongena* L.), potato (*S. tuberosum* L.), black nightshade (*S. nigrum* L.), and apple of Sodom (*S. sodomaeum* L.).

Symptoms

Infection usually occurs on the stem at or above the soil line, but all aboveground parts of the tomato plant can be infected (Fig. 24). The initial symptom of Didymella stem rot is usually a dark-brown, sunken lesion near the soil level (Fig. 25), which may eventually girdle the stem. Next, the lower leaves yellow and wilt, and as the wilting progresses, the plant may die. Numerous black pycnidia, which are visible to the human eye, form on the stem lesions. Fungal spores from the pycnidia may be splashed onto other parts of the plant and cause lesions higher on the stem and in leaf axils.

Infection of tomato fruit is rare in the greenhouse but common outdoors. Fruit infection starts as water-soaked lesions that enlarge rapidly and become black and sunken with concentric rings. Lesions may form anywhere on the fruit but are

Fig. 24. Didymella stem rot on a plant in the field. (Courtesy H. Dankers)

generally found at the stem end. Pycnidia form abundantly on the lesions, sometimes causing the entire fruit to mummify.

Leaf infections begin as small spots and develop into brown lesions with concentric rings. The center of an individual lesion generally becomes lighter in color, and a few pycnidia may develop. Infected leaves may die or individual lesions may drop out, producing a shot-hole effect.

Causal Organism

The causal organism of Didymella stem rot is *Didymella lycopersici* Kleb. (anamorph *Phoma lycopersici* Cooke). The pycnidia are 100–270 µm in diameter, and the pycniospores are 2.5–5.0 × 4.5–17.0 µm, hyaline, and unicellular to uniseptate. The perfect stage was reported in 1944 on tomatoes grown outdoors in England but is rarely found. The perithecia are subglobose; the asci are cylindrical and 9–10 × 70–95 µm; and the ascospores are 5.5–6.5 × 16.0–18.0 µm, spindle shaped, hyaline, and uniseptate.

Growth of the fungus on agar occurs from 4 to 32°C. The optimum temperature is 20°C.

Disease Cycle and Epidemiology

D. lycopersici subsists between crops chiefly on infected plant debris and organic matter in the soil. It also survives on the seed, on stakes used to support plants, and in some solanaceous plants. Pycnidia are numerous on old, dead stems, and when moisture is present, long tendrils of spores exude in a gelatinous matrix. The spores are splashed onto plants by rain or overhead irrigation, and they can be carried to other plants on tools, hands, and clothing.

The fungus penetrates through wounds caused by routine trimming and other activities. Penetration also occurs directly or through stomata. Old plants are more susceptible to infection than young ones. Low levels of nitrogen and phosphorus increase susceptibility of the host.

Infection occurs at temperatures as low as 13°C and as high as 29°C, with an optimum of 20°C. The fungus is killed within 5 min at 44–48°C. A medium level of soil moisture is more favorable for development of the fungus than either extreme.

Management

Several strategies can be used to manage Didymella stem rot. Pathogen-free seed should be used, and infected plant debris (i.e., the main source of inoculum) should be eliminated. Inoculum can also be reduced or eliminated through long rotation, destruction of alternative hosts, and chemical or steam treatment of seedbed soils or mixes and tomato stakes. Soil and stake sterilization by solarization or other means is recommended, as well. When growing tomatoes in the greenhouse, overhead irrigation should be avoided and adequate ventilation should be provided. Effective fungicides can be applied to plants, but no resistant cultivars are available.

Fig. 25. Symptoms of Didymella stem rot on young tomato plants. (Courtesy J. C. Watterson)

Fruit should be harvested carefully when dry to avoid bruising. Harvesting dry fruit also helps avoid developing a film of moisture, which is necessary for invasion by the fungus.

Selected References

Brooks, F. T., and Searle, G. O. 1921. An investigation of some tomato diseases. Trans. Brit. Mycol. Soc. 7:173-197.

Fagg, J., and Fletcher, J. T. 1987. Studies of the epidemiology and control of tomato stem rot caused by *Didymella lycopersici*. Plant Pathol. 36:361-366.

Farr, D. F., Bills, G. F., Chamuris, G. P., and Rossman, A. Y. 1989. Fungi on Plants and Plant Products in the United States. American Phytopathological Society, St. Paul, MN.

Hickman, C. J. 1944. The perithicial stage of *Didymella lycopersici*. Nature 154:708.

Hickman, C. J. 1945. Infection of outdoor tomato crops by *Didymella lycopersici*. J. Pomol. Hort. Sci. 22:69-75.

Morgan-Jones, G., and Burch, K. B. 1988. Studies in the genus *Phoma*. XI. Concerning *Phoma lycopersici*, the anamorph of *Didymella lycopersici*, causal organism of stem canker and fruit rot of tomato. Mycotaxon 32:133-142.

Sheard, E. 1943. A stem rot of tomato caused by *Didymella lycopersici*. Ches. Exp. Res. Stn. Ann. Rep. 29:33-37.

Williams, P. H., Sheard, E., and Read, W. H. 1953. Didymella stem rot of the tomato. J. Hortic. Sci. Biotechnol. 28:278-294.

(Prepared by T. M. Momol, P. Ji, and J. P. Jones)

Early Blight

Early blight occurs wherever tomatoes are grown. In the United States, early blight is damaging in the northeastern, mid-Atlantic, midwestern, and southeastern states but is rarely significant in the Pacific Coast states. The disease occurs in humid climates and in semiarid climates where frequent dews provide sufficient moisture to permit development. When the disease develops as collar rot in plant production houses or seedbeds, it can result in severe plant loss. Uncontrolled, early blight may cause severe defoliation, resulting in reduced number, size, and quality of fruit.

Symptoms

Early blight occurs on the foliage, stem, and fruit of the tomato plant and can cause severe damage during all stages of plant development. Symptoms are first observed in the field as small, brownish-black lesions on the older foliage, which often touches bare ground (Fig. 26). The tissue surrounding a lesion may become yellow, and when spotting is abundant, the entire leaf may become yellow. The lesions enlarge rapidly, and by the time they are 6 mm or larger in diameter, the dark-brown portions may be encircled by concentric rings (Fig. 27). In some instances, when a lesion is located on one of the primary leaf veins, the area beyond the lesion soon dies and becomes brown. In the latter part of the season, the lesions become numerous, and under favorable conditions for disease development, affected plants become defoliated, exposing the fruit to sunscald. Generally, the lesions appear on the older leaves first, and the symptoms progress up the plant during the season.

Stem lesions on seedlings are small, dark, and slightly sunken. They enlarge to form circular or elongated lesions with pronounced concentric rings and light centers (Fig. 28). If stem-infected seedlings are set in the field, the lesions continue to enlarge at the ground line and eventually girdle the plants. These plants often die, and if they do survive, their growth and yield are reduced. This phase of the disease is often referred to as collar rot.

The fruit become infected, generally through the calyx or stem attachment, in either the green or the ripe stage. The fruit lesions attain considerable size, occasionally involving the entire fruit, and they usually form concentric rings (Fig. 29). The diseased areas appear leathery and may be covered by velvety masses of black spores. Infected fruit frequently drop, and losses of 30–50% of immature fruit may occur.

Causal Organism

Traditionally, both early blight of tomato and early blight of potato were considered to be caused by the same species: *Alternaria solani* Sorauer (syns. *A. solani* (Ellis & G. Martin) L. R. Jones & Grout and *Macrosporium solani* Ellis & G. Martin). Recently, however, it was proposed that early blight of tomato is caused by another species: *A. tomatophila* E. G. Simmons. *A. tomatophila* has distinguishable morphological and cultural characteristics, chiefly in the width, length, and number of beaks. *A. tomatophila* is also more virulent on both stems and leaves of tomato than *A. solani,* although both are able to infect tomato and potato (*Solanum tuberosum* L.).

The mycelia are septate and branched and become dark with age. The conidia are beaked, muriform, dark, and borne singly or in chains of two (in pure culture), and they are of variable lengths. The conidia form from buds that arise from the terminal cells of conidiophores. The bodies of mature spores are generally 12–20 × 83–117 µm, although the literature documents conidia up to 300 µm in length. The length of the beak typically exceeds that of the body and may be more than twice as long (75–217 µm). Multiple beaks are common, as well; two, three, and even four beaks may be observed.

In culture, the fungus sporulates poorly or moderately on V8 juice agar, which is a good choice for diagnostic work. A large amount of pale-gray aerial mycelium is also produced. If the mycelium is scraped away, the solid medium beneath is dark gray to black. Scraping also induces much more abundant sporulation. The spores are initially produced in concentric bands, following the photocycle of the room in which they are grown. Darkness is essential for sporulation.

Disease Cycle and Epidemiology

A. tomatophila and *A. solani* survive between crops on infected debris in the soil and on seed. In locations with mild weather conditions, the fungi can survive from season to season on volunteer tomato plants and other solanaceous hosts, such as potato, eggplant (*S. melongena* L.), horsenettle (*S. carolinense* L.), and black nightshade (*S. nigrum* L.).

Primary infection is generally caused by fungi in the soil and occurs during periods of mild (24–29°C), rainy weather.

Fig. 26. Foliage symptoms of early blight. (Courtesy J. P. Jones)

Fig. 27. Concentric rings of early blight lesions on a tomato leaflet. (Courtesy W. R. Stevenson)

Fig. 28. Zonate lesions of early blight on a plant stem. (Courtesy J. P. Jones)

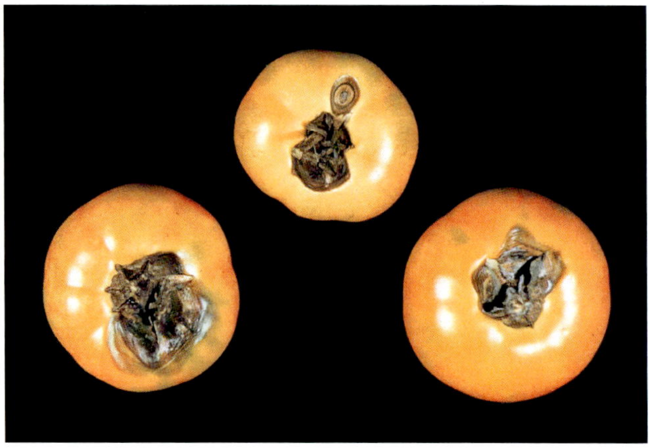

Fig. 29. Zonate lesions of early blight on tomato fruit. (Courtesy J. P. Jones)

However, in warm climates, the disease may be active at higher temperatures. Conidia germinate within 2 h in water at temperatures of 6–34°C and within 35–45 min at the optimum temperature of 28–30°C.

The fungi penetrate directly through the cuticle or through wounds. Under favorable conditions, lesions become visible in 2–3 days. Conidia form when leaf spots are approximately 3 mm in diameter. Heavy dews or frequent rains are necessary for abundant sporulation. The conidia are disseminated by wind.

High soil fertility reduces the severity of early blight of tomato. Older leaves are usually affected first; younger leaves are affected later, as they attain a certain physiological maturity.

Management

Genetic sources of resistance to early blight have been identified within wild species of tomato. Foliar resistance derived from breeding line PI 126445 was used to develop line NC EBR-1. Line C1943, released by the Campbell Institute for Agricultural Research, was used in the development of NC EBR-2, a line with a high level of stem lesion resistance. Both breeding lines were subsequently used in the development of Mountain Supreme, a fresh-market cultivar with moderate resistance to foliar blight and a high level of resistance to the stem lesion phase of the disease. Despite these advances, breeders are still trying to develop cultivars with higher levels of resistance and with earlier maturity.

Several other methods are also used to manage early blight of tomato. Rotation out of tomato for at least 2 years, eradication of weeds and volunteer tomato plants, proper fertilization, and maintaining vigorous crop growth are all necessary for effective management. Covering a greenhouse with ultraviolet (UV)-absorbing vinyl film has been shown to reduce early blight by as much as 50%.

Growers should use pathogen-free seed. Additionally, the soil in seedbeds should be pasteurized with steam or treated with fumigants, or pathogen-free artificial soil mixes should be used. The TOM-CAST computer program is widely used to schedule fungicide spraying of tomato crops.

Selected References

Frazer, J. T. 2002. Two species of *Alternaria* cause early blight of potato (*Solanum tuberosum*) and tomato (*Lycopersicon esculentum*), M.S. thesis. Cornell University, Ithaca, NY.

Gardner, R. G. 1988. NC EBR-1 and NC EBR-2 early blight resistant tomato breeding lines. HortScience 23:779-781.

Gardner, R. G., and Shoemaker, P. B. 1999. "Mountain Supreme" early blight resistant hybrid tomato and its parents, NC EBR-3 and NC eBR-4. HortScience 34:745-746.

Gleason, M. L., MacNab, A. A., Pitblado, R. E., Ricker, M. D., East, D. A., and Latin, R. X. 1995. Disease-warning systems for processing tomatoes in eastern North America: Are we there yet? Plant Dis. 79:113-121.

Moore, W. D. 1942. Some factors affecting the infection of tomato seedlings by *Alternaria solani*. Phytopathology 32:399-403.

Nash, A. F., and Gardner, R. G. 1988. Tomato early blight resistance in a breeding line derived from *Lycopersicon hirsutum* PI 126445. Plant Dis. 72:206-209.

Rotem, J. 1994. The Genus *Alternaria:* Biology, Epidemiology, and Pathogenicity. American Phytopathological Society, St. Paul, MN.

Simmons, E. G. 2000. *Alternaria* themes and variations (244-286): Species on Solanaceae. Mycotaxon 75:1-115.

Thomas, H. R. 1948. Effect of nitrogen, phosphorus, and potassium on susceptibility of tomatoes to *Alternaria solani*. J. Agric. Res. (Washington, D.C.) 76:289-306.

Vakalounakis, D. J. 1991. Control of early blight of greenhouse tomato, caused by *Alternaria solani*, by inhibiting sporulation with ultraviolet-absorbing vinyl film. Plant Dis. 75:795-797.

(Prepared by J. P. Jones;
revised by T. A. Zitter and J. Correll)

Fusarium Crown and Root Rot

Fusarium crown and root rot was first recognized in Japan in 1969 and then in California 2 years later. Today, it has been reported from most tomato-growing countries of the world and is a serious problem in the production of transplants, greenhouse fruit, and field-grown, staked tomatoes. The disease is usually most severe in fumigated or steam-sterilized soil. Yield losses in greenhouse tomato production are often 20–60% and sometimes much higher; in the field, losses of 20–40% have been reported.

Symptoms

Early symptoms of Fusarium crown and root rot include round, brown lesions at the origins of rotted lateral roots. Cotyledons and lower leaves of infected seedlings may yellow and drop prematurely. Later, brown, necrotic lesions of varying sizes develop on taproots or main laterals (Fig. 30). In severe cases, a dry, brown rot of the root cortex may be extensive. Diagnostic chocolate-brown lesions eventually develop on stems from the soil line to 10–30 cm above it. Internally, a reddish-brown or chocolate-brown discoloration in the cortex of the crown extends approximately 5–10 cm (sometimes up to 25 cm) above the stem lesion (Fig. 31).

Infected tomato plants may wilt, especially during the heat of the day. More mature plants may be stunted, and as they begin to bear fruit, their lower leaves turn yellow along the margins of the oldest leaves. The yellowing is followed by the development of necrosis. An abnormal proliferation of adventitious roots may occur above the stem lesions.

Fig. 30. External discoloration of tomato roots, caused by Fusarium crown and root rot. (Courtesy J. P. Jones)

Fig. 31. Internal discoloration of tomato roots, caused by Fusarium crown and root rot. (Courtesy J. P. Jones)

Older plants may die. On dead and dying plants, the fungus produces white mycelium and masses of pinkish-orange spores in necrotic stem lesions.

Causal Agent

The causal agent of Fusarium crown and root rot is the fungus *Fusarium oxysporum* Schltdl.:Fr. f. sp. *radicis-lycopersici* Jarvis & Shoemaker. Unlike *F. oxysporum* f. sp. *lycopersici* (Sacc.) W. C. Snyder & H. N. Hansen, which inhabits the xylem of tomato plants and causes a true wilt, *F. oxysporum* f. sp. *radicis-lycopersici* is a rot-inducing pathogen.

Morphologically, *F. oxysporum* f. sp. *radicis-lycopersici* is indistinguishable from other isolates of *F. oxysporum*. The fungus produces typical microconidia, macroconidia, and chlamydospores. It is unusual among *Fusarium* spp., however, in that the microconidia are airborne and reinfestation and colonization of sterile media occur very quickly.

Disease Cycle and Epidemiology

F. oxysporum f. sp. *radicis-lycopersici* gains access to cortical tissues of roots and hypocotyls through wounds and natural openings made by emerging secondary and adventitious roots. In contrast to Fusarium wilt, which is favored by warm temperatures (25–27°C), crown and root rot is favored by cool temperatures (15–20°C). Because the microconidia (which form in great numbers in necrotic tissue) spread by air currents, the fungus can quickly spread in a greenhouse. Thick-walled chlamydospores enable the fungus to survive in the soil, on wooden stakes, and on other surfaces for more than one cropping season. Colonization of sterile media occurs quite rapidly.

In the field, the fungus is often introduced on infected transplants. Plant-to-plant spread is primarily by root contact; approximately four plants on either side of a diseased plant are infected during one growing season. In the absence of plants, movement of the fungus through the soil is minimal. The fungus can be moved long distances within soil through contaminated shoes, plant stakes, transplant trays, farm machinery, and other equipment. In addition, the fungus is seedborne on tomato seed surfaces. Because spores lodge among the hairs on seeds, they are hard to wet and may survive seed dips in sodium hypochlorite (NaOCl) or hydrochloric acid (1N HCl).

Airborne conidia form in great numbers on stem lesions, and aerial dissemination of conidia is likely an important factor in the development of epidemics. Spores land on all surfaces of tomato plants (including the leaves) and infect tissues in the presence of free moisture. Foliage infection proceeds much more slowly than stem and root colonization and probably plays a small role in development of the disease within a single growing season. Airborne propagules in tomato fields and greenhouses are readily detected (i.e., 1.3–1.6 propagules per square meter every 2 h, in one field study).

F. oxysporum f. sp. *radicis-lycopersici* can also infect and cause small lesions on the roots of many cultivated plants and weeds. In addition to solanaceous crops (except potato (*Solanum tuberosum* L.)), a number of nonsolanaceous plants can be infected by the fungus, including spinach (*Spinacia oleracea* L.), beet (*Beta vulgaris* L.), cabbage (*Brassica oleracea* L.), carrot (*Daucus carota* L.), onion (*Allium* L.), and wheat (*Triticum* L.), along with many legumes and certain cucurbits. Some crops are infected but do not develop symptoms. Regardless, all of these crops may be reservoirs of inoculum. The fungus has also been isolated naturally or experimentally from the roots of a number of weeds, including Brazilian peppertree (*Schinus terebinthifolius* Raddi), carpetweed (*Mollugo verticillata* L.), chickweed (*Stellaria media* (L.) Vill.), corn spurry (*Spergula arvensis* L.), cudweed (*Gnaphalium* L.), curly dock (*Rumex crispus* L.), licorice weed (*Scoparia* L.), narrowleaf plantain (*Plantago lanceolata* L.), redroot pigweed or amaranth (*Amaranthus retroflexus* L.), saltcedar (*Tamarix ramosissima* Ledeb.), shepherd's purse (*Capsella bursa-pastoris* (L.) Medik.), and wild buckwheat (*Fagopyrum cymosum* (Trevir.) Meisn.).

Management

F. oxysporum f. sp. *radicis-lycopersici* is disseminated and survives by a variety of mechanisms, which means control of Fusarium crown and root rot requires an integrated approach to disease management. Effective strategies include the use of pathogen-free seed and planting media, weed control, and other sanitation practices.

Because the use of sterilized soil and potting mixes in greenhouses favors the development of crown and root rot, management strategies in greenhouses are especially important. Once the disease has been introduced into the greenhouse, it is difficult to control. For example, fungicide drenches are generally ineffective. Between tomato crops, pots and transplant trays should be disinfested with steam or a disinfestant, and greenhouses should be cleaned and sanitized in the off season.

Increasing the microbial activity of sterilized media in the greenhouse can reduce activity of *F. oxysporum* f. sp. *radicis-lycopersici*. This can be accomplished by adding selected antagonistic fungi to steamed soil or by inoculating the roots of seedlings with a suspension of saprophytic fungi. The establishment of a microbial community seems to inhibit the saprophytic proliferation of the fungus. In any case, it is clear that the disease is amenable to control by biological means, because it is generally much more severe on tomatoes grown in sterilized soil or potting mix. Another way to enhance microbial activity is to incorporate nontreated field soil, nonsterile peat moss, or a green manure crop into the growing medium. Although lettuce (*Lactuca sativa* L.) is a host of *F. oxysporum* f. sp. *radicis-lycopersici*, it may be a particularly good green manure or rotation crop.

In the field, several strategies can be implemented to minimize damage. Transplants destined for the field should be inspected for the disease and rejected if crown and root rot is identified. When transplants are set in the field, care should be taken to minimize damage and optimal cultural practices should be followed to minimize stress. Following final harvest, crop debris should be rapidly plowed and the field rotated to a nonsusceptible crop. Fumigation with metam sodium works well if optimal incorporation methods are utilized. For staked tomatoes, stakes should be disinfested before reuse or new stakes should be used. Soil solarization, especially in combination with metam sodium, may be particularly effective.

Resistance to Fusarium crown and root rot is available through a single dominant gene (*Frl*) from a tomato relative, Peruvian nightshade (*Solanum peruvianum* L., syn. *Lycopersicon peruvianum* (L.) Mill.). However, the gene has been incorporated into relatively few commercial tomato varieties.

Selected References

Jarvis, W. R. 1988. Fusarium crown and root rot of tomatoes. Phytoprotection 69:49-64.

Marois, J. J., Mitchell, D. J., and Sonoda, R. M. 1981. Biological control of Fusarium crown rot of tomato under field conditions. Phytopathology 71:1257-1260.

McGovern, R. J., Vavrina, C. S., Noling, J. W., Datnoff L. A., and Yonce, H. D. 1998. Evaluation of application methods of metam sodium for management of Fusarium crown and root rot in tomato in southwest Florida. Plant Dis. 82:919-923.

Menzies, J. G., and Jarvis, W. R. 1994. The infestation of tomato seed by *Fusarium oxysporum* f. sp. *radicis-lycopersici*. Plant Pathol. 43:378-386.

Menzies, J. G., Koch, C., and Seywerd, F. 1990. Additions to the host range of *Fusarium oxysporum* f. sp. *radicis-lycopersici*. Plant Dis. 74:569-572.

Rekah, Y., Shtienberg, D., and Katan, J. 1999. Spatial distribution and temporal development of Fusarium crown and root rot of tomato and pathogen dissemination in field soil. Phytopathology 89:831-839.

Rekah, Y., Shtienberg, D., and Katan, J. 2000. Disease development following infection of tomato and basil foliage by airborne conidia of the soilborne pathogens *Fusarium oxysporum* f. sp. *radicis-lycopersici* and *F. oxysporum* f. sp. *basilici*. Phytopathology 90:1322-1329.
Rowe, R. C. 1980. Comparative pathogenicity and host ranges of *Fusarium oxysporum* isolates causing crown and root rot of greenhouse and field-grown tomatoes in North America and Japan. Phytopathology 70:1143-1148.
Sivan, A., Ucko, O., and Chet, I. 1987. Biological control of Fusarium crown rot of tomato by *Trichoderma harzianum* under field conditions. Plant Dis. 71:587-592.

(Prepared by R. M. Davis and A. O. Paulus)

Fusarium Foot Rot

Fusarium foot rot of tomato has been reported from only two countries: Australia, where the disease was first reported in 1975, and the United States, where it was reported in California in 1984. Although the overall effects of foot rot are usually minor, an extensive infection can kill plants. Yield losses of near 5% have been estimated in infested fields in California.

Symptoms

Foot rot is characterized by reddish-brown, cortical lesions on the taproot or large secondary roots, with associated vascular discoloration a few centimeters above and below the lesions (Fig. 32). These lesions may encircle the root. The development of extensive lesions may kill plants, although this is unusual.

Aboveground symptoms of foot rot include interveinal chlorosis and bleaching of the lamina, which resemble symptoms of the viral diseases tomato spotted wilt and alfalfa mosaic (Fig. 33) (see the section Diseases Caused by Viruses, later in Part I). Leaf necrosis often occurs on plants with advanced rot symptoms. Flowers on affected branches may become necrotic, as well. These aboveground symptoms are often restricted to single branches.

Causal Organism

Fusarium foot rot is caused by a specific strain of *Fusarium solani* (Mart.) Sacc.: namely, f. sp. *eumartii* (C. W. Carp.) W. C. Snyder & H. N. Hansen. Aerial mycelium of the fungus on potato dextrose agar (PDA) is abundant and white to cream in color. The color of the colony from the underside of the agar ranges from light tan to burnt orange. With age, a light- to dark-purple pigment diffuses into the agar.

Ovoid to clavate microconidia (3.0 × 9.6 µm) occur on false heads produced on long monophialides. Macroconidia average 3.7 × 32.0 µm in size. In general, macroconidia are stout with thick and mostly parallel walls and have three to seven septa. The basal cell is distinctly foot shaped or notched, whereas the apical cell is not distinctly shaped. Chlamydospores are abundant in 5- to 7-week-old cultures.

Disease Cycle and Epidemiology

The optimum temperature for growth and development of lesions caused by *F. solani* f. sp. *eumartii* isolates from California is 25–30°C. In contrast, lesion development caused by Australian isolates increased as the temperature decreased to 15°C.

Many crops are hosts of the fungus, including bean (*Phaseolus* L.), cotton (*Gossypium* L.), melon (*Cucumis* L.), onion (*Allium* L.), pepper (*Capsicum annuum* L.), safflower (*Carthamus tinctorius* L.), and sunflower (*Helianthus* L.). However, no disease occurs in the field on these hosts. Apparently, the fungus sustains itself on the roots of these plants and probably many other crops and weeds, as well. Thus, once a field has been infested, the fungus likely will remain in the soil.

Management

There are currently no control measures for Fusarium foot rot except to limit the spread of the fungus by soil movement. All commercial tomato cultivars are susceptible to the disease, but their responses differ. Some tolerate the disease to a greater degree than others.

Selected References

Cucuzza, J. D., Watterson, J. C., and Bernhardt, E. A. 1992. Foot rot of tomato caused by *Fusarium solani* in California. Plant Dis. 76:101.
Vawdry, L. L., and Peterson, R. A. 1988. *Fusarium solani,* the cause of foot rot of tomatoes in Central Queensland. Australas. Plant Pathol. 17:24-25.

(Prepared by R. M. Davis and A. O. Paulus)

Fig. 32. Symptoms of Fusarium foot rot on plant roots. (Courtesy Jack Kelly Clark, University of California Statewide IPM Program)

Fig. 33. Aboveground symptoms of Fusarium foot rot. (Courtesy Jack Kelly Clark, University of California Statewide IPM Program)

Fusarium Wilt

Fusarium wilt remains a serious soilborne disease of tomato in almost all areas where tomatoes are grown. First described in England in 1895, it now occurs worldwide, having been reported in more than 40 countries. Researchers have identified three races of the Fusarium wilt pathogen of tomato (races 1, 2, and 3), and all three occur in the United States. The impact on yields can be severe if resistant cultivars are not grown (Fig. 34).

Symptoms

The most common symptoms of Fusarium wilt of tomato include the initial wilting and yellowing of leaves, typically after the plants have flowered. As the plants mature, slight wilting of portions of them can be observed. Chlorotic symptoms begin to appear on one-half of the plant, where all of the leaflets may turn yellow (Fig. 35). As the disease progresses, wilting and yellowing can occur either on one side or the entire plant (Fig. 36).

Fusarium wilt of tomato is considered a warm weather disease. Wilting is more common in the afternoon during the warmest part of the day, and it is not uncommon for plants showing early symptoms of wilting to recover overnight.

Another characteristic symptom of Fusarium wilt is a distinct dark-red to brown discoloration in the vascular tissue at the base of the plant, which is revealed when a cut is made into the xylem tissue (Fig. 37). As symptoms progress, the red-brown vascular discoloration can advance up through the plant and even discolor the vascular bundles at the bases of the petioles.

Causal Organism

Fusarium wilt is caused by the fungal pathogen *Fusarium oxysporum* Schltdl.:Fr. f. sp. *lycopersici* (Sacc.) W. C. Snyder

Fig. 34. Field of tomato plants with a high incidence of Fusarium wilt. (Courtesy J. P. Jones)

Fig. 36. Mature plant showing the yellowing characteristic of Fusarium wilt. (Courtesy J. P. Jones)

Fig. 35. Detached leaf showing unilateral yellowing, caused by Fusarium wilt. (Courtesy J. P. Jones)

Fig. 37. Longitudinal section of a stem with browning of the vascular system, caused by Fusarium wilt. (Courtesy J. P. Jones)

& H. N. Hansen. Cultural characteristics of *F. oxysporum* f. sp. *lycopersici* growing on potato dextrose agar (PDA) include white mycelium, with the underside of the culture typically producing violet, purple, blue, red, orange, or pink pigmentation.

The pathogen produces several types of asexual spores typical for many *Fusarium* spp., including macroconidia, microconidia, and resting chlamydospores. Macroconidia typically are curved, fusoid-shaped, thin-walled spores with three to five septa that have tapered or pointed ends, and they can vary considerably in size (3–5 × 27–60 µm). Microconidia are produced on simple phialides; they can be oval, ellipsoid, straight, or curved and measure 2–4 × 5–12 µm. Macro- and microconidia are hyaline. *F. oxysporum* f. sp. *lycopersici* can produce chlamydospores, which are asexual resting spores. Chlamydospores can be rough or smooth walled, measure approximately 10 µm in diameter, and can form singly, in pairs, or in short chains. The chlamydospores can be terminal or intercalary in the mycelium. Although a sexual stage of *F. oxysporum* f. sp. *lycopersici* is not known, examination of sequence similarities indicates that the pathogen is a member of the *Gibberella fujikuroi* (Sawada) Ito species complex.

The three races of the Fusarium wilt pathogen are known to occur in five distinct vegetative compatibility groups (VCGs) with standardized names: VCGs 0030, 0031, 0032, 0033, and 0035. VCG 0034 was originally reported as *F. oxysporum* f. sp. *lycopersici* but has subsequently been identified as *F. oxysporum* f. sp. *radicis-lycopersici* Jarvis & Shoemaker, the causal agent of Fusarium crown and root rot (see previous section on this disease). Even so, molecular studies have determined that there is a complex relationship among race, VCG, and molecular lineage and that a considerable degree of genetic diversity exists within each race.

Disease Cycle and Epidemiology

F. oxysporum f. sp. *lycopersici* is a soilborne pathogen that can survive for many years, even in the absence of tomato production. It can be transmitted on infected transplants, through the movement of contaminated soil via workers or equipment, and by seed. Seed transmission of the pathogen is uncommon, however, because of industry efforts to grow seed crops on clean ground and to use seed-testing methods to ensure the production of clean seed. *F. oxysporum* f. sp. *lycopersici* can survive as dormant chlamydospores or mycelium in soil or plant debris, and it can be a parasite of many weed hosts, where it colonizes the cortices of roots. The pathogen can colonize tomato roots and then infect the xylem, whereby the pathogen can move passively through the plant's vascular system.

Infection into the xylem is favored by wounds to the roots, including those caused by the root-knot nematode, *Meloidogyne incognita* (Kofoid & White) Chitwood. Secondary spread from plant to plant seldom occurs. Long-distance dispersal of the pathogen can occur by movement of infested soil on equipment, packing boxes, and workers or infected transplants. *F. oxysporum* f. sp. *lycopersici* has also been shown to be aerially disseminated under some environmental conditions.

Management

The most effective means of managing Fusarium wilt of tomato is to use disease-resistant cultivars. Commercial cultivars are available that are resistant to various races of *F. oxysporum* f. sp. *lycopersici,* including races 1, 2, and 3.

Other management practices can help reduce introduction of the pathogen or disease severity, including the use of disease-free transplants, the use of certain fertilizer combinations, the rotation of crops, and the prevention of movement of the pathogen into clean areas on equipment, workers, and infested soil. Although *F. oxysporum* f. sp. *lycopersici* can survive for long periods of time in the soil, crop rotations for 3–5 years are often effective in reducing disease severity. In addition, if the pathogen is present in production fields, disease severity can be reduced on some soil types by using calcium nitrate fertilizers, by avoiding the use of ammonium nitrate fertilizers, and by raising the soil pH to 6.5–7.0. Some bacteria and fungi have also been used successfully as biocontrol agents for Fusarium wilt.

Selected References

Beckman, C. H. 1987. The Nature of Wilt Diseases of Plants. American Phytopathological Society, St. Paul, MN.

Booth, C. 1971. The Genus *Fusarium*. C.A.B. International Mycological Institute, Kew, England.

Cai, G., Gale, L. R., Schneider, R. W., Kistler, H. C., Davis, R. M., Elias, K. S., and Miyao, E. M. 2003. Origin of race 3 of *Fusarium oxysporum* f. sp. *lycopersici* at a single site in California. Phytopathology 93:1014-1022.

Gordon, T. R., and Martyn, R. D. 1997. Evolutionary biology of *Fusarium oxysporum*. Annu. Rev. Phytopathol. 35:111-128.

Jones, J. P., and Woltz, S. S. 1981. Fusarium-incited diseases of tomato and potato and their control. Pages 157-168 in: Fusarium: Diseases, Biology, and Taxonomy. P. E. Nelson, T. A. Toussoun, and R. J. Cook, eds. Pennsylvania State University Press, University Park.

Katan, T., Shlevin, E., and Katan, J. 1997. Sporulation of *Fusarium oxysporum* f. sp. *lycopersici* on stem surfaces of tomato plants and aerial dissemination of inoculum. Phytopathology 87:712-719.

Larkin, R. P., and Fravel, D. R. 1998. Efficacy of various fungal and bacterial biocontrol organisms for control of Fusarium wilt of tomato. Plant Dis. 82:1022-1028.

Walker, J. C. 1971. Fusarium Wilt of Tomato. Monogr. 6. American Phytopathological Society, St. Paul, MN.

(Prepared by J. Correll and J. P. Jones)

Gray Leaf Spot

Gray leaf spot was first noticed in the United States in 1924, and by 1928, it had spread throughout the U.S. state of Florida, causing widespread defoliation. Currently, gray leaf spot remains one of the most destructive diseases of tomato throughout the southeastern United States and the world wherever susceptible cultivars are raised under warm, humid conditions.

Symptoms

Gray leaf spot lesions are limited almost entirely to the leaf blades. Under very favorable conditions, lesions may sometimes develop on the petioles and the more tender parts of the growing stems. In these instances, the spots are linear and parallel with the stem. No fruit symptoms have been observed.

Gray leaf spot first appears as minute, barely visible, brownish-black specks on the lower leaves. The spots are circular to oblong and randomly scattered over both the adaxial (Fig. 38) and abaxial (Fig. 39) leaf surfaces, without any apparent restriction by the veins. The spots enlarge to approximately 2.1 mm in diameter, although on the oldest leaves, near the base of the plant, individual spots may be 4.2 mm or more in diameter. On older leaves, the spots occasionally coalesce, killing large areas of the leaf blade. As the centers of the spots dry out, they often crack in various patterns. It is often at this stage that yellowing of entire leaves becomes most conspicuous, especially if the infection is severe. The leaves then die rapidly, become brown, and drop from the plant.

Serious infections in seedbeds result in marked defoliation without conspicuous yellowing.

Causal Organisms

Gray leaf spot of tomato is caused by three species of *Stemphylium*. The first is *S. solani* G. F. Weber. The hyphae of this fungus are variously branched, septate, and intercellular. Conspicuous and abundant sporulation occurs on various media.

Conidiophores are dark, septate, slightly larger than the infertile hyphae (4–7 × 130–200 μm), and rigid with swollen tips and irregular bases. Conidia are produced acrogenously on simple septate conidiophores. Conidia are muriform, fuscous to very dark, and oblong or rectangular with rounded ends or one round and one somewhat pointed end. Conidia have several longitudinal septa and are constricted near the medial septum, with an average spore size of 22 × 48 μm. No ascerigous stage of *S. solani* is known.

The second fungus that causes gray leaf spot on tomato is *S. lycopersici* (Enjoji) W. Yamam. (syn. *S. floridanum* Hannon & G. F. Weber). The hyphae of this fungus are hyaline to slightly olivaceous, branched, and septate; they range from 5 to 9 μm in diameter. Conidiophores are olivaceous, septate, seldom branched, and rigid; they are 75–300 μm long and 3.0–5.5 μm in diameter with bulbous apexes, one to seven nodular swellings, and irregularly shaped bases. Conidia are produced acrogenously and are olivaceous, muriform with one to six transverse septa, and deeply constricted; they are 7.6–23.0 × 19.9–62.2 μm and have no to many longitudinal septa. No ascerigous stage of *S. lycopersici* is known.

A third *Stemphylium* fungus incites gray leaf spot in Israel. *S. botryosum* Wallr. f. sp. *lycopersici* Rotem, Y. Cohen, & I. Wahl produces the perfect stage, and *Pleospora herbarum* (Pers.:Fr.) Rabenh. occurs in culture but not in nature. The mycelium is branched, septate, and olivaceous green or olivaceous brown, depending on the growth medium. Conidiophores are brown and unbranched, with bulbous apical swellings, each bearing a single conidium. Conidia are echinate and muriform, dark brown to black in color, rectangular in shape, and 9.0–26.6 × 14.0–41.1 μm in size. Key temperatures for growth in pure culture are 5°C minimum, 27°C optimum, and 39°C maximum.

Fig. 38. Gray leaf spot on the adaxial surfaces of tomato leaflets. (Courtesy J. P. Jones)

Fig. 39. Gray leaf spot on the abaxial surfaces of tomato leaflets. (Courtesy J. P. Jones)

Perithecia are 340 μm in diameter. Ascospores (eight per ascus) are mostly straight and oblong or ellipsoid in shape, have seven transverse and two longitudinal septa, and average 17 × 40 μm. They are yellow when young.

Optimum sporulation of the *Stemphylium* pathogens occurs at 23°C, and optimum mycelial growth occurs at 26–29°C. Fungal growth is restricted below pH 5.9 and completely inhibited below pH 4.8. A neutral to slightly alkaline reaction is most conducive to growth, with only a slight inhibition at pH 9.2.

Disease Cycle and Epidemiology

Gray leaf spot begins in seedbeds and transplant houses or on field-transplanted seedlings, often when the plants are in the first true-leaf stage. Cotyledons are frequently but not severely infected.

S. solani and *S. botryosum* f. sp. *lycopersici* remain viable from season to season on infected debris or on tomato plants in southern climates, where they may grow throughout the year. Other sources of inoculum include abandoned fields and seedbeds, volunteer tomato plants, pepper (*Capsicum annuum* L.), gladiolus (*Gladiolus hortulanus* L. H. Bailey), blue lupine (*Lupinus pilosus* L.), horsenettle (*Solanum carolinense* L.), and other wild solanaceous plants. *Stemphylium* spp. are excellent saprophytes and grow readily on necrotic tissue of resistant tomato cultivars.

Much pathogen spread occurs by transplanting diseased seedlings from seedbeds or transplant houses to fields. Conidia can be spread over extensive areas by wind. Spores germinate quickly in the presence of a water film (dew will suffice) and warm temperatures (24–27°C), and they can produce extensive growth during a single night. The pathogens develop rapidly within the host, and symptoms can be distinguished within 2–3 days in growth rooms and 5 days in the field.

In recent years, gray leaf spot has been observed most frequently on cherry and grape tomato cultivars, presumably because they do not have the resistance available in most globe-shaped tomatoes.

Management

Cultivars resistant to gray leaf spot are available and should be used whenever possible. If susceptible cultivars are used, application of effective fungicides on a regular schedule is required.

Selected References

Hannon, C. I., and Weber, G. F. 1955. A leaf spot of tomato caused by *Stemphylium floridanum* sp. nov. Phytopathology 45:11-16.

Rotem, J., Cohen, Y., and Wahl, I. 1966. A new foliage disease in Israel caused by *Stemphylium botryosum*. Can. J. Plant Sci. 46:265-270.

Weber, G. F. 1930. Gray leaf spot of tomato caused by *Stemphylium solani*, sp. nov. Phytopathology 20:513-518.

(Prepared by J. P. Jones and K. L. Pernezny)

Gray Mold

Gray mold probably occurs wherever tomato is grown, because the causal fungus is ubiquitous. The disease occurs in field-planted crops and is commonly observed in greenhouse culture of tomato. It consistently causes minor losses, but occasionally, the disease is a major factor in production, especially in greenhouse systems. Fruit rot can occur both in the field and in postharvest shipments.

Symptoms

Gray mold can occur on all of the aboveground parts of the tomato plant. The most characteristic sign of the disease is the profuse growth of sporophores from necrotic tissue. This growth gives the diseased tissue a fuzzy, gray-brown appearance that may resemble felt (Fig. 40). Spore clouds can be shaken from the sporophores after periods of high humidity. Lesions caused by other fungi, by high salt content of the soil, or by wind may mimic symptoms of gray mold. Gray mold can be distinguished from these other disorders by the presence of sporophores and spores on the surface of the necrotic area.

Lesions on leaflets will progressively expand to include the whole leaflet (Fig. 41), then the petiole (Fig. 42), and finally the stem (Fig. 43). Such lesions may girdle the stem and cause wilting of the plant above the lesion. Stem lesions are the primary cause of plant mortality in greenhouse production systems. Rotted areas may develop on fruit that touch diseased foliage.

Senescent petals are quite susceptible to the fungus. The fungus may grow from the infected petals into the sepals before the petals dehisce, and from there, it may grow into the developing fruit. When infected petals remain attached to fruit, the fungus then grows directly into the fruit. As a result, the disease often occurs at the stem end or blossom end of tomato fruit.

Lesions on fruit are typical of soft rot, and decayed areas are whitish in color. Usually, the skin becomes ruptured in the center of the decayed area but remains unbroken over the rest of the area. Sporophores develop in the broken area but not in the unbroken area. Eventually, the whole fruit becomes affected and mummifies.

The production of halos (called ghost spots) on the fruit is an unusual symptom of gray mold (Fig. 44). This phase of the disease occurs after the spores germinate on the surface of fruit, the germ tubes penetrate, and the mycelium aborts. A small, necrotic fleck appears along with a white halo. The halo is a whitish ring approximately 3–8 mm in diameter. Germ tube

Fig. 40. Diseased tissue with a fuzzy, gray-brown appearance, caused by gray mold. (Courtesy K. L. Ivors)

Fig. 41. Gray mold on a tomato leaflet. (Courtesy M. T. McGrath)

Fig. 42. Gray mold on the leaflets and petiole. (Courtesy M. T. McGrath)

Fig. 43. Stem canker emanating from a diseased petiole, caused by gray mold. (Courtesy M. T. McGrath)

Fig. 44. Halos (or ghost spots) on fruit resulting from an aborted infection by *Botrytis cinerea.* (Courtesy T. A. Zitter)

penetration occurs when the fruit is 1.5–3.0 cm in diameter, but full expression of the disease occurs at the mature-green stage of development. Although no rot ever occurs with ghost spots, the appearance of many halos makes the fruit unsightly and thus unmarketable.

Causal Organism

The gray mold pathogen, *Botrytis cinerea* Pers.:Fr., grows well on a variety of media. It produces conidia on long conidiophores that are dichotomously branched at the tips. Conidia are borne on sterigma at the tip of each branch, and the cluster of spores resembles a bunch of grapes. The conidia are colorless, single celled, and ovoid; they range in size from 7.3 × 8.0 to 9.7 × 11.1 µm. Sclerotia may form in the plant tissue. They are flat, irregular in size, and black to dark brown. Sclerotia may germinate, forming mycelia that produce conidia. Apothecia similar to those produced by *Sclerotinia* spp. have reportedly been produced from sclerotia on rare occasions. Ascospores are formed in the apothecia.

B. cinerea requires relatively cool temperatures (18–23°C) for best growth. As temperatures increase above 24°C, conidial germination decreases. Growth is inhibited at 32°C and above.

Disease Cycle and Epidemiology

There are various sources of primary inoculum for gray mold. *B. cinerea* has a wide host range, so the disease can be present on perennial plants in any geographical area and sporulation can occur when conditions become optimal. Conidia are easily windborne and can be blown from field to field. In addition, the fungus can survive from season to season in the form of sclerotia, which develop on the woody tissues of tomato plants. The fungus is also an efficient saprophyte and can be harbored in organic matter in the soil. The fungus grows from the sclerotia or organic matter in the soil and can infect leaves lying on the ground.

B. cinerea is usually considered a weak parasite. Even so, it can penetrate plant tissue directly through the formation of appressoria, if supplied with a food source. Foliage disease is generally associated with some type of wounding or insect injury. The disease usually starts on senescent foliage. After inoculum increases on this foliage, juvenile tissue may be invaded. Thus, the disease is associated with mature plants that have a dense canopy. This situation is commonly found in greenhouse tomato production, where plants are typically 4–6 m in length and display lush foliar growth. Pruning of leaves and clusters provides wounds that are ideal for the initiation of gray mold. Once the disease begins, it continues to spread by growth of the mycelium through the tissue.

Gray mold begins in relatively cool weather and does not require a prolonged period of high humidity. In fact, the humid conditions within the tomato canopy at night are usually sufficient for the disease to develop. After the disease begins, it continues to develop in rather warm temperatures but at a reduced level. High humidity is common in greenhouse tomato production systems. Once established, gray mold can spread rapidly through an entire greenhouse, resulting in significant economic loss.

Repeated applications of the metal salts of ethylenebis (dithiocarbamate) (EBDC) fungicides may increase the severity of gray mold. This may occur, in part, because of the increased canopy density that results from control of other fungal diseases. However, researchers have demonstrated that the metal salts associated with EBDC fungicides increase tomato plants' susceptibility to gray mold.

Gray mold is particularly severe on plants grown on acidic sandy soils that have high water content.

Management

Several fungicides are effective in the management of gray mold. Regardless, application of fungicides should begin before a dense canopy develops, because covering senescent foliage may be impossible in the center of a dense canopy. This is especially true in greenhouse tomato production. Fungicidal control of gray mold may depend on the presence of residual fungicide in the canopy.

Liming acid soils to increase the calcium content of plants will reduce the disease susceptibility of tomato plants. A calcium to phosphorus ratio of 2:1 or higher in leaf petiole tissues will aid in control.

Selected References

Agrios, G. N. 2005. Plant Pathology. 5th ed. Elsevier Academic Press, Boston.

Ferrer, J. B., and Owen, J. H. 1959. *Botrytis cinerea* the cause of ghost-spot disease of tomato. Phytopathology 49:411-417.

Ingram, D. M., and Meister, C. W. 2006. Managing *Botrytis* gray mold in greenhouse tomatoes using traditional and bio-fungicides. Plant Health Progress. doi:10.1094/PHP-2006-0718-01-RS.

Pepin, H. S., and MacPherson, E. A. 1982. Strains of *Botrytis cinerea* resistant to benomyl and captan in the field. Plant Dis. 66:404-405.

Stall, R. E., Hortenstine, C. C., and Iley, J. R. 1965. Incidence of Botrytis gray mold of tomato in relation to a calcium-phosphorus balance. Phytopathology 55:447-449.

Webb, R. E., Good, J. M., and Danielson, L. L. 1967. Tomato diseases and their control. U.S. Dep. Agric. Handb. 203.

(Prepared by D. M. Ingram and R. E. Stall)

Late Blight

Late blight is a highly destructive disease that affects tomato, potato (*Solanum tuberosum* L.), and solanaceous weeds. This is a disease of historic importance, first described on potato in 1845 and on tomato in 1847. Late blight appeared in both North America and Europe in the 1830s and is thought to have originated in Central America or South America. The disease is responsible for the infamous Irish potato famine of the 1840s.

Late blight can be a serious threat to tomato production around the world, particularly when tomatoes are grown under cool temperatures, sprinkler-irrigated fields, prolonged periods of rain, and fog or high dew conditions. The disease develops most rapidly when the level of humidity is 100% and the temperature is approximately 18–24°C.

The late blight pathogen has been responsible for numerous epidemics of tomato and potato since it was first described. In addition to tomato and potato, this pathogen infects hairy nightshade (*S. sarrachoides* Sendtn.), petunia (*Petunia* Juss.), and bittersweet nightshade (*S. dulcamara* L.), along with other *Solanum* spp.

Symptoms

All aboveground parts of the tomato plant are susceptible to infection from late blight. On leaves, lesions first appear as small, indefinite, water-soaked spots, which may enlarge rapidly into pale-green to brown lesions and cover large areas (Fig. 45). In moist weather, the undersides of small lesions may be covered with a gray to white moldy growth (Fig. 46). Later, only a ring of moldy growth may be observed around the undersides of large, brown lesions. Infected foliage becomes brown, shrivels, and soon dies. Petioles and stems are affected in a similar manner, and the whole plant may die.

Fruit lesions appear as dark, olivaceous, greasy spots, which may continue to enlarge until the entire fruit has been affected (Fig. 47). A thin weft of white mycelium may cover fruit lesions during long periods of rain. Soft rot often follows late blight infection and leads to fruit disintegration. Decaying vines may be recognized by a foul odor that pervades the air.

Causal Organism

The late blight pathogen, *Phytophthora infestans* (Mont.) de Bary, belongs to the kingdom Chromista, phylum Oomycota, class Oomycetes, order Pythiales, and family Pythiaceae. There are two mating types, A1 and A2, and interaction between individuals of opposite mating types stimulates the development of sexual structures (i.e., oogonia and antheridia). Fertilization occurs in the oogonium, which subsequently matures into a thick-walled oospore (24–46 μm in diameter) that can survive between growing seasons. Oospores germinate by means of a germ tube with a terminal sporangium.

In the absence of a second mating type, the pathogen reproduces asexually by forming lemon-shaped, multinucleate, hyaline sporangia (21–23 × 21–38 μm). Production of sporangia requires high moisture and moderate temperatures (10–24°C). Germination of sporangia occurs directly during warm conditions (18–24°C) by production of a single germ tube or indirectly during cool (8–18°C) conditions by production of motile zoospores. Each zoospore may swim for several minutes with the aid of two flagella in a film of water on the plant surface and then encyst and germinate by producing a germ tube that penetrates the host.

The genetic diversity of *P. infestans* has increased dramatically in recent years with the migration of new strains, many of which have increased fitness and resistance to phenylamide fungicides. Another factor in the increased genetic diversity of *P. infestans* is the distribution of the A2 mating type to production areas around the world. Sexual reproduction was initially limited to central Mexico, but the presence of the A1 and A2 mating types worldwide has led to the development of more aggressive and difficult to control populations of the pathogen.

Disease Cycle and Epidemiology

P. infestans survives between cropping seasons on volunteer and abandoned potato and tomato plant materials in fields, cull piles, weed hosts, and home garden potatoes and tomatoes. At this time, oospores are not known to be a major factor in the epidemiology of late blight in the United States, but they are produced in some potato- and tomato-growing regions of the world.

The pathogen is most active during cool, moist weather. Cool nights and warm days provide ideal conditions for late blight development, and during blight-favorable periods (18–22°C), symptoms of new infections can occur in as short a time as 3 days. When the weather is favorable, infection moves so rapidly that affected plants appear as though they have been damaged by frost. Multiple asexual cycles are possible during a single crop production season if weather conditions favor plant infection, pathogen reproduction, and dispersal of inoculum. Temperatures above 30°C are considered unfavorable for late blight development, although the pathogen can survive elevated temperatures.

Management

Infected plant materials serve as inoculum sources for neighboring fields. Cultural control measures include eliminating all potato cull piles in the vicinity of tomato plantings and destroying volunteer potato plants that grow from overwintered tubers. If tomato transplants are used to plant production fields, they must be carefully grown, inspected, and handled to ensure they are free of foliar symptoms at the time of transplanting.

P. infestans is seedborne and has been reported to produce oospores in tomato seeds when mature green fruit are coinoculated with A1 and A2 sporangia. Tomato fruit may also carry oospores, making the transport of infected fruit a potential source of inoculum.

Plant resistance to late blight does not currently play an important role in disease control. Resistance holds promise, however, as an important disease control strategy, and cultivars tolerant to certain races of *P. infestans* are becoming available.

The most effective way to manage late blight disease on tomato is through a fungicide spray program combined with a late blight forecasting system. Several late blight forecasting systems have been developed to improve the timing of field scouting activities and the application of fungicides. These systems have been helpful in identifying the periods when spraying with fungicides will be most helpful, in making the decision to spray, and in ensuring the cost-effective use of fungicides.

The use of fungicides is recommended when the weather favors disease development. Several protectant materials are currently registered and recommended throughout the world. In addition, the registration of fungicides with systemic, curative properties is assisting in late blight control.

Fig. 45. Symptoms of late blight on the adaxial surfaces of leaflets. (Courtesy R. E. Stall)

Fig. 46. Symptoms of late blight on the abaxial surfaces of leaflets. (Courtesy J. P. Jones)

Fig. 47. Dark-brown, greasy, firm lesions on tomato fruit, caused by late blight. (Courtesy W. R. Stevenson)

Selected References

Cohen, Y., and Grinberger, M. 1987. Control of metalaxyl-resistant causal agents of late blight in potato and tomato and downy mildew in cucumber by cymoxanil. Phytopathology 77:1283-1288.

Fry, W. E., and Goodwin, S. B. 1997. Re-emergence of potato and tomato late blight in the United States. Plant Dis. 81:1349-1357.

Fry, W. E., Thurston, H. D., and Stevenson, W. R. 2001. Late blight. Pages 28-30 in: Compendium of Potato Diseases. 2nd ed. W. R. Stevenson, R. Loria, G. D. Franc, and D. P. Weingartner, eds. American Phytopathological Society, St. Paul, MN.

Jones, J. P. 1978. Disease thresholds for downy mildew and target leaf spot of cucurbits and late blight of tomato. Plant Dis. Rep. 62:798-802.

Krause, R. A., Massie, L. B., and Hyre, R. A., 1975. BLITECAST: A computerized forecast of potato blight. Plant Dis. Rep. 59:95-98.

MacKenzie, D. R. 1981. Scheduling fungicide applications for potato late blight with BLITECAST. Plant Dis. 65:394-399.

Rubin, E., Baider, A., and Cohen, Y. 2001. *Phytophthora infestans* produces oospores in fruits and seeds of tomato. Phytopathology 91:1074-1080.

Shattock, R. C. 2002. *Phytophthora infestans:* Populations, pathogenicity and phenylamides. Pest Manag. Sci. 58:944-950.

Sherf, A. F., and MacNab, A. A. 1986. Vegetable Diseases and Their Control. John Wiley & Sons, New York.

Vartanian, V. G., and Endo, R. M. 1985. Overwintering hosts, compatibility types, and races of *Phytophthora infestans* on tomato in southern California. Plant Dis. 69:516-519.

Vartanian, V. G., and Endo, R. M. 1985. Survival of *Phytophthora infestans* in seeds extracted from infected tomato fruits. Phytopathology 75:375-378.

(Prepared by W. R. Stevenson and H. Bolkan)

Leaf Mold

Leaf mold is most severe where tomatoes are grown under conditions of high humidity. Thus, it is primarily a disease of greenhouse-grown tomatoes grown under protected culture; it occurs only occasionally on field-grown tomatoes. The pathogen, which was first described from samples collected in South Carolina in 1883, is now distributed worldwide.

Symptoms

Foliage is generally the only tissue affected by leaf mold. Older leaves are affected first, with symptoms on younger leaves occurring later. Initial symptoms are noticeable on upper leaf surfaces and characterized by pale-green or yellowish spots, which later turn a distinct yellow (Fig. 48). The margins of the spots are not well defined. When infection is severe, the spots coalesce and the foliage is killed. On the lower surfaces of affected leaves, an olive-green mold is associated with the discolored areas (Fig. 49). The mold is more dense and deeper in color toward the center of the discolored area. Leaves curl and wither and may drop from the plant.

In some cases, the fungus attacks the petioles, peduncles, stems, blossoms, and fruit. Blossoms may be killed. Green and mature fruit can develop a black, leathery rot on the stem end. The margins of the rotted areas may be irregular, affecting as much as one-third of the fruit surface.

Causal Organism

The organism that causes leaf mold is *Passalora fulva* (Cooke) U. Braun & Crous (syn. *Fulvia fulva* (Cooke) Cif.). Conidiophores up to 200 μm long protrude through stomatal openings; they are usually unbranched and somewhat constricted at the base and more broad at the tip. Conidia are pale to dark brown, have zero to three septa, contain a thickened hilum, and are 4–10 × 12–47 μm. Macroscopically, the fungus is buff to brown and velvety in appearance.

Disease Cycle and Epidemiology

The fungus that causes leaf mold depends on high relative humidity and moderate temperatures for disease development. Spore germination can occur in free water but only if relative humidity is greater than 85%; thus, below 85% relative humidity, disease does not occur. Disease development is optimal at 22–24°C but may occur at temperatures between 4 and 32°C. Disease rarely occurs below 10.5°C.

The fungus survives as a saprophyte on crop residue and in soil as conidia or sclerotia. The conidia, which are readily disseminated by rain and wind, can survive at least 1 year. Other factors responsible for dispersal of the fungus are tools, workers' clothing, and perhaps insects.

Fig. 48. Leaf mold on the adaxial surface of a tomato leaflet. (Courtesy J. P. Jones)

Fig. 49. Leaf mold on the abaxial surface of a tomato leaflet, with profuse growth of *Passalora fulva.* (Courtesy J. P. Jones)

Seed can be contaminated and serve as a primary source of inoculum.

Management

For both greenhouse and field production, crop residue should be removed and destroyed following crop harvest. In a greenhouse, once the crop residue has been removed, production areas should be cleaned by steaming (i.e., 57°C for at least 6 h). When seed transmission is suspected, hot-water treatment of seeds may help manage the disease. Fungicides also effectively control leaf mold.

Several strategies can help reduce the severity of leaf mold. For instance, long periods of leaf wetness and high relative humidity should be minimized by avoiding the watering of foliage and by circulating air with fans. In addition, night temperatures should be maintained at a higher level than outside temperatures. In field production areas where late crops are severely affected by leaf mold, early planting may reduce severity. Staking and pruning of tomatoes to increase ventilation also helps to control the disease.

Cultivars with resistance to leaf mold have been developed. However, the extreme pathogenic variation in the fungus soon makes these cultivars susceptible to the disease.

Selected References

Gardner, M. W. 1925. Cladosporium leaf mold of tomato: Fruit invasion and seed transmission. J. Agric. Res. 31:519-540.

Jones, J. P. 1973. Evaluation of some fungicide treatments for phytotoxicity to tomato and control of late blight and leafmold. Plant Dis. Rep. 57:612-614.

Kerr, E. A., and Bailey, D. L. 1966. Breeding for resistance to *Cladosporium fulvum* Cke. in tomato. Acta Hortic. 4:145-148.

Rivas, S., and Thomas, C. M. 2005. Molecular interactions between tomato and the leaf mold pathogen *Cladosporium fulvum*. Annu. Rev. Phytopathol. 43:395-436.

Small, T. 1930. The relation of atmospheric temperature and humidity to tomato leaf mould (*Cladosporium fulvum*). Ann. Appl. Biol. 17:71-80.

(Prepared by J. B. Jones and J. P. Jones)

Phoma Rot

Phoma rot has historically been a major problem on tomato in India and in the southern United States, although the disease now appears only sporadically in the United States. Phoma rot may cause foliar and stem damage, but fruit damage during transit is what results in major losses from the disease.

Symptoms

Foliar infection by Phoma rot is typified by small, black spots on both upper and lower leaf surfaces; the spots are irregular and slightly sunken (Fig. 50). Upon expansion, the spots become zonate and range from 6 to 12 mm in diameter. The presence of pycnidia distinguishes Phoma rot from early blight and target spot. The lesions expand and may coalesce, causing the foliage to become chlorotic and die. On stems, dark, elongated lesions develop that have faint concentric rings (Fig. 51). On seedlings, the stems may become completely girdled with lesions.

Both green and ripe fruit are affected. The fungus enters the fruit through growth cracks and stem scars, as well as injuries caused by mechanical means and insect punctures, on occasion. When an injured fruit is invaded, sunken spots develop that are dark brown to black in color (Fig. 52). The lesions expand and affect large sections of the tomato surface. The lesions are leathery and can be distinguished from those caused by other rots by dark areas that contain black, speckled eruptions (i.e.,

Fig. 50. Phoma rot on a tomato leaf. (Courtesy J. P. Jones)

Fig. 51. Phoma rot on tomato stems. (Courtesy L. L. Black)

Fig. 52. Phoma rot on tomato fruit. (Courtesy D. Blancard, INRA)

pycnidia). In transit, fruit lesions may enlarge, causing the fruit to be rejected at market.

Causal Organism

Phoma rot is incited by *Phoma destructiva* var. *destructiva* Plowr. (syn. *Phoma destructiva* Plowr.). This fungus produces mycelia that are septate and darken with age. Pycnidia are subcutaneous and later erumpent; they are subglobose and slightly papillate, have one or two ostioles, are brown or blackish brown in color, and are 30–350 μm in diameter. Conidia are hyaline, single celled, and biguttulate and measure 2.0 × 3.5 μm.

Disease Cycle and Epidemiology

P. destructiva var. *destructiva* survives in debris from tomato and pepper (*Capsicum annuum* L.) plants. During periods of high moisture, masses of spores are released from pycnidia. Spores are spread in the field by splashing rain and droplets of irrigation water.

Infected tomato seed and transplant seedlings can disseminate the fungus long distances. The fungus penetrates the plant through wounds; thus, any damage in the field during and after harvest can result in an increased incidence of Phoma rot.

Management

Transplants should be raised in locations distant from tomato production. Seedbeds must be disinfested and contaminated seed treated to produce pathogen-free plants. Growers who do not produce their own transplants should purchase certified clean transplants. Registered fungicides can be applied when disease is suspected.

To avoid postharvest rot, only dry, smooth fruit should be packed; obviously diseased fruit should be discarded. To minimize fruit injury, crates should not be overfilled.

Selected References

Fournet, J. 1971. Study on the conditions for infection of tomato leaves by *Phoma destructiva* Plowr. Annu. Rev. Phytopath. 3:215-231.

Tisdale, W. B., and Hawkins, S. W. 1937. Experiments for the control of Phoma rot of tomatoes. Fla. Agric. Exp. Stn. Bull. 308.

(Prepared by J. B. Jones and J. P. Jones)

Powdery Mildew

Two important powdery mildew diseases affect field- and greenhouse-grown tomatoes in the United States. One disease is caused by the endophytic powdery mildew pathogen, *Leveillula taurica*. This powdery mildew was first reported on tomato in the United States in 1978 from California's Imperial Valley, but it has subsequently been found throughout North America. This disease is restricted to fairly well-defined subtropical production areas with arid to semiarid conditions during the cropping season.

The second significant powdery mildew disease of U.S. tomatoes is caused by *Oidium neolycopersici*. This disease was found first in Canada in 1994 and then in the United States in 1996; it is now widely distributed throughout North America. Powdery mildew caused by *O. neolycopersici* is favored by more humid environments and can be particularly problematic in greenhouse-grown tomatoes.

A third powdery mildew, caused by *O. lycopersici*, has been reported only from Australia. Initial reports of this fungus in the United States were likely *O. neolycopersici*.

Symptoms

The most common initial symptom of powdery mildew caused by *L. taurica* is the development of small (less than 1 cm), irregularly shaped, light-green to bright-yellow lesions on upper leaf surfaces (Fig. 53). Lesions appear first on older leaves. Under warm conditions, the lesions may expand and become necrotic, often coalescing; entire leaflets may die. Severely affected leaves remain attached to the stem. Sporulation is typically observed on lesions on the undersides of leaves. Under conditions of high humidity, a powdery sporulation may be observed on both upper and lower leaf surfaces.

The most common symptom of powdery mildew caused by *O. neolycopersici* is the appearance of white, powdery colonies on upper leaf surfaces. These colonies result from the production of mycelium and conidia on the leaf surface. As the disease progresses, the colonies often coalesce, forming a dense, powdery sporulation over much of the leaf. Severely affected leaves become chlorotic and then necrotic but often remain attached.

Causal Organisms

As noted earlier, there are three powdery mildew fungal pathogens of tomato: *Leveillula taurica* (Lév.) G. Arnaud (syns. *Oidiopsis sicula* Scalia, *O. taurica* (Lév.) E. S. Salmon); *Oidium neolycopersici* L. Kiss; and *O. lycopersici* Cooke & Massee. Morphological characteristics of *L. taurica* include the production of relatively long, often branched conidiophores and dimorphic conidia (pyriform and cylindrical). Conidia typically are formed singly or in short chains (Fig. 54). Pyriform conidia measure 17–24 × 50–71 μm, whereas cylindrical conidia measure 16–23 × 45–65 μm. The mycelia of *L. taurica* are endophytic. There are several reports of the presence of the perithecia (sexual stage) of *L. taurica,* but it is not common.

Morphological characteristics of the second fungus, *O. neolycopersici,* include the production of erect, long, unbranched conidiophores and a single conidium. However, under conditions of high humidity, conidia can be produced in false (or pseudo) chains of two to six. Ellipsoid–ovoid conidia measure 10–20 × 22–46 μm. The nuclear rDNA internal transcribed spacer (ITS) reference sequence for the neotype specimen of *O. neolycopersici* is deposited in GenBank under accession number AF229019. There are a few documented reports of perithecia of *O. neolycopersici* under greenhouse or field conditions.

Morphological characteristics of *O. lycopersici,* the third powdery mildew pathogen, are similar to those of *O. neo-*

Fig. 53. Symptoms of powdery mildew on leaves, caused by *Leveillula taurica*. (Courtesy R. L. Forster)

lycopersici, but conidia are formed in true chains of three to five. The nuclear rDNA ITS reference sequence for the neotype specimen of *O. lycopersici* is deposited in GenBank under accession number AF229021.

Disease Cycle and Epidemiology

Both *L. taurica* and *O. neolycopersici* have a broad host range and are able to infect a number of crop and weed species. However, it is not known if there are biotypes within either species that might have host specialization. Both are believed to represent species complexes composed of many different strains. It is likely that at least some weed species can serve as sources of primary inoculum for infecting tomato. Also, cross-infectivity between other solanaceous crop hosts and tomato has been demonstrated for both species. Conidia of both pathogens can germinate over a wide temperature range (10–35°C) and likely cause infection on tomato under conditions of high humidity. Infection can occur with little or no free moisture under high humidity. Secondary infection also readily occurs under conditions of high humidity. Conidia are readily windborne. Infection by *L. taurica* typically occurs during growing seasons with prolonged dry periods (little to no rainfall).

Early symptoms of powdery mildew are often difficult to observe. Even so, observing symptoms is critical for implementing fungicide-based management practices.

Management

Some tomato cultivars have resistance (or tolerance) to powdery mildew, but most of these cultivars are not well adapted for commercial production. The most common management practice involves the use of a wide range of registered fungicides, inorganic bicarbonate salts, sulfur, cupric hydroxide, and paraffinic oils. Some biological control agents—such as *Ampelomyces quisqualis* Ces. ex Schltdl., a hyperparasite—have also been shown effective in managing the disease.

Selected References

Arredondo, C. R., Davis, R. M., Rizzo, D. M., and Stahmer, R. 1996. First report of powdery mildew of tomato in California caused by *Oidium* sp. Plant Dis. 80:1303.

Belanger, R. R., and Jarvis, W. R. 1994. Occurrence of powdery mildew (*Erysiphe* sp.) on greenhouse tomatoes in Canada. Plant Dis. 78:640.

Braun, U., Cook, R. T. A., Inman, A. J., and Shin, H.-D. 2002. The taxonomy of the powdery mildew fungi. Pages 13-55 in: The Powdery Mildews: A Comprehensive Treatise. R. R. Bélanger, W. R. Bushnell, A. J. Dik, and T. L. W. Carver, eds. American Phytopathological Society, St. Paul, MN.

Correll, J. C., Gordon, T. R., and Elliott, V. J. 1987. Host range, specificity, and biometrical measurements of *Leveillula taurica* in California. Plant Dis. 71:248-251.

Karasevicz, D. M., and Zitter, T. A. 1996. Powdery mildew occurrence on greenhouse tomato plants in New York. Plant Dis. 80:709.

Kiss, L., Cook, R. T. A., Saenz, G. S., Cunnington, J. H., Takamasu, S., Pascoe, I., Bardin, M., Nicot, P. C., Sato, Y., and Rossman, A. Y. 2001. Identification of two powdery mildew fungi, *Oidium neolycopersici* sp. nov. and *O. lycopersici,* infecting tomato in different parts of the world. Mycol. Res. 105:684-697.

LaMondia, J. A., Smith, V. L., and Douglas, S. M. 1999. Host range of *Oidium lycopersicum* on selected solanaceous species in Connecticut. Plant Dis. 83:341-344.

(Prepared by J. Correll)

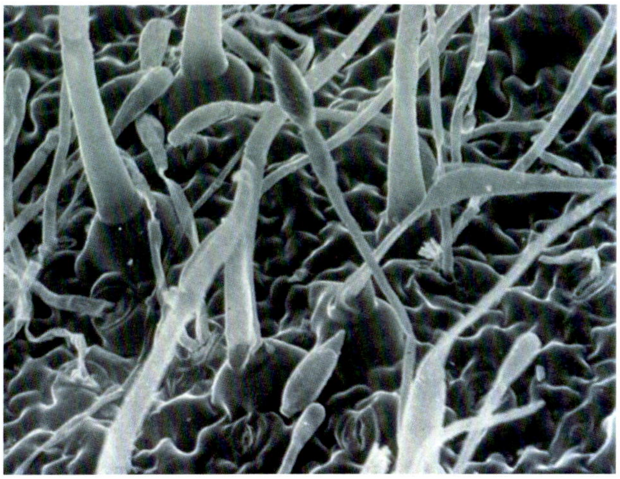

Fig. 54. Conidia and conidiophores of *Leveillula taurica* on a tomato leaf surface (scanning electron micrograph). (Courtesy S. V. Thomson)

Diseases Caused by *Pythium* Species

Several species of *Pythium* can infect tomato plants, causing pre- and postemergence damping-off or stem rot. The diseases occur on tomato in both greenhouse and field plantings and cause plant losses and uneven stand establishment. Under wet weather conditions and in poorly drained areas of fields, certain *Pythium* spp. can cause a fruit rot called watery rot or cottony leak of green or ripe fruit.

Symptoms

Infection of tomato by *Pythium* spp. can occur at any stage of seed germination or seedling development. Infected seeds develop a soft, mushy rot before the radicle emerges, which causes a reduction in seedling emergence. The most common symptom associated with infection by *Pythium* spp. is preemergence damping-off, which occurs during the early stages of germination. Germinating seedlings develop dark-brown or black, water-soaked lesions that spread rapidly.

Seedlings infected by postemergence damping-off develop dark-colored, water-soaked lesions on the root, which extend up the basal stem to or above the soil line. When dark-colored, soft lesions develop around a major portion of the stem or throughout the stem, the seedling falls over, desiccates, and dies. *P. aphanidermatum* and *P. myriotylum* sometimes attack plants up to 10 cm tall, causing dark-brown or black, water-soaked lesions that may extend 2–4 cm above the soil line (Fig. 55). Plants infected by these species usually wilt and die. A less severe root infection may result in stunted growth. *P. myriotylum* causes an aerial blight on leaves, petioles, and stems on tomato seedlings in the field. This pathogen is associated with higher

Fig. 55. Stem lesions, caused by *Pythium aphanidermatum.* (Courtesy J. P. Jones)

than average rainfall; white, cottony mycelial growth may be visible on plant parts under humid and moist conditions.

The fruit rot caused by *Pythium* spp. usually begins with the formation of small, water-soaked lesions on mature-green or ripe fruit in contact with or in proximity to the soil. Within 72 h, the rot engulfs the entire fruit, the epidermis is ruptured, and the fruit collapses, releasing its watery contents. Under high moisture conditions—such as when dew is present in the early morning in the field or in a moist chamber—a dense growth of white, cottony mycelium may appear on the lesion surface.

Causal Organisms

Pythium spp. are fungal-like organisms commonly referred to as water molds or oomycetes. Tomato seedling disease and stem rot are caused by several *Pythium* species: *P. aphanidermatum* (Edson) Fitzp., *P. myriotylum* Drechsler, *P. arrhenomanes* Drechsler, *P. ultimum* Trow, and *P. debaryanum* R. Hesse. *P. aphanidermatum* is more commonly associated with seedling disease than the other species, especially under moderate to high temperature conditions. *P. aphanidermatum* and *P. ultimum* have also been reported to cause fruit rot.

All *Pythium* spp. produce white, rapidly growing, coenocytic mycelia. *Pythium* spp. reproduce asexually by producing sporangia of various sizes and shapes, depending on the species. Some species, such as *P. aphanidermatum* and *P. myriotylum*, produce lobulate sporangia that arise from inflated, lobed hyphae (Fig. 56). Other species, such as *P. ultimum* and *P. debaryanum*, produce spherical sporangia in a terminal or intercalary fashion from mycelia. Depending on the species and the environmental conditions, the sporangia may germinate directly to form a germ tube or produce a sporangial vesicle for discharge of zoospores. Zoospores swim freely for a time and then encyst, producing cell walls. Cysts germinate and form germ tubes that infect plant tissues.

The *Pythium* spp. that infect tomato are homothallic. Thick-walled oospores arise within the oogonia after fertilization of the oosphere (egg) by nuclei from the attached antheridia. The morphology of the oogonium and antheridium and the number of antheridia attached to each oogonium vary among species. Sporangial and antheridial–oogonial characteristics are useful in the identification of *Pythium* spp. The morphology of the asexual and sexual stages of various *Pythium* spp. has been described by J. T. Middleton.

A mature oospore germinates after a period of dormancy to form a germ tube or a sporangium. The sporangium may germinate directly as a germ tube or release zoospores by means of a sporangial vesicle.

The optimum temperatures for organism growth and disease development also vary among *Pythium* spp. Some species, such as *P. aphanidermatum* and *P. myriotylum*, develop best at high temperatures (30°C or above), whereas other species, such as *P. ultimum* and *P. debaryanum*, favor low temperatures (below 20°C).

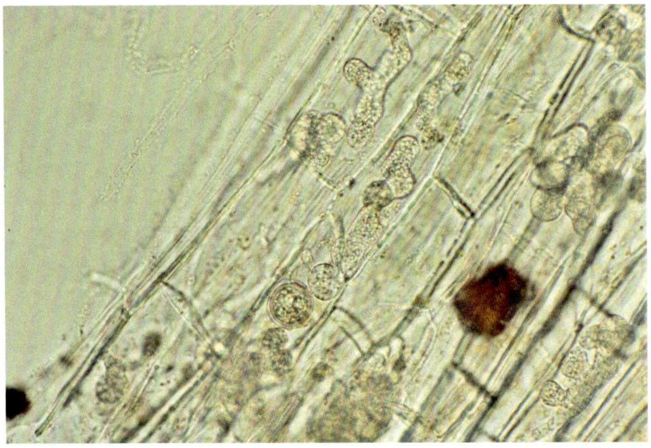

Fig. 56. Inflated, lobed sporangium and oospores of *Pythium aphanidermatum* on an infected stem (photomicrograph). (Courtesy A. W. Engelhard)

Disease Cycle and Epidemiology

Pythium spp. are good soil saprophytes and can grow as vegetative mycelia in the soil indefinitely on various types of organic substrates. Growth is influenced by a number of soil factors, such as the nature of the nutrient base, the temperature, the pH, the moisture level, and the activity of associated organisms. Sporangia and zoospores are produced when conditions are optimal, particularly the availability of free moisture. Nearly saturated soil stimulates mycelial vegetative growth and asexual reproduction. In the absence of a host, the thick-walled oospore provides for long-term survival.

Propagules capable of infecting the host include vegetative mycelia, germinating sporangia, germinating cysts, and germinating oospores. Some propagules, particularly zoospores, may be chemically attracted (i.e., chemotaxis) to parts of tomato plants, such as the roots. Infective hyphae are capable of direct penetration, but wounds enhance the penetration and infection process. Susceptibility to diseases caused by *Pythium* spp. is greater among plants that are highly succulent or etiolated because of suboptimum growth conditions, such as too low or too high temperature, excessive moisture, poor-quality light, and unbalanced nutrition (particularly resulting from excessive nitrogen).

Tomato fruit growing in very wet soil conditions can be invaded by certain *Pythium* spp. when the fruit touch the soil or soil is splashed onto the fruit surface. *Pythium* spp. cause rapid tissue destruction on both vegetative parts and tomato fruit, which is related to strong pectolytic and cellulolytic enzyme activity.

Management

Diseases caused by *Pythium* spp. can be managed by both cultural and chemical practices. Seed should be coated with fungicide. Transplants should be produced under optimal temperature, moisture, and nutritional conditions, avoiding excessive watering and providing adequate air circulation, plant spacing, light, and ventilation.

In the greenhouse, the growing medium should be sterilized and the containers disinfested before reuse. Disinfestation can be performed using hot water or steam or by placing items in a 10% sodium hypochlorite solution for 30 min. Workers should clean their hands and tools before handling healthy plants.

In field plantings, losses from diseases caused by *Pythium* spp. and other soilborne pathogens can be reduced through soil treatment with a broad-spectrum fumigant or solarization, under certain environmental conditions. Seeds and transplants should be planted in well-drained soil, avoiding low-lying, poorly drained areas. Fungicide can be applied in bands in beds by mixing it with transplant water, or it can be applied during fertigation. Applications of fungicide can begin as soon as the first symptoms of damping-off are observed. In addition to oomycetes, several fungi can cause damping-off, and the causal agent must be identified prior to selecting a fungicide treatment. Some biological control agents have been shown to suppress disease when used in treating seeds or applied in fields.

Pythium fruit rot can probably best be controlled by avoiding poorly drained areas and by using measures such as staking and covering beds with mulch to prevent the fruit from contacting the soil. Fungicide applications may also help to suppress this disease.

Selected References

Dixon, G. R. 1981. Vegetable Crop Diseases. Avi, Westport, CT.

MacNab, A. A., and Sherf, A. F. 1986. Vegetable Diseases and Their Control. 2nd ed. John Wiley & Sons, New York.

Middleton, J. T. 1943. The taxonomy, host range, and geographical distribution of the genus *Pythium*. Mem. Torrey Bot. Club 20:1-171.

Pearson, R. C., and Hall, D. H. 1973. Ripe fruit rot of tomato caused by *Pythium ultimum* and *Pythium aphanidermatum*. Plant Dis. Rep. 57:1066-1069.

Roberts, P. D., Urs, R., and McGovern, R. J. 1999. First report of aerial blight caused by *Pythium myriotylum* on tomato in Florida. Plant Dis. 83:301.1.

van der Plaats-Niterink, A. J. 1981. Monograph of the genus *Pythium*. Studies in Mycology 21. Centraalbureau voor Schimmelcultures, Baarn, the Netherlands.

(Prepared by R. D. French-Monar, P. D. Roberts, and S. M. McCarter)

Diseases Caused by *Rhizoctonia solani*

Rhizoctonia diseases of tomato—including damping-off, root rot, basal stem rot (foot rot), stem canker, and fruit rot—are prevalent throughout the world. Pre- and postemergence damping-off of young seedlings occur in both greenhouse and field plantings. Root rot is most severe when plants are under stress or when roots have been injured (e.g., by root-knot nematodes). Aboveground stem cankers sometimes develop at pruning wounds. Fruit rot is a serious problem in warm, wet regions of the world and is a limiting factor in the production of processing tomatoes in the southern United States.

Symptoms

Germinating seedlings can be killed by Rhizoctonia disease before or soon after they emerge from the soil (Fig. 57). Before emergence, seedlings rot after the pathogen kills the growing tips. Germinating seedlings that are infected have lesions that are tan to reddish brown in color. After emergence, seedlings have lesions that are brown, reddish brown, or nearly black; the lesions form on the plant near the soil line. The young stem becomes constricted and soft, and the plant topples over and dies. Older seedlings are also susceptible, but damage is restricted mostly to the cortical regions as the plant becomes more resistant. Under certain conditions, symptoms may occur on older or mature plants.

Root symptoms appear either as distinct, dark-colored lesions or as a more general reddish or brown rot. Root rot can occur on plants growing under adverse soil and environmental conditions or when roots are damaged (e.g., by root-knot nematodes).

Fig. 57. Damping-off of tomato seedlings, caused by *Rhizoctonia solani*. (Courtesy J. P. Jones)

A basal stem canker (sometimes called foot rot) can develop on growing or mature plants under conditions that are favorable for fungal development but unfavorable for plant growth. The canker is characterized by sunken brown or reddish-brown lesions at or slightly below the soil line. The lesions may enlarge to include most of the basal part of the plant and may even extend into the root system.

Sunken stem cankers sometimes develop at the bases of branches. Symptoms include lesions with alternating bands of dark/red brown and tan/light brown and sharply defined margins. These lesions result from the growth of the pathogen through the midribs of leaves in contact with the soil. Stem lesions that are brown and shrunken and filled with coarse, brown mycelium have been observed at pruning sites, presumably from the splashing of inoculum or the transmission of airborne basidiospores.

Fruit in contact with the soil in warm, wet areas develop a brown rot in which lesions often have alternating light- and dark-colored, concentric bands (Fig. 58). Green fruit are sometimes infected, but ripe fruit are more susceptible. The rot may be fairly firm at first, but the fruit rapidly become soft and mushy as the pathogen advances and secondary organisms invade. If fruit with incipient infections are harvested and shipped, the fungus may grow on fruit in transit, causing a brown "nest" rot. Under moist conditions, the brown, coarse mycelium may appear on the surface of any infected plant part. Mycelial growth is most obvious on fleshy parts, such as fruit.

Causal Organism

The causal organism, *Rhizoctonia solani* J. G. Kühn, is a complex species consisting of strains that differ in host range, pathogenicity, cultural characteristics, and response to the environment. The species is commonly divided into genetically different anastomosis groups defined by pairing an isolate with tester strains and observing hyphal fusion. Some anastomosis groups have been subdivided based on frequency of hyphal anastomosis, cultural characteristics, polymorphism in the ribosomal RNA (rRNA) gene, morphology, and pathogenicity.

The fungus grows readily on laboratory media, such as potato dextrose agar. Although minimum, optimum, and maximum temperatures for growth of individual isolates may vary, most grow satisfactorily in the laboratory at 25–30°C.

Isolates vary greatly in appearance in culture. Some are almost white, but most are various shades of brown, especially as they get older. Also, some strains grow flat, while others produce a considerable amount of aerial mycelium. Loosely formed

Fig. 58. Fruit rot, caused by *Rhizoctonia solani*. (Courtesy J. P. Jones)

brown to black sclerotia may appear in culture. When observed microscopically, the mycelium appears hyaline or light tan and has characteristic right-angle branching, often with a slight constriction and a septum near each branch (Fig. 59).

R. solani occurs ubiquitously, causing some type of disease on almost every cultivated plant. The organism has excellent saprophytic ability, but the degree varies among isolates. The pathogen does not produce any asexual spores but sometimes produces a basidial sexual stage (*Thanatephorus cucumeris* (A. B. Frank) Donk) in culture and in association with its host. Basidiospore inoculum has been implicated in the formation of aerial stem cankers on tomato. A basidial hymenium layer may develop on tomato fruit in contact with the soil.

Disease Cycle and Epidemiology

R. solani is a common soil inhabitant, surviving as actively growing mycelium, resting mycelium, and sometimes sclerotia. The fungus can colonize most dead plant material. The fungus's survival and inoculum potential are influenced by soil factors such as temperature, moisture, and pH, as well as the competitive activity of associated microorganisms.

Isolates of *R. solani* differ in survival capacity. Propagules of the fungus can be disseminated through movement of contaminated plant parts and soil. Isolates also differ in the optimum temperatures at which they cause disease. Disease is usually most severe under temperature conditions that are adverse to the host, such as unusually low temperatures (15–18°C). However, the fruit rot phase is severe under midsummer conditions in the southeastern United States. The pathogen is most active when the soil is moderately moist; dry and waterlogged soil conditions tend to inhibit fungal development.

The fungus readily invades plants through wounds but can also invade young, succulent tissue directly. *R. solani* also produces mycelial mats that enhance penetration of plant parts. The pathogen causes rapid tissue destruction by strong enzymatic action.

Management

Several measures can be used to manage the seed rot and seedling disease phases caused by *R. solani*. The seed used for greenhouse and field plantings should be treated with a fungicide or fungicide mixture to control *R. solani* and other seedborne and soilborne pathogens. Tomato plants grown with good-quality seed under optimum temperature, moisture, and nutritional conditions will be less susceptible to seedling disease than those grown under suboptimal conditions.

In the greenhouse, the planting medium should be sterilized, and in field plantings, the soil should be fumigated with a broad-spectrum fumigant. These measures will help control *R. solani* and other soilborne fungal pathogens. Biological control agents, applied in the greenhouse or in the field, may also reduce the impact of diseases caused by *R. solani*. Maintaining optimum growing conditions in the field—such as irrigating and fertilizing adequately and preventing plant injury (especially by nematodes)—will reduce the incidence of root rot and basal stem rot caused by *R. solani*.

The most satisfactory way to prevent losses from fruit rot is to prevent direct contact of the fruit with the soil when grown under warm, moist conditions. Staking and using various types of film mulches to cover the soil surface are also effective management practices. Certain fungicides may be applied to reduce losses caused by fruit rot.

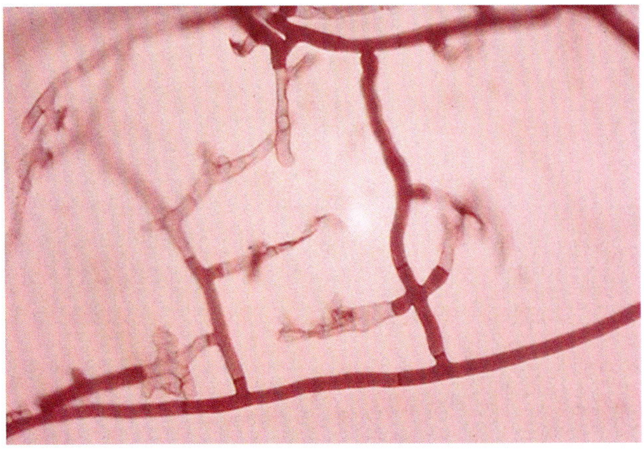

Fig. 59. Right-angle branching of mycelium of *Rhizoctonia solani* (photomicrograph). (Courtesy S. M. McCarter)

Selected References

Conover, R. A. 1949. Rhizoctonia canker of tomato. Phytopathology 39:950-951.

Cubeta, M. A., Echandi, E., Abernethy, T., and Vilgalys, R. 1991. Characterization of anastomosis groups of binucleate *Rhizoctonia* species using restriction analysis of amplified ribosomal RNA gene. Phytopathology 81:1395-1400.

Golden, J. K., and Van Gundy, S. D. 1975. A disease complex of okra and tomato involving the nematode, *Meloidogyne incognita,* and the soil-inhabiting fungus, *Rhizoctonia solani.* Phytopathology 65:265-273.

Parmeter, J. R., Jr. 1970. *Rhizoctonia solani:* Biology and Pathology. University of California Press, Berkeley.

Small, T. 1927. Rhizoctonia "foot rot" of the tomato. Ann. Appl. Biol. 14:290-295.

Sneh, B., Burpee, L., and Ogoshi, A. 1991. Identification of *Rhizoctonia* species. American Phytopathological Society, St. Paul, MN.

(Prepared by P. D. Roberts, R. D. French-Monar, and S. M. McCarter)

Rhizoctonia Foliar Blight

Rhizoctonia foliar blight of tomato was recently identified in Japan and North America, and there are undocumented reports of this disease occurring in Latin America. The causal relationship of *Rhizoctonia solani* anastomosis group 3 (AG-3) and this disease on tomato was not demonstrated by fulfillment of Koch's postulates until 2009, although foliar diseases caused by *R. solani* AG-3 and teleomorph *Thanatephorus cucumeris* have been reported from other solanaceous hosts, including eggplant (*Solanum melongena* L.), potato (*S. tuberosum* L.), and tobacco (*Nicotiana* L.).

Symptoms

Symptoms observed in the field on plants naturally infected with Rhizoctonia foliar blight include brown, necrotic lesions and blighted leaves (Fig. 60). Extensive necrosis can be observed on the adaxial leaf surface (Fig. 61). White mycelial growth on the undersides of infected leaves is also observed (Fig. 62).

Causal Organism

The soil fungus *Rhizoctonia solani* J. G. Kühn exists as a heterogeneous species whose current classification is based on hyphal anastomosis (fusion) to established anastomosis groups (AGs). Rhizoctonia foliar blight on tomato is caused by *R. solani* anastomosis group 3 (AG-3) (teleomorph *Thanatephorus cucumeris* (A. B. Frank) Donk). The identity of the fungus is based on hyphal anastomosis criteria and sequence analysis of the internal transcribed spacer (ITS) region of the ribosomal DNA (rDNA) repeat. Isolates of *R. solani* AG-3 from tomato are phylogenetically related to isolates of the same AG from potato more so than tobacco, although their disease biology (epidemiology) is more similar to that of the pathogen that causes target spot of tobacco.

Hymenia may be white to various shades of brown and resupinate with or without aerial hyphae. Hyphae are 5–13 µm wide, hyaline to brown in color, without clamp connections, and multinucleate; they typically have slight constrictions at branch points, with dolipore septa (Fig. 63). Aggregations of monilioid cells form loose, cream- to brown-colored sclerotia on plant tissue and in culture. Holobasidia of the teleomorph are formed discontinuously on short, vertically branching hyphae arising from a basal hyphal layer. Basidia are 8–10 × 15 µm, hyaline, and barrel shaped or subcylindrical, and they typically have four sterigmata (Fig. 64). Sterigmata are 2–4 × 5–13 µm and typically broader at the base than at the tip. Basidiospores are 4–5 × 4–7 µm, smooth, hyaline, and oblong to ellipsoid; they can be unilaterally flattened and are prominently apiculate.

Disease Cycle and Epidemiology

Little is known about the disease cycle of Rhizoctonia foliar blight. The disease was initially identified in the United States in 2005 from a commercial tomato field (cultivar Mountain Fresh Plus) in Rowan County, North Carolina, and then subsequently identified a year later in four fields in two additional North Carolina counties (Henderson and Polk). These fields had no history of fungicide use, and the disease was observed during warm, humid weather conditions with extended periods of leaf wetness. The economic impact and importance of Rhizoctonia foliar blight disease is currently not known.

When foliar isolates of *R. solani* were used to infest soil in greenhouse inoculations, root necrosis developed on both tomato and eggplant seedlings. Basidia and basidiospores were

Fig. 62. Symptoms of Rhizoctonia foliar blight on abaxial leaf surfaces. (Courtesy K. L. Ivors)

Fig. 60. Necrotic lesions and mycelia on a tomato leaf, caused by Rhizoctonia foliar blight. (Courtesy K. L. Ivors)

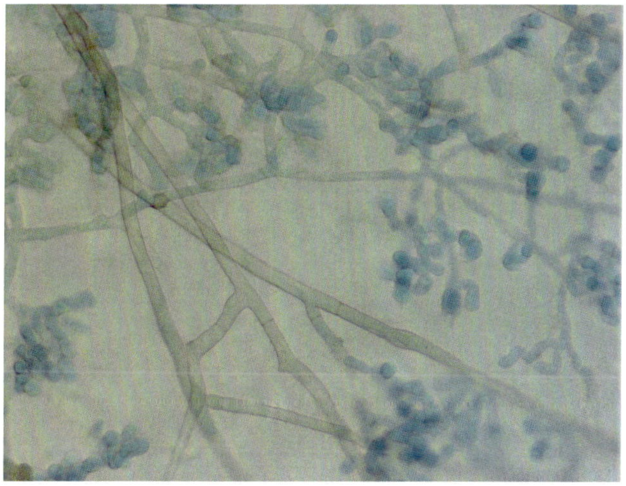

Fig. 63. Mycelia (stained with lactophenol blue) from the abaxial side of an infected tomato leaf (photomicrograph). (Courtesy K. L. Ivors)

Fig. 61. Symptoms of Rhizoctonia foliar blight on adaxial leaf surfaces. (Courtesy K. L. Ivors)

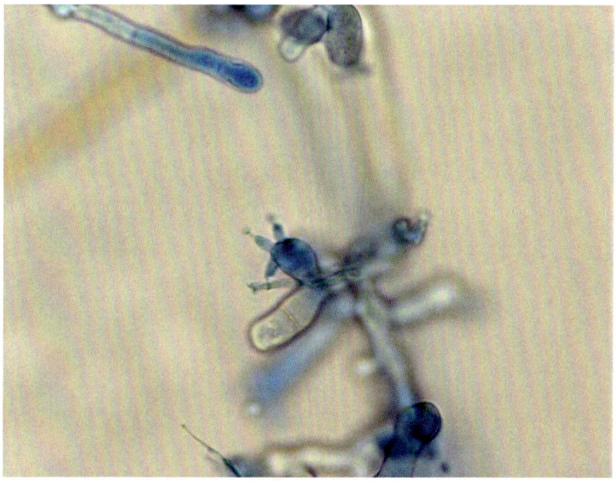

Fig. 64. Basidia with sterigmata (stained with lactophenol blue) on a tomato stem (photomicrogaph). (Courtesy K. L. Ivors)

observed in tape mounts taken from infected tomato stem tissue after infesting soil with oat grains colonized with mycelium of the fungus in growth chamber experiments (Figs. 61 and 62).

Since disease symptoms in the field were observed on leaves in the middle and upper plant canopy, the primary inoculum is assumed to be aerially dispersed basidiospores. No disease symptoms have been observed on tomato leaves lower in the canopy—a pattern likely to result from splash dispersal of fungal propagules from soil.

Management

Rhizoctonia foliar blight has been identified only on tomato plants not exposed to foliar fungicides. This indicates that some fungicides, such as products containing mancozeb and chlorothalonil, may be effective at preventing this pathogen from colonizing leaf surfaces and causing infection.

Selected References

Bartz, F. E., Cubeta, M. A., Toda, T., Naito, S. and Ivors, K. L. 2010. An in planta method for assessing the role of basidiospores in Rhizoctonia foliar disease of tomato. Plant Dis. 94:515-520.

Date, H., Yagi, S., Okamoke, Y., and Oniki, N. 1984. On the leaf blight of tomatoes by *Thanatephorus cucumeris* (Frank) Donk (*Rhizoctonia solani*). Ann. Phytopathol. Soc. Jpn. 50:399.

Ivors, K. L., Bartz, F. E., Toda, T., Naito, S. and Cubeta, M. A. 2009. First report of tomato foliar blight caused by *Rhizoctonia solani* AG-3 basidiospore infection in North America. (Abstr.) Phytopathology 99:S57.

Johnk, J. S., Jones, R. K., Shew, H. D., Carling, D. E. 1993. Characterization of populations of *Rhizoctonia solani* AG-3 from potato and tobacco. Phytopathology 83:854-858.

Parmeter, J. R., Jr. 1970. *Rhizoctonia solani:* Biology and Pathology. University of California Press, Berkeley.

Sneh, B., Jabaji-Hare, S., Neate, S., and Dijst, G. 1996. *Rhizoctonia* Species: Taxonomy, Molecular Biology, Ecology, Pathology, and Disease Control. Kluwer Academic, Dordrecht, the Netherlands.

(Prepared by F. Bartz, M. Cubeta, and K. L. Ivors)

Septoria Leaf Spot

Septoria leaf spot is one of the most destructive diseases of tomato foliage. It is widely distributed throughout the world, with reports coming from Africa, Asia, Australia, Europe, North America, and South America. The disease occurs in most U.S. states where tomatoes are grown and is particularly severe in areas where wet, humid conditions persist for extended periods. In seasons of moderate temperature and abundant rainfall, the disease reduces yield and often destroys so much of the foliage that either the fruit fails to mature properly or sunscald becomes a major problem.

Symptoms

Symptoms of Septoria leaf spot usually appear on the lower leaves after the first fruit sets. Lesions also appear on the stems, petioles, and calyx. Lesions are circular (approximately 1–3 mm in diameter) and have dark-brown margins and tan to gray centers dotted with black pycnidia (Figs. 65 and 66). When conditions are ideal for enlargement and the tissue is succulent, lesions may approach 5 mm in diameter. A narrow, yellow halo is often associated with leaf lesions. Stem, petiole, and calyx lesions are generally smaller and darker brown than those on leaves, and pycnidia may or may not be produced.

The disease spreads upward, from oldest to youngest growth. If the leaf lesions are numerous, some infected leaves will turn slightly yellow and then brown, and they will wither. Fruit infection is rare.

Causal Organism

The fungal pathogen, *Septoria lycopersici* Speg., produces numerous black pycnidia in the centers of mature lesions. The pycnidia produce hyaline, filiform conidia up to 10 cells long. The conidia average 3.2 × 67.0 µm in size. No sexual stage has been reported for this species, but at least two physiological races have been identified.

Disease Cycle and Epidemiology

S. lycopersici overwinters on infected tomato debris or debris of solanaceous weed hosts, such as horsenettle (*Solanum carolinense* L.) The fungus can also survive on equipment such as plant stakes and cages. Seed infection can aid in the long-distance spread of the pathogen, but seed produced in arid areas is likely to be pathogen free. Once conidia have been introduced to a planting area, they are spread primarily by rainfall. Workers, equipment, and insects moving through wet foliage can also spread the fungus from plant to plant.

Germinating conidia penetrate leaf tissue through the stomata. Symptoms can be observed within 6 days after inoculation, whereas pycnidia do not appear until approximately 14 days after inoculation. Older leaves seem more susceptible to infection than younger leaves when the inoculum density is high.

Fig. 65. Symptoms of Septoria leaf spot. (Courtesy W. R. Stevenson)

Fig. 66. Pycnidia on a leaflet, characteristic of Septoria leaf spot. (Courtesy W. R. Stevenson)

Temperatures of 20–25°C are optimum for infection, symptom expression, and development of pycnidia. Long periods of high relative humidity, moderate temperatures, and dew constitute ideal conditions for disease development.

Management

Currently grown tomato cultivars are susceptible to Septoria leaf spot, although breeding offers the potential for including resistance as part of future disease management programs. Several protectant fungicides are currently labeled for use on tomato for Septoria leaf spot control. To achieve effective control, these products must be applied at regular intervals and achieve adequate coverage during the growing season, particularly when environmental conditions favor infection and dispersal of conidia.

A variety of cultural controls can also be used to manage Septoria leaf spot on tomato. Crops should be rotated away from tomato for at least 1–2 years, and susceptible weeds (such as horsenettle) should be controlled in tomato and rotation crops. In addition, crop debris should be removed from planting areas (gardens and small plantings) or incorporated immediately after harvest. Careful attention must be paid to the timing of irrigation to reduce the length of time during which plants remain wet, and plants should be staked to improve air circulation. Plastic or organic mulches should be used to reduce contact between the foliage and soil, and workers and equipment should be kept out of fields with wet foliage.

Selected References

Barksdale, T. H. 1982. Control of an epidemic of Septoria leaf spot of tomato by resistance. Plant Dis. 66:239-240.

Barksdale, T. H., and Stoner, A. K. 1978. Resistance in tomato to *Septoria lycopersici*. Plant Dis. Rep. 62:844-847.

Elmer, W. H., and Ferrandino, F. J. 1995. Influence of spore density, leaf age, temperature, and dew periods on Septoria leaf spot of tomato. Plant Dis. 79:287-290.

Malnati, W. D. 1993. Possibilities of including genetic resistance to control tomato Septoria leaf spot. Fitopatol. Bras. 18:469-477.

Parker, S. K., Gleason, M. L., and Nutter, F. W., Jr. 1995. Influence of rain events on spatial distribution of Septoria leaf spot of tomato. Plant Dis. 79:148-152.

Sherf, A. F., and MacNab, A. A. 1986. Vegetable Diseases and Their Control. John Wiley & Sons, New York.

(Prepared by W. R. Stevenson)

Southern Blight

Southern blight was first reported on tomato, but it also occurs on hundreds of other species of plants, including many economically important vegetable, ornamental, and agronomic crops. The disease has several other common names, including southern wilt, southern stem rot, and Sclerotium stem rot. Southern blight occurs worldwide but is most important in tropical and subtropical regions.

Symptoms

Symptoms of southern blight usually appear on plant parts in contact with or near the soil. Plants of any age are susceptible when environmental conditions are suitable for development of the disease.

The most common symptom is a rot of the stem, in which a brown to black lesion develops near the soil line. The lesion develops rapidly to completely girdle the stem, resulting in a sudden and permanent wilt of all aboveground plant parts (Fig. 67). Young plants may fall over at the soil line. Under moist conditions, an abundance of robust, white mycelium develops on the lesion and sometimes extends several centimeters up the stem of a mature plant. After a few days, tan to reddish-brown, spherical sclerotia (average diameter 1–2 mm) appear on the mycelial mat (Fig. 68). Brown lesions with surface mycelium may appear on foliage in contact with the soil or on lower leaves splashed with inoculum from the soil.

The fungus that causes southern blight quickly penetrates the epidermis of fruit in contact with infested soil. The infection site first appears sunken and slightly yellow, with a ruptured epidermis. The lesion becomes water soaked and soft and is often star shaped. The fruit collapses within 3–4 days of infection. White mycelium and developing sclerotia quickly fill the lesion cavity (Fig. 69).

Fig. 67. Field-grown tomato plants affected by southern blight. (Courtesy J. P. Jones)

Fig. 68. Mycelium and sclerotia of *Sclerotium rolfsii* on the plant stem. (Courtesy J. C. Watterson)

Fig. 69. Mycelium and sclerotia of *Sclerotium rolfsii* on tomato fruit. (Courtesy S. M. McCarter)

Causal Organism

Southern blight is caused by the fungus *Sclerotium rolfsii* Sacc. *S. rolfsii* grows rapidly on common laboratory media, such as potato dextrose agar (PDA). The white mycelium of the fungus grows outward in a fan shape. Sclerotia may differ in size, color, or morphology among different isolates. Sclerotia begin forming on the surface of the mycelial mat after 5–6 days of growth at an optimum temperature (27–30°C). The round sclerotia form as fuzzy, white mycelia, which turn brown when mature. The mature sclerotium measures 1–2 mm and is covered with a hard, protective rind.

A basidial perfect stage (*Athelia rolfsii* (Curzi) Tu & Kimbr.) has been reported. Two or four thin-walled, colorless basidiospores are produced and borne on short spines at the ends of slightly enlarged short threads.

Diagnostic signs of *S. rolfsii* include the characteristic white mycelial fans and tan to reddish-brown sclerotia extending from infected tissues. The sclerotia sometimes extend into the soil in close proximity to the infected tissue.

Disease Cycle and Epidemiology

The fungus can survive for years as sclerotia in soil and host debris. The sclerotia disseminate by the movement of infested soil or plant material. Environmental conditions that are favorable for the fungus and disease development are high temperatures (25–35°C), aerobic and moist conditions, and acidic soil. Germination of sclerotia occurs at pH 2–5 and is inhibited at pH higher than 7.

Infection by *S. rolfsii* is usually limited to plant parts in contact with the soil. The fungus is also highly saprophytic and capable of producing abundant mycelial growth on a variety of host substrates. It often uses organic matter as a substrate for the production of oxalic acid and enzymes that result in host tissue disintegration.

Management

Management of southern blight is difficult when inoculum levels are high and environmental conditions are conducive to the disease. Avoiding the disease by selecting fields that are free of *S. rolfsii* is the most successful method of control. Crop rotation with nonsusceptible grass crops—including corn (*Zea mays* L.) and small grains, such as wheat (*Triticum* L.) and barley (*Hordeum* L.)—reduces the incidence of disease by reducing inoculum levels. Because of the pathogen's wide host range, crop rotation must be done with a nonhost crop, such as grass, corn, or wheat.

The use of effective fungicides in the transplanting solution has proven economical. The fumigation of soils with broad-spectrum chemicals reduces disease incidence, but this strategy is limited by economic considerations. Certain fertilization regimes, such as the use of high calcium levels and ammonium-type fertilizers, have provided some control under low disease pressure. Use of a physical barrier (e.g., aluminum foil, plastic) to protect plant stems at the soil line is sometimes recommended for homeowners.

Sanitation methods and cultural control practices may also help to manage southern blight. Diseased plants should be rogued and weed populations reduced or eliminated. Injuring plants during cultivation should be avoided as much as possible, and adopting wider plant spacings can help reduce infection, since a dense canopy increases disease incidence. Modifying the planting date can help reduce disease incidence, as well, if planting is timed to avoid wet, warm periods. Soil pH can be raised by liming. Disease incidence is lower when ammonium, rather than nitrate, fertilizers are applied.

S. rolfsii is highly aerobic, which means the inoculum level can be reduced by burying infected plant debris and sclerotia via deep plowing (at least 20 cm) with a mold-board extension that inverts the soil. Buried soil must not be brought back to the surface during the growing season. The use of plastic mulch and row covers reduces disease severity and incidence by providing a barrier between the plants and the soil. The addition of oat or straw compost or straw to the soil has also been shown to reduce disease incidence.

These soil amendments may increase populations of antagonistic soil microorganisms. A number of commercially available biological control agents have been shown to reduce damage from *S. rolfsii*. Soil solarization combined with the introduction of *Trichoderma harzianum* Rifai has been shown more effective in reducing disease incidence than either treatment alone.

Selected References

Aycock, R. 1966. Stem rot and other diseases caused by *Sclerotium rolfsii*. N.C. Agric. Exp. Stn. Tech. Bull. 174.

Baxter, L. W., Jr., Witcher, W., Fagan, S. G., and Owen, M. G. 1977. Control of southern blight by physical means. Plant Dis. Rep. 61:341-342.

Beute, M. K., and Rodriguez-Kabana, R. 1981. Effects of soil moisture, temperature, and field environment on survival of *Sclerotium rolfsii* in Alabama and North Carolina. Phytopathology 71:1293-1296.

Brown, J. E., Stevens, C., Osborn, M. C., and Bryce, H. M. 1989. Black plastic mulch and spunbonded polyester row cover as method of southern blight control in bell pepper. Plant Dis. 73:931-932.

Gurkin, R. S., and Jenkins, S. F. 1985. Influence of cultural practices, fungicides, and inoculum placement on southern blight and Rhizoctonia crown rot of carrot. Plant Dis. 69:477-481.

Jenkins, S. F., and Averre, C. W. 1986. Problems and progress in integrated control of southern blight of vegetables. Plant Dis. 70:614-619.

Mihail, J. D., and Alcorn, S. M. 1984. Effects of soil solarization on *Macrophomina phaseolina* and *Sclerotium rolfsii*. Plant Dis. 68:156-159.

Mordue, J. E. M. 1974. *Corticium rolfsii*. Descriptions of Pathogenic Fungi and Bacteria, no. 410. C.A.B. International Mycological Institute, Kew, England.

Peeples, J. L., Curl, E. A. and Rodriguez-Kabana, R. 1976. Effect of the herbicide EPTC on the biocontrol activity of *Trichoderma viride* against *Sclerotium rolfsii*. Plant Dis. Rep. 60:1050-1054.

Punja, Z. K. 1985. The biology, ecology, and control of *Sclerotium rolfsii*. Annu. Rev. Phytopathol. 23:97-127.

Weber, G. F. 1931. Blight of carrots caused by *Sclerotium rolfsii*, with geographic distribution and host range of the fungus. Phytopathology 21:1129-1140.

(Prepared by P. D. Roberts, R. D. French-Monar, and S. M. McCarter)

Target Spot

Target spot of tomato has been reported on both greenhouse and field tomatoes in more than 25 countries—including Brazil, India, Japan, Nigeria, the Philippines, and the United States—and in both the Pacific and the Caribbean Basins. This field disease is restricted to crops in tropical and subtropical regions during extended periods of high humidity. In the United States, reports of losses from target spot in field tomatoes are restricted to the Southeast. Since the first observation of target spot in Florida in 1972, the disease has increased in both severity and occurrence, to the point that it is now one of the most damaging tomato diseases, especially in the southern half of the state.

Symptoms

The initial symptom of target spot is the appearance of small (pinpoint-size), water-soaked lesions on leaves' upper surfaces. The lesions gradually increase in size and become circular and pale brown. Individual lesions develop conspicuous yellow halos (Fig. 70). In highly susceptible leaflets, the lesions are sometimes so numerous that they are difficult to count. The coalescence of two or more lesions in a leaflet may result in a rapid collapse of tissue, but the leaflet will remain attached to the petiole.

Lesions on the petioles and stems are brown and oblong to elongate and resemble lesions on leaves. Pinpoint lesions rapidly increase in size and may girdle the petioles and stems, leading to leaflet collapse. The breakdown of the petioles and blades is sometimes quite rapid, resembling symptoms of a blight similar to bacterial spot or speck. Tomato spot is sometimes overlooked early in an epidemic because it begins deep in the canopy and "melts out" of the tomato crop starting from the inside. Sporulation is greater on the undersides of leaves than on the uppersides.

Lesions of young fruit first appear as dark, sunken, pinpoint, brown spots. These lesions can be confused with symptoms of abiotic problems, such as sand injury. The legions may enlarge and develop into craters (Fig. 71).

On ripe fruit, tomato spot is characterized by large, circular lesions with pale-brown centers that crack in the middles. The fungus often sporulates on the surfaces of large fruit lesions. In some tomato cultivars, fruit lesions are reduced in size or absent but may be invaded by secondary pathogens as the fruit increases in size.

Causal Organism

Corynespora cassiicola (Berk. & M. A. Curtis) C. T. Wei is easily cultured on potato dextrose agar (PDA), producing pale-brown to dark-olive or gray colonies with sporulation in 7–10 days under ambient light. Sporulation can easily be induced from stem and leaf tissue within 24 h of placement in a moisture chamber. Conidial length (40–200 µm), width (9–22 µm), and pseudoseptation (4–20), as well as conidiophore length (110–850 µm) and width (4–11 µm), exhibit considerable morphological plasticity that is dependent on humidity, light, temperature, and substrate. Conidia are straight or curved, obdurate or cylindrical, and smooth subhyaline to pale, olivaceous brown; they occur singly or in short chains of two to six on the conidiophore tip.

On tomato tissue, key diagnostic features are indeterminate conidiophores (Fig. 72) and apically budding, pseudoseptate (i.e., walls of the individual septa do not extend all the way to the outer walls) conidia with pronounced basal hila (Fig. 73).

Fig. 70. Target spot lesions on leaves, caused by *Corynespora cassiicola.* (Courtesy J. P. Jones)

Disease Cycle and Epidemiology

C. cassiicola has been found on more than 250 genera in 77 plant families, including hydrangea (*Hydrangea* L.), soybean (*Glycine max* (L.) Merr.), cucumber (*Cucumis sativus* L.), papaya (*Carica* L.), rubber tree (*Hevea brasiliensis* (Willd. ex A. Juss.) Müll. Arg.), and common weeds. There are also numerous subspecies with more limited host ranges. Individual isolates from tomato vary in virulence and host specificity but have been shown to infect cucumber.

Development of tomato spot is favored by temperatures of 20–28°C, but infection can occur at temperatures of 16–32°C. Long periods (16–44 h) of high moisture are necessary for optimum disease development.

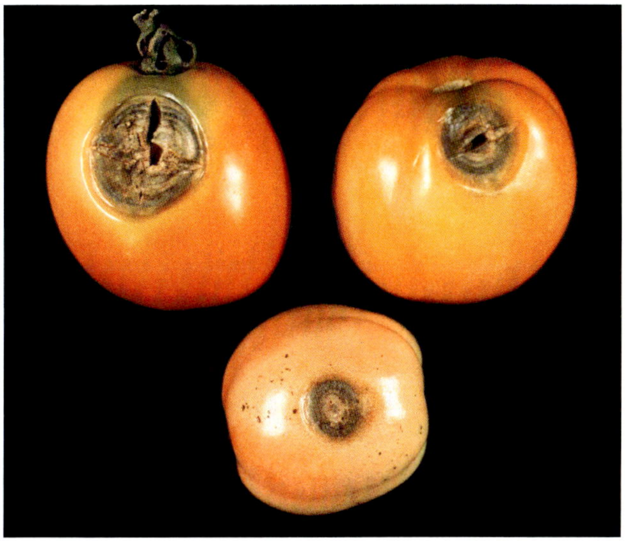

Fig. 71. Target spot lesions on fruit, caused by *Corynespora cassiicola.* (Courtesy J. P. Jones)

Fig. 72. Conidiophore of *Corynespora cassiicola.* (Note the attached conidia.) (Courtesy G. Vallad)

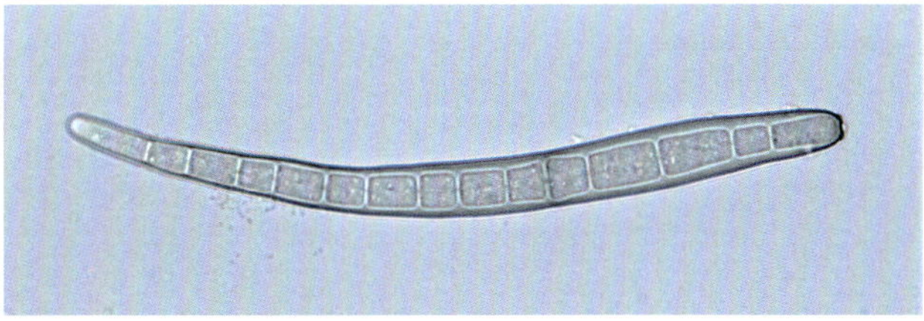

Fig. 73. Conidium of *Corynespora cassiicola.* (Courtesy G. Vallad)

Management

Some resistance to tomato spot has been found in cultivars screened in fields in southern Nigeria and in greenhouses in the U.S. state of Wisconsin, but no resistant commercial cultivars are available. Tomato plants should be sprayed with efficacious fungicides, starting before symptoms of disease are apparent. Fungicides must be used in rotation as part of a resistance management program.

Selected References

Blazquez, C. H. 1967. Corynespora leaf spot of cucumber. Proc. Fla. State Hortic. Soc. 80:177-182.

Blazquez, C. H. 1977. Blight of tomatoes caused by *Corynespora cassiicola*. Plant Dis. Rep. 61:1002-1006.

Deighton, F. C. 1936. A preliminary list of fungi and diseases of plants from Sierra Leone. Kew. Bull. 1936:397-424.

Gussow, H. T. 1906. Uber eine neue krankheit an gurken in England (*Corynespora mazei,* Gussow gen et spec. nov.) (In German.) Z. Pflanzenkr. 16:10-13.

Jones, J. P., and Jones, J. B. 1984. Target spot of tomato. Epidemiology and control. Proc. Fla. State Hortic. Soc. 97:216-218.

Jones, J. P., and Jones, J. B. 1988. Field control of target spot and bacterial speck of tomato. Proc. Fla. State Hortic. Soc. 101:358-361.

Pernezny, K., Datnoff, L. E., Mueller, T., and Collins, J. 1996. Losses in fresh-market tomato production in Florida due to target spot and bacterial spot and the benefits of protectant fungicides. Plant Dis. 80:559-563.

Pernezny, K., Stoffella, P., Collins, J., Carroll, A., and Beany, A. 2002. Control of target spot of tomato with fungicides, systemic acquired resistance activators, and a biocontrol agent. Plant Prot. Sci. 38:81-88.

Schlub, R. L., Smith, L. J., Datnoff, L. E., and Pernezny, K. 2009. An overview of target spot of tomato caused by *Corynespora cassiicola*. Acta Hortic. 808:25-28.

Smith, L. J., Datnoff, L. E., Pernezny, K., and Schlub, R. L. 2009. Phylogenetic and pathogenic characterization of *Corynespora cassiicola*. Acta Hortic. 808:51-56.

Volin, R. B., Pohronezny, K., and Simone, G. W. 1989. Severe spotting of fresh market tomato fruit incited by *Corynespora cassiicola* after storm-related injury. Plant Dis. 73:1018-1019.

(Prepared by K. L. Pernezny, C. H. Blazquez, L. J. Smith, and R. L. Schlub)

Verticillium Wilt

Phytopathogenic *Verticillium* spp. are polyphagous and cause wilt disease in approximately 200 plant species, including many vegetables. Verticillium wilt occurs worldwide in all tomato-growing regions. However, symptom severity and yield reductions vary widely across locations and years.

Symptoms

Symptoms of Verticillium wilt can be easily confused with those of Fusarium wilt and other wilt diseases. Symptoms of Verticillium wilt may also be so subtle as to suggest abiotic problems, such as suboptimal irrigation. Visible symptoms of disease become evident only in the later stages of infection.

Often, the first indicators of possible Verticillium wilt are epinasty and a diurnal wilting pattern (Fig. 74), in which plants show mild to moderate wilting during the warmest part of the day but recover at night. As the disease advances, marginal and interveinal chlorosis develop on lower leaflets initially (which may also show characteristic V-shaped lesions) (Fig. 75) and progresses acropetally. In highly susceptible plants, stunting, defoliation, and premature death may occur. Vascular discoloration is evident when stems of infected plants are cut open longitudinally, especially in the lower parts of the plants. There is often no vascular browning in the petioles or pith.

Causal Organisms

Verticillium wilt is caused by two soilborne fungi: *Verticillium albo-atrum* Reinke & Berthier and *V. dahliae* Kleb. The latter pathogen appears to be most common, especially in the major tomato-growing areas of the U.S. Southeast and California, along with Canada and some European countries.

Both species are relatively slow growing in culture (Fig. 76) and may be obscured by more rapidly growing fungi. On an agar medium such as potato dextrose (PDA), these *Verticillium* fungi produce ellipsoidal, hyaline conidia (2.0–4.0 × 3.5–10.0 µm) that are unicellular and uninucleate and borne in clusters in mucilaginous liquid droplets on verticillate conidiophores. In a liquid environment, the fungi exhibit dimorphic growth, producing both spores and mycelia. After a 2- to 3-week incu-

Fig. 74. Verticillium wilt affecting an entire tomato plant. (Courtesy J. P. Jones)

Fig. 75. The V-shaped lesions characteristic of Verticillium wilt. (Courtesy J. P. Jones)

bation period on agar media, cultures of *V. albo-atrum* appear dark because of the formation of the resting mycelia as aggregations of torulose, melanized hyphal cells.

For many years, *V. dahliae* isolates were classified as microslerotial-forming strains of *V. albo-atrum*. Growth of *V. dahliae* on agar is similar to that of *V. albo-atrum,* except that cultures of the former turn black because of the abundant production of microsclerotia (15–50 × 80–120 μm) as grapelike clusters of spherical, thick-walled, melanized cells (Fig. 36, page 28). These propagules form best at cool temperatures (10–20°C) and are extremely tolerant to environmental stresses. The formation of microsclerotia in senescing plant tissue, as well as in culture, is of key taxonomic importance.

Recent molecular studies have demonstrated that both *V. dahliae* and *V. albo-atrum* are genetically diverse. DNA-based assays have been developed that can be used to differentiate the two fungi from one another and from two closely related species: *V. nigrescens* Pethybr. and *V. tricorpus* I. Isaac.

Disease Cycle and Epidemiology

Verticillium wilt is a disease of relatively cool weather. Substantial damage has occurred in Florida during winter months when the daytime high temperature has averaged 20–24°C. The disease also seems to be more severe in neutral to alkaline soils. Both *V. dahliae* and *V. albo-atrum* are poor soil competitors but can survive for extended periods of time as dark, resting mycelium or microsclerotia in infected plant debris. Although *V. albo-atrum* does not overwinter well in certain regions (such as eastern Canada), *V. dahliae* is particularly hardy and can maintain a high inoculum potential in soil in the absence of a susceptible host for several years.

Colonization studies with susceptible crops have shed light on the intimate interaction between the host and *V. dahliae.* While the entire root system is susceptible, only colonization of the root tip or within the root elongation zone leads to successful vascular infection and disease. Wounding of roots during cultivation, in the formation of secondary roots, or as a result of nematode feeding has not been shown to increase susceptibility. However, a number of studies have demonstrated a synergistic enhancement of disease severity in the presence of certain nematode species.

During much of the disease cycle of Verticillium wilt, *V. dahliae* is confined to the xylem. In the later stages of disease, the fungus moves out of the xylem as infected plant tissues senesce and undergoes a limited phase of saprophytic growth, in which it produces the resting structures that constitute the primary source of disease inoculum in the field.

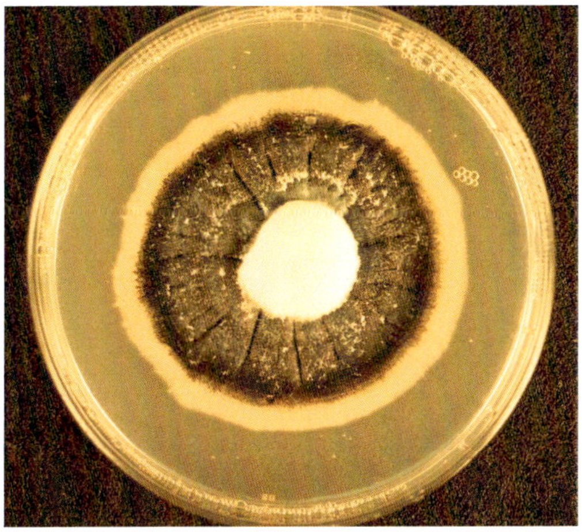

Fig. 76. *Verticillium dahliae* growing in culture. (Courtesy K. L. Pernezny)

Management

Although Verticillium wilt is a monocyclic disease, the longevity of the pathogens, combined with intensive cropping, can lead to a substantial buildup of pathogen populations in the soil. The incidence of wilt is definitely less in areas with long rotations between susceptible crops. However, the benefits of rotation are limited by the pathogens' wide host ranges. (In addition to many crop plants, nightshade (*Solanum* L.) and other common weeds are good hosts.) In the case of *V. dahliae,* the benefits of rotation are further limited by the extensive length of time the fungus may survive in the soil.

In some cases, removing and destroying plant debris will reduce the amount of inoculum liberated into the soil. Soil solarization and fumigation with broad-spectrum methyl isothiocyanate compounds are also frequently used to reduce inoculum levels, although sometimes with only limited success.

When possible, resistant cultivars should be grown. These cultivars carry the *Ve* genes that confer resistance against race 1 strains but not race 2 strains. However, the widespread use of *Ve* cultivars has caused a slow shift from race 1 to race 2 strains in local populations.

Selected References

Bender, C. G., and Shoemaker, P. B. 1984. Prevalence of Verticillium wilt of tomato and virulence of *Verticillium dahliae* race 1 and race 2 isolates in western North Carolina. Plant Dis. 68:305-309.

Bewley, W. F. 1922. Sleepy disease of the tomato. Ann. Appl. Biol. 9:116-133.

Bowers, J. H., Nameth, S. T., Riedel, R. M., and Rowe, R. C. 1996. Infection and colonization of potato roots by *Verticillium dahliae* as affected by *Pratylenchus penetrans* and *P. crenatus.* Phytopathology 86:614-621.

Conover, R. A. 1959. Verticillium wilt of tomato in Dade County, Florida. Proc. Fla. State Hortic. Soc. 72:199-201.

Dobinson, K. F., Tenuta, G. K., and Lazarovits, G. 1996. Occurrence of race 2 of *Verticillium dahliae* in processing tomato fields in southwestern Ontario. Can. J. Plant Pathol. 18:55-58.

Domsch, K. H., Gams, W., and Anderson, T. H. 1980. Compendium of Soil Fungi. Vol. 1. Academic Press, London.

Green, R. J., Jr. 1969. Survival and inoculum potential of conidia and microsclerotia of *Verticillium albo-atrum* in soil. Phytopathology 59.874-876.

Grogan, R. G., Ioannou, N., Schneider, R. W., Sall, M. A., and Kimble, K. A. 1979. Verticillium wilt on resistant tomato cultivars in California: Virulence of isolates from plants and soil and relationship of inoculum density to disease incidence. Phytopathology 69:1176-1180.

Harrington, M. A., and Dobinson, K. E. 2000. Influences of cropping practices on *Verticillium dahliae* populations in commercial processing tomato fields in Ontario. Phytopathology 90:1011-1017.

Huisman, O. C., and Ashworth, L. J., Jr. 1974. Quantitative assessment of *Verticillium albo-atrum* in field soils: Procedural and substrate improvements. Phytopathology 64:1043-1044.

Jones, J. P., and Crill, P. 1973. The effect of *Verticillium* on resistant, tolerant, and susceptible tomato varieties. Plant Dis. Rep. 57:122-124.

Jones, J. P., and Woltz, S. S. 1972. Effect of soil pH and micronutrient amendments on Verticillium and Fusarium wilt of tomatoes. Plant Dis. Rep. 56:151-153.

Jorge, P. E., Green, R. J., Jr., and Chaney, W. R. 1992. Inoculation with *Fusarium* and *Verticillium* to increase resistance in *Fusarium*-resistant tomato. Plant Dis. 76:340-343.

Nachmias, A., Buchner, V., Tsror, L., Burstein, Y., and Keen, N. 1987. Differential phytotoxicity of peptides from culture fluids of *Verticillium dahliae* races 1 and 2 and their relationship to pathogenicity of the fungi on tomato. Phytopathology 77:506-510.

Nadakavudaren, M. J., and Horner, C. E. 1959. An alcohol agar medium selective for determining *Verticillium* microsclerotia in soil. Phytopathology 49:527-528.

Nazar, R. N., Hu, X., Schmidt, J., Culham, D., and Robb, J. 1991. Potential use of PCR-amplified ribosomal intergenic sequences in the detection and differentiation of Verticillium wilt pathogens. Physiol. Mol. Plant Pathol. 39:1-11.

Robb, J., Moukhamedov, R., Hu, X., Platt, H., and Nazar, R. N. 1993. Putative subgroups of *Verticillium albo-atrum* distinguishable by PCR-based assays. Physiol. Mol. Plant Pathol. 43:423-436.

Vallad, G. E., and Subbarao, K. V. 2008. Colonization of resistant and susceptible lettuce cultivars by a green fluorescent protein-tagged isolate of *Verticillium dahliae*. Phytopathology 98:871-885.

Volin, R. B., and Bryan, H. H. 1975. 'Flora-Dade,' a fresh market tomato for south Florida with resistance to Verticillium wilt. Proc. Fla. State Hortic. Soc. 88:218-221.

Wilhelm, S. 1955. Longevity of the Verticillium wilt fungus in the laboratory and field. Phytopathology 45:180-181.

(Prepared by K. L. Pernezny, G. Vallad, and K. F. Dobinson)

White Mold

White mold (also known as Sclerotinia stem rot and timber rot) can be a problem on tomato whenever cool conditions prevail during crop development. The causal organism has a wide host range, causing disease on more than 360 plant species. White mold diseases of economic importance occur on bean (*Phaseolus* L.), cabbage (*Brassica oleracea* L.), potato (*Solanum tuberosum* L.), lettuce (*Lactuca sativa* L.), and sunflower (*Helianthus* L.).

Symptoms

White mold is usually first noticed on tomato about the time of flowering. Infection begins in leaf axils or in stem joints where flower petals have fallen and lodged. Water-soaked areas develop at these sites. Stems are subsequently invaded and become soft, and large portions of invaded tissue may eventually die. Infected stems turn a bleached, light gray and look like animal bones dried in the sun (Fig. 77).

Signs of the fungus that causes white mold are apparent in most cases and are extremely useful in making the correct field diagnosis of the disease. During cool, moist weather, white, cottony mycelium is often evident on diseased stems. Sclerotia are seen on mycelial mats 7–10 days after infection. Sclerotia readily form inside stems, as well, assuming an elongated shape that is directed toward the stem cavity (Fig. 78). Some sclerotia are almost tubular. The fungus can also enter plants at the soil line, especially if senescent tissue is present.

Mycelial growth accounts for some plant-to-plant spread, especially in the case of lush, overlapping tomato growth. Large portions of a field can become diseased, resulting in circular areas of dead plants.

Fig. 77. Tomato plant infected by white mold (also known as Sclerotinia stem rot). (Courtesy K. L. Pernezny)

Infected fruit may also be found. Typically, such fruit are gray and break down rapidly in a watery rot. A ring of sclerotia often develops around the calyx.

Causal Organism

Sclerotinia sclerotiorum (Lib.) de Bary is the causal agent of white mold, but at times, *S. minor* Jagger can also cause white mold on tomato. Sclerotia are hard and variable in shape and measure 3–7 × 3–10 mm; they are black outside with a white interior. (Old specimens may be black throughout.) Each sclerotium produces one to several apothecia (Fig. 79). These sporocarps are normally cup shaped but are sometimes expanded and flat with a central dimple. They are usually white, yellow, or some shade of brown. Asci are cylindrical or clavate in shape and up to 10 × 130 µm in size; they contain eight spores. Ascospores are nonseptate, uniseriate, hyaline, and elliptical and measure 4.65 × 9.00–13.00 µm. Cylindrical, hyaline paraphyses are present on the hymenium.

Disease Cycle and Epidemiology

White mold is a disease of cool, moist weather. It is particularly severe at temperatures of 15–21°C. High humidity and free moisture are important in the disease cycle.

The inoculum for initiation of an epidemic of tomato white mold originates primarily from ascospores. Sclerotia on the surface or in the top 2–3 cm of soil are responsible primarily for airborne inoculum. Sclerotia require a holding period under cool, moist conditions before they are preconditioned to undergo carpogenic germination. Temperatures of 11–15°C are particularly favorable for the formation of apothecial initials. During fluctuating weather conditions, ascospores are ejected from asci and carried by the wind to potential host tomatoes.

Ascospores require an exogenous energy source to initiate infection. Most of the time, the source is fallen flower petals. Once sufficient fungal growth has taken place on the dead or senescent tissue, healthy stems are invaded. *S. sclerotiorum* may even colonize tomato pollen as an initial organic substrate. A long period (16–72 h) of continuous wetness is required for both ascospore infection of senescent tissue and subsequent in-

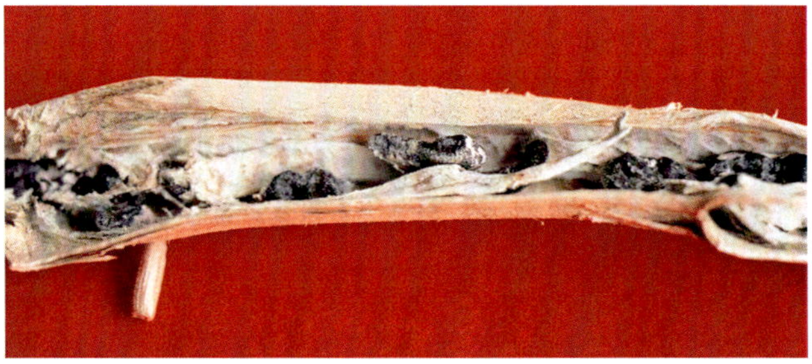

Fig. 78. Sclerotia of *Sclerotinia sclerotiorum.* (Courtesy K. L. Pernezny)

Fig. 79. Apothecia of *Sclerotinia sclerotiorum.* (Courtesy P. B. Adams)

fection of living plant parts. Even lesion expansion seems dependent on free moisture.

Prebloom infection sometimes occurs in tomato. In these cases, an exogenous energy source, such as frost-damaged tissue or mechanically damaged stems, serves as the substrate for initiation of disease. The relative importance of prebloom infection to the epidemiology of tomato white mold is not well understood. Any condition that contributes to poor air circulation and retention of moisture will likely aggravate the disease. Thus, disease intensity can be expected to be higher in low-lying areas, in fields adjacent to heavily wooded lots, and in crops with lush vegetative growth.

A secondary cycle probably is not important. However, plant-to-plant spread does occur by direct contact of healthy plants with mycelial mats on infected hosts.

Management

Practical management of tomato white mold is based primarily on preventative application of appropriate fungicides prior to colonization of senescent and dead tissue by *Sclerotinia* spp. Application of a broad-spectrum fumigant has been shown effective in destroying sclerotia in the soil. However, in the commonly used full-bed, plastic mulch culture of fresh-market tomatoes, there are considerable amounts of unfumigated soil between beds.

Management of crop growth to prevent formation of a dense canopy will help control white mold, and deep plowing to bury sclerotia at sufficient depths will also aid in control. Flooding of fields for 4–5 weeks will reduce the number of viable sclerotia; however, this is not practical for most tomato growers and may in fact spread other pathogens. Encouraging results have been found with some biocontrol agents, particularly mycoparasites.

Selected References

Abawi, G. S., and Grogan, R. G. 1979. Epidemiology of diseases caused by *Sclerotinia* species. Phytopathology 69:899-904.

Gerlogh, M., Whipps, J. M., Budge, S. P., and Goosen-van de Geijin, H. M. 1996. Efficiency of isolates of *Coniothyrium minitans* as mycoparasites of *Sclerotinia sclerotiorum, Sclerotium cepivorum,* and *Botrytis cinerea* on tomato stem pieces. Eur. J. Plant Pathol. 102:787-793.

Huang, H. C., Kokko, E. G., and Erickson, R. S. 1997. Infection of alfalfa pollen by *Sclerotinia sclerotiorum*. Phytoparasitica 25:17-24.

Kohn, L. M. 1979. Delimitation of the economically important plant pathogenic *Sclerotinia* species. Phytopathology 69:881-886.

Moore, W. D. 1949. Flooding as a means of destroying the sclerotia of *Sclerotinia sclerotiorum*. Phytopathology 39:920-927.

Purdy, L. H. 1979. *Sclerotinia sclerotiorum:* History, diseases and symptomatology, host range, geographical distribution, and impact. Phytopathology 69:875-880.

Purdy, L. H., and Bardin, R. 1953. Mode of infection of tomato plants by the ascospores of *Sclerotinia sclerotiorum*. Plant Dis. Rep. 37:361-362.

(Prepared by K. L. Pernezny)

Zonate Leaf Spot

Zonate leaf spot is primarily a foliar disease of tomato, but it has been observed infecting stems and fruit. The disease occurs sporadically in the field but sometimes causes significant defoliation and yield reductions in localized areas. The disease was first described on maple (*Acer* L.) in 1947 and has subsequently been reported on 73 species belonging to 36 families of vascular plants, including woody perennials and annual succulents. On tomato, zonate leaf spot has a wide geographic distribution in the eastern United States, ranging from New York and Pennsylvania to Florida.

Symptoms

Zonate leaf spot derives its name from the development of large, circular leaf lesions with distinct concentric rings (Fig. 80). Initially, the leaf lesions are light tan and the concentric rings are symmetrical, appearing like "bull's eyes." Early lesions are 5–10 mm in diameter and continue to expand by adding more rings; lesions can reach 30 mm in diameter. Eventually, the lesions coalesce, causing the leaves to senesce.

Leaf lesions caused by zonate leaf spot can be distinguished from those caused by early blight (*Alternaria solani* Sorauer) in several ways. The concentric rings that characterize zonate leaf spot are more symmetrical than those caused by early blight. In addition, no distinct chlorosis is associated with the lesions caused by zonate leaf spot, and the lesions are not localized near the bottom of the tomato canopy.

Lesions can also form on the stems of plants infected with zonate leaf spot. Fruit infection appears to result from the expansion of lesions on the sepals and peduncle.

Following a period of high humidity, large, pyramid-shaped reproductive structures form on older leaf lesions. These structures (referred to as fruiting bodies) are visible with a powerful hand lens or under low magnification with a stereo/dissecting microscope. Sclerotia are occasionally observed on the peduncle end of an infected fruit.

Causal Organism

Zonate leaf spot is caused by *Hinomyces moricola* (I. Hino) Narumi & Y. Harada (syns. *Cristulariella moricola* (I. Hino) Redhead and *C. pyramidalis* Waterman & R. P. Marshall). The fungus is identified by the production of pyramid-shaped fruiting bodies that measure 300–400 μm in length (Fig. 81). These hyaline, multicellular, asexual reproductive structures are produced on the apex of a septate hypha. They detach easily and function as reproductive propagules. On nutrient media, germinated propagules produce a mycelial mat that ultimately bears phialides and masses of globose, single-celled phialocondia. Sclerotia are eventually produced on media and infected leaves after abscission. Sclerotia are black and superficial, 1–4 mm in length, and pulvinate to fusiform in shape.

The teleomorph of *H. moricola* has been produced in vitro after a cold treatment of 4–6 months at 4°C and subsequent incubation at 15°C with a 14 h photoperiod. It has been designated

Fig. 80. Concentric rings characteristic of zonate leaf spot. (Courtesy T. A. Zitter)

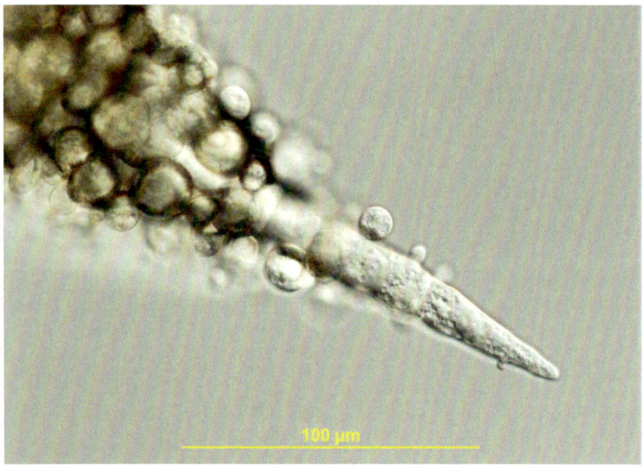

Fig. 81. Pyramid-shaped fruiting body of the fungus that causes zonate leaf spot (*Cristulariella* sp.). (Courtesy Bruce Watt, University of Maine, Bugwood.org)

as *Grovesinia pyramidalis* M. N. Cline, J. L. Crane, & S. D. Cline. The apothecia are macroscopically similar to those of *Sclerotinia sclerotiorum* (Lib.) de Bary but have smaller asci and ascospores.

Disease Cycle and Epidemiology

The broad and diverse host range of *H. moricola* figures prominently into the epidemiology of zonate leaf spot in tomato production fields. The initial inoculum originates from infected trees and shrubs in wooded and riparian areas adjacent to tomato fields. The disease is spread by the pyramid-shaped reproductive propagules, which are readily dislodged from infected leaves. However, because of the relatively large size of the reproductive propagules, aerial dissemination is limited. Thus, symptoms of zonate leaf spot appear initially along the borders or edges of fields in close proximity to the wooded or riparian area. Over time, a distinct disease gradient extends from this area to the interior of the tomato field.

In the field, infection of tomato plants progresses from the upper portion of the vine and spreads downward. The disease cycle is favored by extended periods of high humidity and rainfall and wide fluctuations in temperature. In the greenhouse, sporulation has been shown to be inhibited at 37°C on leaves of pecan (*Carya illinoinensis* (Wangenh.) K. Koch). In vitro, mycelial growth on V8 juice agar ceased and colonies died at 30°C.

Management

Information is lacking on effective fungicides for management of zonate leaf spot on tomato. On other crops, various fungicides have been effective in controlling zonate leaf spot.

Selected References

Cline, M. N., Crane, J. L., and Cline, S. D. 1983. The teleomorph of *Cristulariella moricola*. Mycologia 75:988-994.

Dillard, H. R., Cobb, A. C., Carman, W. C., and Brewer, R. 1995. First report of zonate leaf spot on tomato, caused by *Cristulariella moricola,* in New York. Plant Dis. 79:319.

Redhead, S. A. 1975. The genus *Cristulariella*. Can. J. Bot. 53:700-707.

Trolinger, J. C., Elliott, E. S., and Young, R. J. 1978. Host range of *Cristulariella pyramidalis*. Plant Dis. Rep. 62:710-714.

(Prepared by D. O. Chellemi)

Diseases Caused by Bacteria

Major Bacterial Diseases

Bacterial Canker

Bacterial canker, first described in 1909 in the U.S. state of Michigan, is now distributed worldwide. While canker epidemics occur only sporadically, they can be devastating, resulting in up to 100% yield loss. Bacterial canker is considered one of the most important diseases of tomatoes produced in greenhouses. Both direct-seeded crops and those grown from transplants can be affected, and the pathogen spreads readily during clipping and pruning. Bacterial canker can also occur on pepper (*Capsicum annuum* L.), but it is economically significant only on tomato.

Symptoms

Systemic wilting is the most conspicuous symptom of bacterial canker. Early symptoms include one-sided wilting of leaves (Fig. 82) and downward turning of lower leaves. Leaf petioles often remain turgid as leaflets wilt and become distorted. Young plants sometimes undergo a rapid and complete wilt, whereas older plants show wilting on the lower leaves first, with the upper leaves remaining turgid until complete collapse occurs (Fig. 83). If the pathogen gains entry through a wounded branch tip after a terminal bud is clipped, symptoms can begin in the upper portion of the plant and rapidly spread to the lower leaves, killing the plant.

Systemic wilting caused by bacterial canker can be mistaken for that caused by bacterial wilt (*Ralstonia solanacearum* (Smith) Yabuuchi et al.). However, the bacterial streaming from canker-affected tissue is far less than that from tissue affected by bacterial wilt. (A yellow ooze can sometimes be produced by squeezing the cut end of a canker-infected stem.)

Adventitious roots may develop on the stems, and on occasion, the stems may develop external, discolored streaks, which may split open to reveal brown cankers (Figs. 84 and 85). The vascular tissues of affected stems develop light-yellow to brown

Fig. 82. Unilateral wilting of leaves, symptomatic of bacterial canker. (Courtesy J. C. Watterson)

streaks, which later darken to reddish brown; this discoloration is most noticeable at the nodes. The stem pith eventually becomes discolored and mealy.

Marginal browning of leaflets may begin at the tips of the lobes and then spread to include the entire margins (Fig. 86). This browning is sometimes accompanied by a yellow margin, in which case it is often referred to as "firing." Entire leaflets or leaves may wither as the browning advances, but symptoms that begin as marginal browning seldom progress to a general vascular wilt. On young plants in the greenhouse, small, creamy-white, blisterlike spots can appear on leaflets (Fig. 87). These spots can become encircled by narrow, dark rings of necrotic tissue, and affected leaflets can eventually turn yellow. Similar white spots can also develop on stems in the field in some years and locations. Dark, sunken leaflet veins and stem streaks can also develop on greenhouse-grown seedlings (Fig. 88).

External fruit symptoms, referred to as bird's-eye spots, are 3–6 mm in diameter and have tan or brown centers surrounded by creamy-white halos (Fig. 89). Although bird's-eye spots are often described as the most reliable symptom for field diagnosis, fruit spots caused by the bacterial spot pathogen can be whitish when they first appear, and canker fruit spots can turn brown with age, mimicking those of bacterial spot. Furthermore, fruit symptoms do not always accompany canker outbreaks. Vascular infection of fruit sometimes appears as yellowing or browning of the vascular strands (Fig. 90).

In greenhouse-produced tomatoes, leaf symptoms appear as interveinal, pale-green to yellow patches that become necrotic (Fig. 91); the leaves eventually die. Wilting generally begins with the lower leaves and moves up the plant, or it may start at the point of inoculation. Wilting may also be more pronounced on one side of the plant than the other. Light- to dark-brown streaks develop on the stems of infected plants. Bird's-eye spots

Fig. 83. Complete collapse of plants in field, caused by bacterial canker. (Courtesy D. Fulbright)

Fig. 84. A stem crack that may develop into a canker. (Courtesy R. X. Latin)

Fig. 85. A stem canker at a late stage of development. (Courtesy R. M. Davis)

Fig. 86. Marginal browning of leaflets, caused by bacterial canker. (Courtesy R. X. Latin)

Fig. 87. White spots on the leaflet of a greenhouse-grown seedling, associated with bacterial canker. (Courtesy R. Chang)

are rarely observed on fruit, but the fruit surface may appear netted or marbled (Fig. 92).

Causal Organism

Clavibacter michiganensis subsp. *michiganensis* (Smith) Davis et al. is a non-spore-forming, non-acid-fast, gram-positive, aerobic bacterium that is not motile or encapsulated. Cells from plant material are characteristically rod shaped but may become pleomorphic, minute, coccoid, or club shaped under certain growth conditions.

Fig. 88. Greenhouse-grown seedling with darkened, sunken leaf veins and stem streaks. (Courtesy R. Chang)

Fig. 89. External symptoms of bacterial canker: bird's-eye spots on fruit. (Courtesy R. X. Latin)

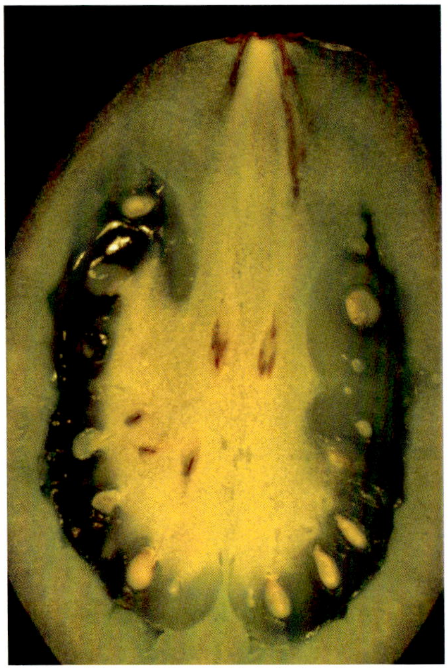

Fig. 90. Vascular infection of fruit, caused by bacterial canker. (Courtesy R. Chang)

Fig. 91. Greenhouse-grown tomato plants with chlorotic (top) and necrotic (bottom) leaf symptoms of bacterial canker. (Top: Courtesy S. A. Miller; Bottom: Courtesy K. Tifft)

Fig. 92. Fruit netting or marbling of greenhouse-grown tomatoes. (Courtesy K. Tifft)

Colonies on nutrient agar are characteristically yellow and reach a diameter of 2–3 mm in 5 days at 25°C. They are smooth and have entire margins and a butyrous consistency. The appearance of colonies on semiselective media varies, depending on the medium.

C. michiganensis subsp. *michiganensis* causes a hypersensitive response in four o'clock plants (*Mirabilis jalapa* L.), also known as marvel of Peru. Serological assays based on monoclonal antibodies, as well as ordinary and real-time polymerase chain reaction (PCR), are used routinely to detect and identify the pathogen. Repetitive extragenic palindromic (REP) PCR has also been used to distinguish at least four groups within *C. michiganensis* subsp. *michiganensis,* but the biological significance of these groups is not yet clear.

Disease Cycle and Epidemiology

Sources of inoculum for bacterial canker include seeds, plant debris, weed hosts, and volunteer tomatoes, along with contaminated wooden stakes, flats, and benches. Under conditions that are highly favorable for spread of the pathogen, such as the clipping of transplants, as few as one infected seed in 10,000 can give rise to a devastating canker epidemic. Although plants infected from seed transmission may wilt and die or fail to set fruit, many plants that develop from infected seed display no symptoms.

Secondary spread of the pathogen can occur by splashing water, contaminated equipment, workers' hands, and pruning and clipping of transplants. In hydroponic culture, secondary spread can also occur from the roots of infected plants to those of healthy plants. In the field, secondary spread often results in foliar symptoms, fruit spotting, or both. Marginal firing of leaflets results from the ingress of bacteria in guttation droplets through hydathodes, whereas fruit spots result from invasion through flowers or the bases of trichomes on young fruit.

Yield losses in the field typically result from primary infection at the seedling stage; secondary spread often does not harm yield or fruit quality. However, in greenhouse production systems, secondary spread can significantly reduce yield. During transplant production, secondary spread occurs by clipping, grafting, and topping or by splashing water. Plants may wilt and die within several weeks after transplanting.

In some production regions, *C. michiganensis* subsp. *michiganensis* can survive for more than 2 years in infected tomato stems; burial of crop debris reduces survival time. The pathogen can survive for at least 1 year in plant tissue in rockwool used in hydroponic production.

Management

To manage bacterial canker, emphasis should be placed on avoiding the pathogen by using clean seed and transplants. Seed should be tested for the presence of *C. michiganensis* subsp. *michiganensis* according to well-accepted methods that use large samples of seed (e.g., 10,000 seeds per sample). To minimize infection, seed lots should be treated with dilute solutions of hydrochloric acid, calcium hypochlorite, sodium hypochlorite, heated cupric acetate and acetic acid, or hot water.

For field transplant production, essential sanitation practices include sterilizing growth media and pots. Seedlings should be watered only enough to promote healthy plant growth, and foliage should dry before nightfall. Before wooden stakes are reused, they should be steamed or washed with a disinfestant; wooden greenhouse benches should also be treated with disinfestant.

Solanaceous weeds should be eliminated in and near tomato production fields. Debris from diseased fields should be plowed under promptly, and crop rotation to a nonhost crop for 3–4 years is desirable.

In hydroponic systems, strict sanitation practices are required throughout the production cycle, from propagation through harvest and cleanout. These practices include providing dedicated boots and clean clothes for greenhouse use, handwashing before and after entry into the greenhouse, using disinfestant footbaths, sanitizing water, and disinfesting tools. Tomato or pepper cull piles and solanaceous weeds should never be allowed near tomato greenhouses.

Spraying with copper compounds or copper plus mancozeb during greenhouse transplant production has been shown to protect yield and quality of processing tomatoes against bacterial canker in Michigan, but in production fields of processing tomatoes, posttransplant spraying has generally been shown ineffective in suppressing the disease. However, on fresh-market tomatoes in the wetter climate of North Carolina, spraying with copper compounds has effectively suppressed the foliar blight phase of bacterial canker in production fields.

Selected References

Bach, H.-J., Jessen, I., Schloter, M., and Munch, J. C. 2003. A TaqMan-PCR protocol for quantification and differentiation of phytopathogenic *Clavibacter michiganensis* subspecies. J. Microbiol. Methods 52:85-91.

Carlton, W. M., Braun, E. J., and Gleason, M. L. 1998. Ingress of *Clavibacter michiganensis* subsp. *michiganensis* into tomato leaves through hydathodes. Phytopathology 88:525-529.

Chang, R. J., Ries, S. M., and Pataky, J. K. 1991. Dissemination of *Clavibacter michiganensis* subsp. *michiganensis* by practices used to produce tomato transplants. Phytopathology 81:1276-1281.

Chang, R. J., Ries, S. M., and Pataky, J. K. 1992. Local sources of *Clavibacter michiganensis* subsp. *michiganensis* in the development of bacterial canker on tomatoes. Phytopathology 82:553-560.

Dreier, J., Bermpohl, A., and Eichenlaub, R. 1995. Southern hybridization and PCR for specific detection of phytopathogenic *Clavibacter michiganensis* subsp. *michiganensis*. Phytopathology 85:462-468.

Fatmi, M., and Schaad, N. W. 1988. Semiselective agar medium for isolation of *Clavibacter michiganense* subsp. *michiganense* from tomato seed. Phytopathology 78:121-126.

Gitaitis, R. D. 1990. Induction of a hypersensitivelike reaction in four-o'clock by *Clavibacter michiganensis* subsp. *michiganensis*. Plant Dis. 74:58-60.

Gleason, M. L., Braun, E. J., Carlton, W. M., and Peterson, R. H. 1991. Survival and dissemination of *Clavibacter michiganensis* subsp. *michiganensis* in tomatoes. Phytopathology 81:1519-1523.

Gleason, M. L., Gitaitis, R. D., and Ricker, M. D. 1993. Recent progress in understanding and controlling bacterial canker of tomato in eastern North America. Plant Dis. 77:1069-1076.

Hausbeck, M. K., Bell, J., Medina-Mora, C., Podolsky, R., and Fulbright, D. W. 2000. Effect of bactericides on population sizes and spread of *Clavibacter michiganensis* subsp. *michiganensis* on tomatoes in the greenhouse and on disease development and crop yield in the field. Phytopathology 90:38-44.

Louws, F. J., Bell, J., Medina-Mora, C. M., Smart, C. D., Opgenorth, D., Ishimaru, C. A., Hausbeck, M. K., de Bruijn, F. J., and Fulbright, D. W. 1998. rep-PCR-mediated genomic fingerprinting: A rapid and effective method to identify *Clavibacter michiganensis*. Phytopathology 88:862-868.

Medina-Mora, C. M., Hausbeck, M. K., and Fulbright, D. W. 2001. Bird's eye lesions of tomato fruit produced by aerosol and direct application of *Clavibacter michiganensis* subsp. *michiganensis*. Plant Dis. 85:88-91.

Ricker, M. D., and Riedel, R. M. 1993. Effect of secondary spread of *Clavibacter michiganensis* subsp. *michiganensis* on yield of northern processing tomatoes. Plant Dis. 77:364-366.

Santos, M. S., Cruz, L., Norskov, P., and Rasmussen, O. F. 1997. A rapid and sensitive detection of *Clavibacter michiganensis* subsp. *michiganensis* in tomato seeds by polymerase chain reaction. Seed Sci. Technol. 25:581-584.

Strider, D. L. 1969. Bacterial canker of tomato caused by *Corynebacterium michiganense:* A literature review and bibliography. N.C. Agric. Exp. Stn. Tech. Bull. 193.

Werner, N. A., Fulbright, D. W., Podolsky, R., Bell, J., and Hausbeck, M. K. 2002. Limiting populations and spread of *Clavibacter michiganensis* subsp. *michiganensis* on seedling tomatoes in the greenhouse. Plant Dis. 86:535-542.

(Prepared by M. L. Gleason, R. D. Gitaitis, and S. A. Miller)

Bacterial Speck

Bacterial speck and its causal agent were originally described and identified in both Taiwan and the United States in 1933. Since that time, the disease has become prevalent in many tomato-growing regions worldwide. Bacterial speck is favored by cool temperatures and high moisture conditions. Leaf blight may result in stunting of plants and reduced fruit size. Fruit spots will reduce the marketable yield of fresh-market varieties, but moderate spotting can be tolerated in the processing tomato industry, since the lesions are superficial and do not interfere with peeling the fruit.

Symptoms

On leaflets, bacterial speck produces lesions that are round and dark brown to black. The lesions extend through the entire leaf but are most prominent on the lower surface. Spots may coalesce, killing large areas of tissue (Fig. 93). A halo around the lesion is lacking in the early stages of the disease but develops with time (Fig. 94).

Stems, petioles, peduncles, pedicels, and sepals are also affected by this disease (Fig. 95). Lesions on these plant parts are oval to elongated in shape.

On fruit, dark lesions or specks develop that are minute in size (rarely larger than 1 mm in diameter) (Fig. 96). The lesions are slightly raised or flat when first observed, and in some instances, the spots are sunken. Dark-green halos may form around lesions on fruit.

Causal Organism

The organism that causes bacterial speck is *Pseudomonas syringae* van Hall pv. *tomato* (Okabe) Young, Dye, & Wilkie. This bacterium is a strictly aerobic, gram-negative rod measuring 0.69–0.97 × 1.80–2.80 μm. It produces a diffusible, green pigment that fluoresces when exposed to ultraviolet (UV) light on medium B of King et al.

The LOPAT profile of *P. syringae* pv. *tomato* is (1) levan positive, (2) oxidase negative, (3) pectolytic on sodium polypectate medium at pH 4.6 (though does not rot potato), (4) arginine dihydrolase negative, and (5) positive for the hypersensitive reaction in tobacco (*Nicotiana* L.) leaves. The bacterium is positive

Fig. 93. Field-grown plants with blighted leaves, caused by bacterial speck. (Courtesy J. P. Jones)

Fig. 95. Field-grown plants with necrosis of petioles, caused by bacterial speck. (Courtesy J. B. Jones)

Fig. 94. Leaf spots with distinct yellow halos, caused by bacterial speck. (Courtesy J. B. Jones)

Fig. 96. Fruit lesions on field-grown plants, produced by bacterial speck. (Courtesy A. A. MacNab)

for β-glucosidase activity, and it utilizes D-mannitol, inositol, trigonelline, D-quinate, D-sorbitol, and D-(–)-tartrate but not L-(+)-tartrate, erythritol, and adonitol as sole carbon sources.

P. syringae pv. *tomato* is also ice-nucleation negative—a characteristic that can be used to differentiate it from *P. syringae* subsp. *syringae,* which also occurs on tomato but is of little significance after the seedling stage. Polymerase chain reaction (PCR) assay primers COR1/COR2, derived from the coronatine gene cluster, and MM5/MM6, derived from the $hrpZ_{Pst}$ gene, are available for detection and diagnosis of the pathogen.

There are two known races of *P. syringae* pv. *tomato,* designated 0 and 1. They can be determined based on the differential reaction on the tomato cultivar Ontario 7710.

Disease Cycle and Epidemiology

P. syringae pv. *tomato* is seedborne. The importance of seed for dissemination of the bacterium is not fully understood but may help explain increasing reports of bacterial speck worldwide. In seedling production greenhouses, a high plant population density and overhead watering favor spread of the bacterium, which may develop as an epiphytic population on seedlings and thus remain undetected. Once seedlings have been transplanted into open fields, rain, overhead irrigation, tools, machinery, and workers spread the bacterium.

P. syringae pv. *tomato* overwinters at least one season in crop residue in temperate climates, although it may not oversummer in tropical and subtropical regions. Volunteer tomatoes are a viable source of initial inoculum, but survival of the pathogen on weeds has not been consistently demonstrated. Survival of the bacterium in untreated soil is short term (less than 30 days). Both high humidity and cool temperatures (18–24°C) favor the development of bacterial speck.

Management

Only high-quality seed free of detectable populations of the pathogen should be planted. If seed has not been treated by the seed supplier, it should be treated by the grower with sodium hypochlorite, hydrochloric acid, or hot water. Resistant tomato cultivars should be used, if available.

In the transplant production greenhouse, overwatering seedlings and handling them more than necessary should both be avoided. Workers should wash their hands thoroughly after handling plants, and trays, benches, tools, and greenhouse structures should be washed and sanitized between seedling crops. If the disease is detected in the greenhouse, diseased plants and neighboring trays of seedlings should be removed and destroyed. Seedlings may be treated preventatively with copper-based bactericide products combined with a fungicide containing mancozeb, especially if copper-resistant strains are present.

Seedlings should be transplanted into well-drained fields that are free of volunteer tomatoes and weeds. Weekly applications of copper-based fungicides tank mixed with mancozeb should be considered during rainy periods when cool temperatures prevail. Application of plant activators, such as acibenzolar-S-methyl, has also been shown to reduce bacterial speck severity. Overhead irrigation should be avoided, if possible, and workers should stay out of fields when plants are wet. The tools used for pruning and suckering should be disinfested regularly.

At the end of the season, fields should be plowed or disk harrowed to bury crop debris. A minimum 2-year rotation with other crops should be practiced before returning to tomato.

Selected References

Braun-Kiewnick, A., and Sands, D. C. 2001. II. Gram-negative bacteria: C. *Pseudomonas.* Pages 84-120 in: Laboratory Guide for the Identification of Plant Pathogenic Bacteria. 3rd ed. N. W. Schaad, J. B. Jones, and W. Chun, eds. American Phytopathological Society, St. Paul, MN.

Bryan, M. K. 1933. Bacterial speck of tomatoes. Phytopathology 23:897-904.

Cuppels, D. A., Louws, F. J., and Ainsworth, T. 2006. Development and evaluation of PCR-based diagnostic assays for the bacterial spot and bacterial speck pathogens of tomato. Plant Dis. 90:451-458.

Goode, M. J., and Sasser, M. 1980. Prevention—the key to controlling bacterial spot and bacterial speck of tomato. Plant Dis. 64:831-834.

Jardine, D. J., Stephens, C. T. and Fulbright, D. W. 1988. Potential sources of initial inoculum for bacterial speck in early planted tomato crops in Michigan: Debris and volunteers from previous crops. Plant Dis. 72:246-249.

King, A. M. Q., Lefkowitz, E. J., Adams M. J., and Carstens, E. B. 2012. Virus Taxonomy: Ninth Report of the International Committee on Taxonomy of Viruses. Elsevier, London.

Louws, F. J., Wilson, M., Campbell, H. L., Cuppels, D. A., Jones, J. B., Shoemaker, P. B., Sahin, F., and Miller, S. A. 2001. Field control of bacterial spot and bacterial speck of tomato using a plant activator. Plant Dis. 85:481-488.

McCarter, S. M., Jones, J. B., Gitaitis, R. D., and Smitley, D. R. 1983. Survival of *Pseudomonas syringae* pv. *tomato* in association with tomato seed, soil, host tissue, and epiphytic weed hosts in Georgia. Phytopathology 73:1393-1398.

Miller, S. A., and Lewis Ivey, M. L. 2005. Hot water and chlorine treatment of vegetable seeds to eradicate bacterial plant pathogens. Ohio State Univ. Ext. Fact. HYG-3085-05. Available online at http://ohioline.osu.edu/hyg-fact/3000/3085.html

Schneider, R. W., and Grogan, R. G. 1977. Bacterial speck of tomato: Sources of inoculum and establishment of a resident population. Phytopathology 67:388-394.

Zaccardelli, M., Spasiano, A., Bazzi, C., and Merighi, M. 2005. Identification and *in planta* detection of *Pseudomonas syringae* pv. *tomato* using PCR amplification of $hrpZ_{Pst}$. Eur. J. Plant Pathol. 111:85-90.

(Prepared by S. A. Miller and J. B. Jones)

Bacterial Spot

Bacterial spot is present wherever tomatoes are grown. This disease was observed in the United States and South Africa in 1912 and 1914, respectively, and it was identified in South Africa in 1921. In North America, the disease is most severe in Florida and other states in the southeastern United States, but significant losses also occur in the U.S. Northeast and Midwest and in Ontario, Canada. Crop losses result from both the yield reduction that occurs from defoliation and the severe fruit spotting that renders the fruit unsalable. On processing tomatoes, the presence of deep spots prevents complete peeling of the fruit.

Symptoms

Bacterial spot affects all aboveground plant parts. On leaves, stems, and fruit spurs, the spots are generally brown and circular. Lesions rarely develop to more than 3 mm in diameter. The spots are water soaked when observed during a rainy period or when dew is present (Fig. 97).

On leaflets, the spots caused by bacterial spot can easily be confused with those caused by early blight, Septoria leaf spot, gray leaf spot, target spot, or bacterial speck. Bacterial spot lesions do not have the concentric zones that characterize target spot and early blight lesions, and compared to gray leaf spot lesions, bacterial spot lesions are generally darker in color and less uniformly distributed. With target spot and early blight, prominent halos often surround pinpoint lesions, but that is not the case with bacterial spot lesions. The chlorosis that surrounds bacterial spot lesions may be lighter and more diffuse than that observed for early blight and target spot lesions.

When the bacterium *Xanthomonas perforans* is present, leaf spots often have a shot-hole appearance (Fig. 98). Under conditions optimal for disease development, spots on the leaves, petioles, and rachis coalesce to form long, dark streaks. A general yellowing may occur on leaflets that have many lesions. Blighting of the foliage occurs with the coalescing of lesions (Fig. 99), and plants become huddled in appearance because of severe epinasty. Often, the dead foliage remains on the plant, giving it a scorched appearance.

On fruit, lesions begin as minute, slightly raised blisters. As the spots increase in size, they become brown, scablike, and slightly raised (Fig. 100); they may also become raised around the margin but sunken in the middle. A developing lesion may have a faint to prominent white halo, which eventually disappears. Bacterial spot lesions feel slightly rough to the touch compared to those of bacterial speck, which are generally smaller and smoother.

Causal Organisms

Bacterial spot is incited by four species of *Xanthomonas: X. vesicatoria* (ex Doidge) Vauterin et al., *X. euvesicatoria* Jones et al., *X. gardneri* (ex Šutic) Jones et al., and *X. perforans* Jones et al. The four species can easily be differentiated using a polymerase chain reaction (PCR) technique in which unique banding patterns are produced for each group following restriction of the PCR product with specific restriction enzymes.

The bacteria are motile, strictly aerobic, gram-negative rods; they measure 0.7–1.0 × 2.0–2.4 µm and possess a single polar flagellum. On nutrient agar, the bacteria grow relatively slowly, and the colonies are circular, wet, shining, Naples yellow, and entire in appearance. The bacterium produces acid but no gas from arabinose, glucose, sucrose, galactose, trehalose, cellobiose, and fructose. Xanthomonadins are produced by the bacteria.

Strains of *X. euvesicatoria* also cause bacterial spot of pepper (*Capsicum annuum* L.), a potentially severe disease.

Previously, the bacterial spot pathogens were considered races of *X. campestris* pv. *vesicatoria* (Doidge) Dye. Designated T1–T4, the four races were differentiated based on their reactions on three tomato genotypes (Hawaii 7981, Hawaii 7998, and Bonny Best) and on *Lycopersicon pennellii* (Correll) D'Arcy LA716.

Disease Cycle and Epidemiology

Seed serve as a medium for survival and dissemination of the *Xanthomonas* bacteria. Most fresh-market and processing tomatoes grown in North America are established in the field from transplanted seedlings produced in a greenhouse. The greenhouse environment is often conducive for development of bacterial spot, with high temperatures, high relative humidity, a large and dense plant population, and overhead irrigation, which facilitates water-splash transmission of bacteria from plant to plant. The pathogens may survive epiphytically on tomato seedlings without causing symptoms during transplant production, or they may cause typical leaf symptoms.

Once in the field, plants harboring a bacterial spot pathogen develop symptoms under environmentally conducive condi-

Fig. 97. Bacterial spot lesions on leaflets, with significant water soaking. (Courtesy G. E. Vallad)

Fig. 99. Coalescing of lesions on leaves, resulting in a blighted appearance. (Courtesy J. B. Jones)

Fig. 98. Leaf spots on leaflets with a shot-hole appearance, associated with *Xanthomonas perforans*. (Courtesy J. B. Jones)

Fig. 100. Bacterial spot lesions on tomato fruit. (Courtesy K. L. Pernezny)

tions. In northern regions, the bacteria generally do not survive more than 1–2 years on plant debris in the soil, but in tropical and subtropical regions, they may persist for many months. In addition to surviving on plant debris, the bacterial spot organisms can survive on tomato and pepper volunteers.

Disease development is favored by temperatures of 24–30°C and by high precipitation. The bacteria are disseminated within a field by wind-driven rain droplets, handling of plants (e.g., suckering, clipping, training, harvesting, etc.), and aerosols. The bacteria penetrate the plants through stomates and wounds created by wind-driven sand, insect punctures, or mechanical means.

Management

The most effective management strategy is to plant high-quality seeds free of detectable populations of the *Xanthomonas* pathogens. Seeds should be treated with sodium hypochlorite, hydrochloric acid, or hot water to reduce the potential for seedling infection.

In the transplant production greenhouse, efforts should be made to minimize overwatering and handling of seedlings. In addition, trays, benches, tools, and greenhouse structures should be washed and sanitized between seedling crops, and workers should wash their hands thoroughly after handling plants. If the disease is detected in the greenhouse, diseased plants and neighboring trays of seedlings should be removed and destroyed.

Seedlings may be treated preventatively with bacteriophages or copper-based bactericide products combined with a fungicide containing mancozeb. Seedlings should be transplanted into fields free of volunteer tomatoes and weeds. Overhead irrigation should be avoided, if possible, and plants should be handled on a minimal basis when wet. Tools used for pruning and suckering should be disinfested frequently.

Weekly applications of copper-based fungicides tank mixed with mancozeb should be considered during rainy periods when warm temperatures prevail, especially when copper-resistant strains of the pathogens are present. Bacteriophages specific to the bacterial spot pathogens have been shown to reduce disease severity and to result in increased yield. Plant activators, such as acibenzolar-S-methyl, also have been shown to reduce bacterial spot severity.

At the end of the season, fields should be plowed or disk harrowed to bury crop debris. A 1-year rotation with another crop before returning to tomato is recommended. Volunteer tomatoes should be eliminated from rotational crops.

Selected References

Astua-Monge, G., Minsavage, G. V., Stall, R. E., Vallejos, C. E., Davis, M. J., and Jones, J. B. 2000. *Xv4-vrxv4:* A new gene-for-gene interaction identified between *Xanthomonas campestris* pv. *vesicatoria* race T3 and the wild tomato relative *Lycopersicon pennellii*. Mol. Plant-Microbe Interact. 13:1346-1355.

Conover, R. A., and Gerhold, N. R. 1981. Mixtures of copper and maneb or mancozeb for control of bacterial spot of tomato and their compatibility for control of fungus diseases. Proc. Fla. State Hortic. Soc. 94:154-156.

Flaherty, J. E., Jones, J. B., Harbaugh, B. K., Somodi, G. C., and Jackson, L. E. 2000. Control of bacterial spot on tomato in the greenhouse and field with h-mutant bacteriophages. HortScience 35:882-884.

Gardner, M. W., and Kendrick, J. B. 1921. Bacterial spot of tomato. J. Agric. Res. (Washington, D.C.) 21:123-256.

Jones, J. B., Lacy, G. H., Bouzar, H., Stall, R. E., and Schaad, N. W. 2004. Reclassification of the xanthomonads associated with bacterial spot disease of tomato and pepper. Syst. Appl. Microbiol. 27:755-762.

Jones, J. B., Pohronezny, K. L., Stall, R. E., and Jones, J. P. 1986. Survival of *Xanthomonas campestris* pv. *vesicatoria* in Florida on tomato crop residue, weeds, seeds, and volunteer tomato plants. Phytopathology 76:430-434.

McInnes, T. B., Gitaitis, R. D., McCarter, S. M., Jaworski, C. A., and Phatak, S. C. 1988. Airborne dispersal of bacteria in tomato and pepper transplant fields. Plant Dis. 72:575-579.

Obradovic, A., Jones, J. B., Momol, T. M., Balogh, B., and Olson, S. M. 2004. Management of tomato bacterial spot in the field by foliar applications of bacteriophages and SAR inducers. Plant Dis. 88:736-740.

Obradovic, A., Mavridis, A., Rudolph, K., Janse, J. D., Arsenijevic, M., Jones, J. B., Minsavage, G. V., and Wang, J.-F. 2004. Characterization and PCR-based typing of *Xanthomonas campestris* pv. *vesicatoria* from peppers and tomatoes in Serbia. Eur. J. Plant Pathol. 110:285-292.

Peterson, G. H. 1963. Survival of *Xanthomonas vesicatoria* in soil and diseased tomato plants. Phytopathology 53:765-767.

Pohronezny, K. L., and Volin, R. B. 1983. The effect of bacterial spot on yield and quality of fresh market tomatoes. HortScience 18:69-70.

(Prepared by J. B. Jones and S. A. Miller)

Bacterial Stem Rot

Bacterial stem rot is considered a relatively minor disease of tomato, but major losses sometimes occur. The disease occurs in both the greenhouse and the field, and it is most likely to develop when plants are pruned. Thus, the highest incidence of bacterial stem rot occurs on staked and trellised tomatoes.

Symptoms

The initial symptom of bacterial stem rot is a wilting of tomato plants at the time of first fruit harvest or later in the crop. The wilted plants remain in this condition for many days. When cross sections of stems are made, the pith has usually disintegrated, causing a hollow stem (Fig. 101). The stem appears wet and slimy from bacterial growth in the pith area. The development of a hollow stem can also be determined by pinching the stem.

The bacterium sometimes develops in the bark of the stem, causing the bark to turn black and slough easily from the stem. Vascular discoloration does not extend appreciably above or below the point of the hollow stem.

Causal Organism

Bacterial stem rot is caused by *Pectobacterium carotovorum* (Jones) Waldee (syn. *Erwinia carotovora* subsp. *carotovora* (Jones) Dye). This bacterium also causes soft rot of many

Fig. 101. Internal and external symptoms of bacterial stem rot. (Courtesy T. A. Zitter)

vegetables, including tomato fruit, before and after harvest (see Bacterial Soft Rot, later in Part I).

Isolation of *P. carotovorum* from stem rot lesions can be difficult because of the presence of many saprophytic bacteria in rotted tissues. However, the bacterium grows well on many laboratory media. Colonies of *P. carotovorum* appear approximately 48 h after isolation and are translucent, round, smooth, and slimy. The optimal temperature for growth is 25–30°C. *P. carotovorum* can be selectively isolated on a sodium polypectate medium. Colonies on this medium form a pit, which can be used as an identification characteristic. The bacterium grows anaerobically as well as aerobically and is gram negative; it has peritrichous flagella and is rapidly motile.

A rapidly growing soft rot develops in inoculated tomato fruit. The bacterium is negative for phosphatase activity.

Disease Cycle and Epidemiology

P. carotovorum is probably ubiquitous in tomato fields. It has been found in the gut of insect larvae and may also be spread from decaying vegetation to plants by adult insects. Infected fruit invaded by fruitworms are usually the source of a large amount of inoculum.

The bacterium must be present on a fresh wound for invasion; if a wound heals before inoculation, then disease will not develop. The removal of suckers and leaves is the most common cause of wounds prior to invasion by *P. carotovorum*.

Disease will not occur on healthy, green leaves, but the bacterium may develop in senescent leaves. Stems may be invaded from senescent leaves through the vascular system. High relative humidity is essential for disease development.

In greenhouses, the disease sometimes develops in a nonrandom fashion. *P. carotovorum* can be prevalent on plants under gutters in greenhouses, where high moisture conditions are common.

Management

Although *P. carotovorum* is probably present in all crops, it does not reach populations that are epidemiologically significant except in diseased plants. Thus, the spread of the bacterium from diseased plants to healthy plants through cultural practices can be avoided. Bacterial stem rot can be prevented by proper sanitation.

Selected References

Cuppels, D., and Kelman, A. 1974. Evaluation of selective media for isolation of soft-rot bacteria from soil and plant tissue. Phytopathology 64:468-475.

Dhanvantari, B. N., and Dirks, V. A. 1987. Bacterial stem rot of greenhouse tomato: Etiology, spatial distribution, and the effect of high humidity. Phytopathology 77:1457-1463.

Speights, D. E., Halliwell, R. S., Horne, C. W., and Hughes, A. B. 1967. A bacterial stem rot of greenhouse-grown tomato plants. Phytopathology 57:902-904.

(Prepared by R. E. Stall)

Bacterial Wilt

Bacterial wilt is a serious disease of tomato and many other economically important crops. It is widely distributed and causes serious crop losses, especially in tropical and subtropical regions and in some warm temperate regions of the world. In the United States, bacterial wilt occurs on tomato mostly in the southern states.

Symptoms

Bacterial wilt is first observed as a flaccid condition that develops on one or more of the youngest leaves. Under favorable environmental conditions, a rapid and complete wilt soon develops (Fig. 102). Advanced stages of the disease may occur 2–3 days after the appearance of initial symptoms (Fig. 103).

Adventitious roots may appear on the stem of an infected plant. These roots are most pronounced when bacterial wilt develops slowly under conditions that are not optimal for disease development. Low temperature, low strain virulence, and host resistance are factors that promote the formation of adventitious roots. Leaf epinasty also can occur when the disease develops slowly.

When the stem of a tomato plant in the early stages of bacterial wilt is cut in transverse or longitudinal section, the vascular system appears yellow or light brown; it becomes a darker brown as the disease progresses. When the plant is completely wilted, the pith and cortex are also brown. Massive invasion of the cortex may result in the appearance of water-soaked lesions on the external surface of the stem. When an infected stem is cut crosswise, tiny drops of dirty-white or yellowish, viscous ooze exude from the severed vascular bundles.

Bacterial wilt can easily be distinguished from diseases caused by fungal vascular pathogens by suspending a clean section of a diseased stem in water. A white, milky stream of bacterial cells and slime will flow from xylem elements of the infected plant in 3–5 min. If the stem is severely infected, the water will become completely milky in 10–15 min (Fig. 104).

Belowground symptoms of bacterial wilt appear as various degrees of root decay, depending on the stage of disease devel-

Fig. 102. Tomato plants with sudden wilt, characteristic of bacterial wilt. (Courtesy S. M. Olson)

Fig. 103. Tomato plants with wilted and necrotic leaves, caused by bacterial wilt. (Courtesy T. M. Momol)

opment. Initially, one or a few roots may show a brown rot, but as the disease progresses and the plant becomes permanently wilted, the entire root system may show a brown rot.

Causal Organism

Bacterial wilt is caused by *Ralstonia solanacearum* (Smith) Yabuuchi et al. This bacterium was known for many years as *Pseudomonas solanacearum* (Smith) Smith. It was then transferred to the genus *Burkholderia* and then more recently to the genus *Ralstonia,* which was established to accommodate *R. solanacearum* and the closely related species within the ribosomal RNA (rRNA) homology group II: *R. pickettii* and *R. eutropha*. The binomial *R. solanacearum* is currently the widely accepted name for the bacterial wilt pathogen.

R. solanacearum is a gram-negative rod, 0.5–0.7 × 1.5–2.5 µm in size, and motile; it is aerobic and oxidase and catalase positive, accumulating poly-β-hydroxybutyrate. Colonies are nonfluorescent on complex media. Most strains of the bacterium produce nitrite from nitrate, except those of biovar 2. The bacterium grows in 0.5% and 1.0% but not 2.0% sodium chloride (NaCl) broth.

In general, virulent colonies of *R. solanacearum* are fluidal and irregular. Each has a characteristic light-pink center and whitish periphery on triphenyl tetrazolium chloride (TTC) medium. Avirulent mutants lacking production of extracellular polysaccharide appear small, round, red, and butyrous. TTC medium (also known as Kelman's medium) is easy to prepare and provides a consistent colony appearance, which makes it the most widely used medium to distinguish *R. solanacearum* colonies on pure culture. Semiselective SMSA medium can be used to detect *R. solanacearum* from water and soil samples, and quick serological tests are available to identify *R. solanacearum* from wilted plants or pure culture. Species- and race-3-specific PCR assays are available if sensitive detection is needed.

R. solanacearum can be stored in sterile distilled water at room temperature for years without significant loss of virulence. However, the organism is best stored when frozen at –80°C in a glycerol suspension.

Fig. 104. White, milky ooze of bacterial cells in clear water, produced by a plant infected with bacterial wilt. (Courtesy D. B. Langston, University of Georgia, Plant Pathology Extension, Bugwood.org)

R. solanacearum is an extremely complex and diverse bacterial species, which enables it to be pathogenic to several hundred plant species belonging to more than 50 families. At the subspecies level, the pathogen was initially divided into three races based on the ability to infect different plants. Race 1 (which has a wide host range) is known as the solanaceous race, race 2 as the *Musa* race, and race 3 (a quarantined pathogen listed as a Select Agent in the United States) as the potato race. Races 4 (the ginger race) and 5 (the mulberry race) were later proposed. The identification of races to differentiate strains of *R. solanacearum* has become less practical, however, with the rapid increase in the number of identified host plants.

Strains of *R. solanacearum* have also been found to have different biochemical characteristics. A. C. Hayward divided the species into different biovars based on the ability of the bacterial isolates to utilize or oxidize three hexose alcohols and three disaccharides.

The use of modern molecular techniques will certainly provide a better understanding of the diversity of *R. solanacearum*. Based on restriction fragment length polymorphism (RFLP) analysis, the bacterium has been separated into two broad RFLP groups: division I includes isolates of biovars 3, 4, and 5, and division II includes isolates of biovars 1 and 2. A hierarchical scheme based on DNA sequence and PCR analysis has been proposed that taxonomically subdivides the *R. solanacearum* species complex into phylotypes and sequevars.

Disease Cycle and Epidemiology

Bacterial wilt of tomato is caused by biovars 1, 2, 3, and 4 of *R. solanacearum*. Biovars 1, 3, and 4 belong predominantly to race 1 and have a wide host range, which guarantees long-term survival of the pathogen in soil in the absence of a susceptible crop. The pathogen can survive in the rhizospheres of nonhost plants, including weeds. Soil factors also influence the survival of the bacterium. For example, bacterial wilt is an important disease of tomato in Florida, but it rarely occurs in calcareous soils with high pH. Moderate pH and moderate to high temperatures are associated with longer bacterial survival in soil. Soils that suppress bacterial wilt promote desiccation of the pathogen or microbial activity that favors antagonistic microorganisms.

Infested soil and surface water (including irrigation water) are the primary sources of inoculum. It is not unusual to find bacterial wilt in the first crop in recently cleared land in tropical and subtropical regions. Disease-free areas can become infested through infected planting material (i.e., tomato transplants) and contaminated irrigation water, stakes, and machinery. Workers that contact surface contamination from infested fields and other sources may also infest disease-free fields. Transmission by seed is not considered important in tomato, even though *R. solanacearum* has been reported to be seed transmitted in tomato, eggplant (*Solanum melongena* L.), and peanut (*Arachis hypogaea* L.). Within the field, the most significant method of pathogen dissemination is contaminated irrigation water flowing through wilted plants.

R. solanacearum can infect the undisturbed roots of susceptible hosts through microscopic wounds caused by the emergence of lateral roots. Transplanting, nematodes, insects, and agricultural equipment are other common causes of root wounding that allow the bacterium to enter the plant. After infecting the plant, the bacterium colonizes the cortex and makes its way toward the xylem vessel, from where it rapidly spreads in the plant. Bacterial masses prevent waterflow from the roots to the leaves, resulting in plant wilting.

The severity of bacterial wilt depends on soil temperature, soil moisture, soil type (which influences soil moisture and microbial populations), host susceptibility, and strain virulence. High soil temperature (28–35°C) and moisture levels are the main factors associated with high incidence and severity. Under these conditions, as a plant wilts, bacteria are released into the soil from the roots, which can infect nearby plants.

Management

Bacterial wilt is caused by a diverse soilborne pathogen with a wide host range and can persist in the soil for a long time in association with weed and crop hosts. Thus, the disease is very difficult to manage once it has become established in the field. No single management strategy effectively prevents losses caused by bacterial wilt. In the southeastern United States, the prevalent race 1, biovar 1 isolates of *R. solanacearum* have caused significant losses in tomato and other crops. The management information provided here is applicable mostly to race 1 of *R. solanacearum*.

The first steps in excluding primary inoculum are to use pathogen-free soil, irrigation water, and transplants. Irrigation water from contaminated ponds must be treated with chlorine or peroxygen products.

In the United States and elsewhere, some tomato cultivars have been found resistant to bacterial wilt. These cultivars have failed across locations primarily because of variations in environmental factors and bacterial strains. Grafting a susceptible tomato cultivar onto resistant rootstock has been used successfully in Japan. In Florida, applications of a plant activator in combination with the use of moderately resistant cultivars (e.g., FL7514 and BHN 466) resulted in higher tomato yield and lower bacterial wilt incidence compared to the untreated control.

Cultural practices, if judiciously used, can also reduce disease incidence. Crop rotation and planting cover crops of nonsusceptible plants can both reduce soilborne populations of the bacterium. The rotation period and the effective break crops should be identified locally for economic reasons, although grasses should always be preferred. In Florida, planting a summer cover crop of sorghum–sudangrass (*Sorghum bicolor* (L.) Moench × *S. bicolor* (L.) Moench subsp. *drummondii* (Nees ex Steud.) de Wet & Harlan) and a winter cover crop of rye (*Secale* L.) reduced population of *R. solanacearum* significantly. Corn (*Zea mays* L.) is another crop that can be used in rotation.

Fields should not be overirrigated because excess soil moisture favors disease buildup, and planting dates should be shifted to cooler periods of the year to help avoid disease development. Higher pH and calcium do not favor development of bacterial wilt. Liming is used routinely in Florida to reduce the disease. Soil amendments such as S-H mixture (which contains agricultural and industrial wastes, including bagasse, rice husk, oyster shell powder, urea, potassium nitrate, calcium super phosphate, and mineral ash) and urea–calcium oxide (CaO) reduced wilt incidence in some locations, but more research and demonstrations are required to provide their suitability in commercial tomato production.

Soil fumigation with chloropicrin has been reported to provide limited success when combined with other control methods. Thymol, a plant-derived volatile chemical, was shown to reduce disease incidence and increase yield in several field experiments. Further research is required to find economical and practical applications of thymol in commercial production.

Selected References

Anith, K. N., Momol, T. M., Kloepper, J. W., Marois, J. J., Olson, S. M., and Jones, J. B. 2004. Efficacy of plant growth-promoting rhizobacteria, acibenzolar-*S*-methyl, and soil amendment for integrated management of bacterial wilt on tomato. Plant Dis. 88:669-673.

Buddenhagen, I. W., and Kelman, A. 1964. Biological and physiological aspects of bacterial wilt caused by *Pseudomonas solanacearum*. Annu. Rev. Phytopathol. 2:203-230.

Elphinstone, J. G., Hennessy, J., Wilson, J. K., and Stead, D. E. 1996. Sensitivity of different methods for the detection of *Ralstonia solanacearum* in potato tuber extracts. EPPO Bull. 26:663-678.

Fegan, M., and Prior, P. 2005. How complex is the "*Ralstonia solanacearum* species complex"? Pages 449-461 in: Bacterial Wilt Disease and the *Ralstonia solanacearum* Species Complex. C. Allen, P. Prior, and A. C. Hayward, eds. American Phytopathological Society, St. Paul, MN.

Hayward, A. C. 1964. Characteristics of *Pseudomonas solanacearum*. J. Appl. Bacteriol. 27:265-277.

Hayward, A. C. 1991. Biology and epidemiology of bacterial wilt caused by *Pseudomonas solanacearum*. Annu. Rev. Phytopathol. 29:65-87.

Kelman, A. 1953. The bacterial wilt caused by *Pseudomonas solanacearum*. N.C. Agric. Exp. Stn. Tech. Bull. 99.

Kelman, A. 1954. The relationship of pathogenicity in *Pseudomonas solanacearum* to colony appearance on a tetrazolium medium. Phytopathology 44:693-695.

Rivard, C., and Louws, F. 2008. Grafting to manage soilborne diseases in tomato production. HortScience 43:2104-2111.

Yabuuchi, E., Kosako, Y., Yano, I., Hotta, H., and Nishiuchi, Y. 1995. Transfer of two *Burkholderia* and an *Alcaligenes* species to *Ralstonia* gen. Nov.: proposal of *Ralstonia pickettii* (Ralston, Palleroni and Doudoroff 1973) comb. Nov., *Ralstonia solanacearum* (Smith 1896) comb. Nov. and *Ralstonia eutropha* (Davis 1969) comb. Nov. Microbiol. Immunol. 39:897-904.

(Prepared by T. M. Momol, P. Ji, and S. M. McCarter)

Tomato Pith Necrosis

Tomato pith necrosis has been reported extensively in the United States and Europe since 1978. It has traditionally been a serious disease of tomato in greenhouses and high tunnels, but in recent years, it has also affected crops in fields. The disease may be most severe—causing almost total plant loss—in greenhouses and high tunnel cultures, where high levels of relative humidity and stress conditions persist.

Symptoms

The initial symptom of tomato pith necrosis is chlorosis on young leaves, followed by yellowing and wilting of the lower leaves. In cases of high severity, chlorosis and wilting occur on the top of the plant and are associated with necrosis of the lower stem (Fig. 105).

Diseased plants exhibit irregular dark-brown/black lesions on the surfaces of stems, peduncles, and leaf petioles (Fig. 106). Stem nodes and branches coming from infected plants are enlarged and darkened (Fig. 107). Adventitious roots on the cracked stem and brown discoloration of the pith (Fig. 108) may also develop, sometimes without external symptoms (Fig. 109). The vascular system may contain a brown discoloration; this is frequently found in plants that do not have external lesions (Fig. 110). The affected areas appear firm on the outside, but longitudinal cuts made into the stem reveal a hollow or chambered

Fig. 105. Chlorosis and wilting, associated with tomato pith necrosis. (Courtesy H. Saygili, F. Sahin, and Y. Aysan)

pith. Profuse development of adventitious roots is associated with areas where the pith is affected.

On fruit, symptoms include initially small, round, black spots, which eventually become scabby in appearance and surrounded with soft rot (Fig. 111). When the lower stem is affected, the plant sometimes dies.

Although the initial symptoms of pith necrosis appear destructive, the disease does not progress. In fact, the condition often becomes less severe and almost undetectable in the field.

Causal Organisms

Several *Pseudomonas* spp. cause pith necrosis on tomato worldwide. Known species include *Pseudomonas corrugata* Roberts & Scarlett, *P. mediterranea* Catara et al., *P. cichorii* (Swingle) Stapp, *P. viridiflava* (Burkholder) Dowson, and *P. fluorescens* Migula. Several unidentified *Pseudomonas* spp. cause the disease, as well.

All of the *Pseudomonas* spp. that cause pith necrosis on tomato are gram-negative rods with one or more polar flagella. In addition, all of the bacteria are strictly aerobic and induce a hypersensitive reaction in tobacco (*Nicotiana* L.). LOPAT testing is useful for distinguishing all of these species (Table 3). *P. corrugata* and *P. mediterranea* can also be differentiated by the assimilation of meso-tartrate, 2-ketogluconate, and histamine. *P. mediterranea* utilizes meso-tartrate, 2-ketogluconate, and histamine, whereas *P. corrugata* does not. *P. corrugata*

Fig. 106. Stem cankers, caused by tomato pith necrosis. (Courtesy H. Saygili, F. Sahin, and Y. Aysan)

Fig. 107. Enlarged stem nodes, a symptom of tomato pith necrosis. (Courtesy H. Saygili, F. Sahin, and Y. Aysan)

Fig. 108. Brown discoloration of the pith, caused by tomato pith necrosis. (Courtesy H. Saygili, F. Sahin, and Y. Aysan)

Fig. 109. External symptoms of tomato pith necrosis. Note the discoloration **(A)** and adventitious roots **(B)**. (Courtesy H. Saygili, F. Sahin, and Y. Aysan)

Fig. 110. Discolored and hollow pith, caused by tomato pith necrosis. (Courtesy H. Saygili, F. Sahin, and Y. Aysan)

Fig. 111. Fruit lesions associated with tomato pith necrosis. (Courtesy M. McGrath)

and *P. mediterranea* are nonfluorescent, but *P. cichorii*, *P. viridiflava*, and other unknown *Pseudomonas* species are fluorescent. On yeast dextrose calcium carbonate agar, *P. corrugata* strains produce two different types of colonies. One type is yellowish and may have a greenish center; it produces a yellow to yellow-green diffusible pigment. The second type is round and raised and usually has a wrinkled appearance.

The maximum temperature for the sustainability of all of the *Pseudomonas* spp. is approximately 37°C.

Disease Cycle and Epidemiology

The pathogens that cause tomato pith necrosis are opportunistic and require an impaired host and favorable environment to incite disease. The bacteria enter plants through wounds and natural openings. The disease frequently occurs when the first fruit set is close to mature green. The spread of the bacteria in greenhouses and fields can occur via contaminated equipment during pruning and harvesting. Bacteria are also easily splashed from plant to plant by irrigation water. High levels of nitrogen fertilization, extensive irrigation, high humidity, and extreme temperature differences between day and night have also been shown to increase host plants' susceptibility. The distribution of the disease in the field usually appears random.

The information available on the epidemiology of pith necrosis on tomato is incomplete. *P. viridiflava* is ubiquitous and causes diseases of a number of vegetable, ornamental, and fruit crops, in addition to tomato. The bacterium may survive epiphytically on the roots and aboveground parts of host plants and in association with infected debris and seeds. Soil, infected debris, seeds, and alternative hosts are probable sources of inoculum.

P. cichorii survives in plant residue in the soil and also numerous hosts. In addition to tomato, this bacterium causes diseases of tobacco; several ornamental plants, such as chrysanthemum (*Chrysanthemum* × *morifolium* Ramat. (pro. sp.)); and a number of vegetables, including lettuce (*Lactuca sativa* L.), chicory (*Cichorium intybus* L.), endive (*Cichorium endiva* L.), cabbage (*Brassica oleracea* L.), cauliflower (*B. oleracea* L. var. *botrytis* L.), and celery (*Apium* L.).

P. corrugata is known to be a pathogen of pepper (*Capsicum annuum* L.) and chrysanthemum, in addition to tomato. It may also cause small, necrotic lesions on alfalfa (*Medicago sativa* L.) and has been isolated from apparently healthy alfalfa roots.

Management

Bactericides have not been developed for management of pith necrosis. Since the pathogens are opportunistic, effective management requires the use of good cultural practices, such as proper sanitation, use of pathogen-free seed and seedlings, proper nutrition, good soil drainage, and good ventilation in greenhouses.

TABLE 3. LOPAT Characteristics of *Pseudomonas* Species Causing Pith Necrosis of Tomato

Pathogen	F/NF[a]	L[b]	O	P	A	T
Pseudomonas corrugata	NF	–	+	–	+	+
Pseudomonas mediterranea	NF	–	+	–	+	+
Pseudomonas cichorii	F	–	+	–	–	+
Pseudomonas viridiflava	F	–	–	+	–	+
Unidentified *Pseudomonas* sp.[c]	F	+	+	–	–	+
Unidentified *Pseudomonas* sp.[d]	F	+	+	–	+	–

[a] F = Fluorescent; NF = Nonfluorescent.
[b] LOPAT characteristics: L = Levan production; O = Oxidase reaction; P = Pectolitic activity on potato slices; A = Arginine dihydrolase activity; T = Hypersensitive reaction on tobacco leaves.
[c] Based on Dhanvantari, 1990.
[d] Based on Sutra et al., 1997.

It is also extremely important to minimize the environmental conditions that enhance development of the disease. For example, excessive nitrogen and high humidity should be avoided, especially in glasshouse and high tunnel culture, and diseased plant debris should be removed from the growing area as promptly as possible. Overhead irrigation and splashing of water should be avoided to minimize leaf wetness, and handling of wet foliage should be avoided, as well.

Selected References

Alivizatos, A. S. 1984. Aetiology of tomato pith necrosis in Greece. Pages 55-57 in: Proceedings of Second Working Group of *Pseudomonas syringae* Pathovars. C. G. Panagopoulous, P. G. Psallidas, and A. S. Alivizatos, eds. Hellenic Phytopathological Society, Athens, Greece.

Aysan, Y., Yildiz, N., and Yücel, F. 2004. Identification of *Pseudomonas viridiflava* on tomato by traditional methods and enzyme-linked immunosorbent assay. Phytoparasitica 32:146-153.

Bazzi, C., Piazza, C., Mazzucchi, U. 1984. Survival in the field of *Pseudomonas cichorii* (Swingle) Stapp, causal agent of lettuce varnish spot. Phytopathol. Z. 111:251-258.

Carroll, N. B., Exhanti, E., and Shoemaker, P. B. 1992. Pith necrosis of tomato in western North Carolina: Etiology and influence of cultural practices on its incidence and severity. N.C. Agric. Res. Serv. Tech. Bull. 300.

Catara, V., Sutra, L., Morineau, A., Achouak, W., Christen, R., and Gardan, L. 2002. Phenotypic and genomic evidence for the revision of *Pseudomonas corrugata* and proposal of *Pseudomonas mediterranea* sp. nov. Int. J. Syst. Evol. Microbiol. 52:1749-1758.

Dhanvantari, B. N. 1990. Stem necrosis of greenhouse tomato caused by novel *Pseudomonas* sp. Plant Dis. 74:124-127.

Jones, J. B., Jones, J. P., Stall, R. E., and Miller, J. W. 1983. Occurrence of stem necrosis on field-grown tomatoes incited by *Pseudomonas corrugata* in Florida. Plant Dis. 67:425-426.

Lai, M., Opgenorth, D. C., and White, J. B. 1983. Occurrence of *Pseudomonas corrugata* on tomato in California. Plant Dis. 67:110-112.

Lukezic, F. L. 1979. *Pseudomonas corrugata,* a pathogen of tomato, isolated from symptomless alfalfa roots. Phytopathology 69:27-31.

Lukezic, F. L., Levine, R. G., and MacNab, A. A. 1983. *Pseudomonas viridiflava* associated with stem necrosis of mature tomato plants. (Abstr.) Phytopathology 73:370.

Mariano, R. L. R., and McCarter, S. M. 1993. Epiphytic survival of *Pseudomonas viridiflava* on tomato and selected weed species. Microb. Ecol. 26:47-58.

Saygili, H., Aysan, Y., Sahin, F., Ustun, N., and Mirik, M. 2004. Occurrence of pith necrosis caused by *Pseudomonas flourescens* on tomato plants in Turkey. Plant Pathol. 53:803.

Scarlett, C. A., Fletcher, J. T., Roberts, P., and Lelliott, R. A. 1978. Tomato pith necrosis caused by *Pseudomonas corrugata* n. sp. Ann. Appl. Biol. 88:105-114.

Sutra, L., Siverio, F., Lopez, M. M., Hunault, G., Bollet, C., and Gardan, L. 1997. Taxonomy of *Pseudomonas* strains isolated from tomato pith necrosis, emended description of *Pseudomonas cor-*

rugata and proposal of three unnamed fluorescent *Pseudomonas* genomospecies. Int. J. Syst. Evol. Microbiol. 47:1020-1033.
Wielke, J. P., and Dye, D. W. 1974. *Pseudomonas cichorii* causing tomato and celery diseases in New Zealand. N.Z. J. Agric. Res. 17:123-130.
Yildiz, N., Aysan, Y., Sahin, F., and Çınar, Ö. 2004. Potential inoculum sources of tomato stem and pith necrosis caused by *Pseudomonas viridiflava* in the Eastern Mediterranean Region of Turkey. J. Plant Dis. Prot. 111:380-387.

(Prepared by H. Saygili, F. Sahin, Y. Aysan, J. B. Jones, and T. A. Zitter)

Minor Bacterial Diseases

Tomato Root Mat

In the 1970s, protected soil and straw bed crops of cucumber (*Cucumis sativus* L.) were affected by a root disorder known as root mat. The disease had disappeared by the end of the decade, but a similar disorder emerged in the United Kingdom in 1993—the year by which all U.K. cucumber production had switched to a hydroponic system of propagation (usually in rockwool). Since 1997, cases of a similar disorder have also been observed in rockwool-grown hydroponic tomato crops. Significant financial losses were estimated during a severe outbreak of tomato root mat. The disease has also been reported in tomato crops in France and Spain.

Symptoms

The predominant symptom of tomato root mat is extensive root proliferation within the propagation cube and across the rockwool slab surface. Infected roots are thicker than normal and have minimal branching, and they can be seen growing up out of the cube and slab. Swelling of the cube is another commonly observed symptom. In the first cases of root mat in tomato in the late 1990s, the level of root proliferation was generally thought to be less than that observed in affected cucumber crops. However, extensive root proliferation (Fig. 112) has occurred, matching or exceeding the level observed in severe cucumber outbreaks.

Tomato root mat affects the fruit, as well. Investigations in a severely infected nursery indicated that fruit size was reduced in a crop of cherry tomato (cv. Favorita). Fruit diameter was typically 2–3 mm smaller, resulting in a higher than usual proportion of unmarketable fruit (i.e., less than 28 mm in diameter). The number of fruit did not appear to be affected, although overall yield was reduced by 15%.

Some crops affected by tomato root mat have also been more susceptible to other pathogens, such as *Pythium* spp.

Causal Organism

Root mat in tomato is caused by strains of *Agrobacterium radiobacter* (Beijerinck & van Delden) Conn, biovar 1, that harbor a root-inducing (Ri) plasmid (pRi). *A. radiobacter* is the same bacterium that causes the disease in cucumber.

Root mat symptoms are induced following transfer and expression of a segment of pRi DNA (T-DNA) into the plant cell genome. The T-DNA of pRi in transformed plant tissue encodes genes that induce cell proliferation and the production of unusual amino acid derivatives or sugar-phosphodiesters. (These compounds are termed opines.) The majority of pathogenic *Agrobacterium* strains isolated from affected crops harbor a monomorphic cucumopine type pRi, although occasionally, other unknown opine types of pRi are isolated.

Subsequent to identifying rhizogenic *A. radiobacter* strains as the causal organism of root mat disease, other non-*Agrobacterium* strains, including one from tomato, have been isolated that also possess a pRi. These strains are from the genera *Rhizobium, Sinorhizobium,* and *Ochrobactrum* and induced typical root mat symptoms on cucumber cotyledons in an in vitro pathogenicity test. However, when the strains were inoculated onto whole cucumber plants, only a pRi-harboring *Rhizobium* strain induced symptoms. This finding indicates that bacterial strains acquiring a pRi may not necessarily become pathogenic.

Disease Cycle and Epidemiology

Once present in a nursery, *A. radiobacter* is difficult to eradicate, despite the use of extensive disinfesting programs between growing seasons and new growing media. Symptom expression has been shown to occur within 4–6 weeks after a single inoculation. The pathogen has been found within the rhizospheres of several weed species growing in or around nurseries.

Avirulent *Agrobacterium* strains without pRi are abundant in nursery environments. These strains, in addition to the other bacteria described previously, can receive copies of pRi from rhizogenic strains and thus acquire pathogenicity.

Management

Once infection has occurred, the effects of root mat are irreversible because of the transformation of affected root tissues.

Fig. 112. Typical symptoms of tomato root mat on hydroponically grown tomatoes. Note the extensive root proliferation. (Courtesy S. Weller)

No tomato cultivars have yet been identified with an inherent resistance to the disease.

Most horticultural disinfectants have action against pathogenic strains of the bacterium; sodium hypochlorite-based disinfectants are especially effective. No known chemical treatment can be applied to plants to prevent root mat, although circumstantial reports suggest that the addition of chlorine to irrigation water may suppress symptoms.

Root mat disappeared from soil-grown cucumber crops in the 1970s and has also been eradicated from tomato grown in organic soil beds amended between growing seasons with compost. These developments suggest that some form of biological control may develop within the soil system and prevent or delay infection. Recreating this biological control in hydroponic crops may be possible following the isolation of suitable organisms.

Selected References

Davioud, E., Petit A., Tate, M. E., Ryder, M. H., and Tempe, J. 1988. Cucumopine—A new T-DNA encoded opine in hairy root and crown gall. Phytochemistry 27:2429-2433.

Weller, S. A., Stead, D. E., O'Neill, T. M., Hargreaves, D., and McPherson, G. M. 2000. Rhizogenic *Agrobacterium* biovar 1 strains and cucumber root mat in the UK. Plant Pathol. 49:43-50.

Weller, S. A., Stead, D. E., O'Neill, T. M., and Morley, P. S. 2000. Root mat of tomato caused by rhizogenic strains of *Agrobacterium* biovar 1 in the UK. Plant Pathol. 49:799.

Weller, S. A., Stead, D. E., and Young, J. P. W. 2004. Acquisition of an *Agrobacterium* Ri plasmid, and pathogenicity, by other *α-Proteobacteria* in cucumber and tomato crops affected by root mat. Appl. Environ. Microbiol. 70:2779-2785.

Weller, S. A., Stead, D. E., and Young, J. P. W. 2005. Induction of root-mat symptoms on cucumber plants by *Rhizobium*, but not by *Ochrobactrum* or *Sinorhizobium*, harbouring a cucumopine Ri plasmid. Plant Pathol. 54:799-805.

Weller, S. A., Stead, D. E., and Young, J. P. W. 2006. Recurrent outbreaks of root mat in cucumber and tomato are associated with a monomorphic, cucumopine, Ri-plasmid harboured by various *Alphaproteobacteria*. FEMS Microbiol. Lett. 258:136-143.

(Prepared by S. Weller and D. Stead)

Syringae Leaf Spot

Correct identification of bacterial leaf spot diseases of tomato transplants is difficult if diagnosis is based solely on symptoms. For most of these diseases, early symptoms include the development of water-soaked, pinpoint necrotic areas, which may or may not be surrounded by yellow halos. The diagnostic difficulty is compounded by the diversity of tomato cultivars and bacterial strains, which may also affect symptom expression.

Correctly diagnosing syringae leaf spot and distinguishing it from other bacterial leaf spot diseases is critical because syringae leaf spot is relatively innocuous. Consequently, in certification programs, transplants with only syringae leaf spot should be allowed to retain their "disease-free" certification, whereas plants with bacterial speck or bacterial spot should be placed under quarantine. The transplant-growing region of the southeastern United States has the greatest incidence of syringae leaf spot; thus, specific and more costly laboratory analysis in plant certification programs is necessary in this area.

Symptoms

Symptoms of syringae leaf spot range from small, brown lesions without halos to black lesions with bright-yellow halos (characteristic of bacterial speck) (Fig. 113). The bacterium can occasionally be isolated from necrotic leaf margins, extensively blighted areas, and tissues that have suffered apparent frost damage. Symptom development is more severe on some tomato cultivars than others, and symptoms are extremely difficult to reproduce in the greenhouse. Disease development is favored by high humidity and wounding.

Causal Organism

The organism that causes syringae leaf spot is *Pseudomonas syringae* pv. *syringae* van Hall, a non-spore-forming, gram-negative, aerobic, rod-shaped bacterium. It is fluorescent on King's medium B, is positive for levan production, induces a hypersensitive reaction in tobacco (*Nicotiana* L.), and is negative for oxidase, arginine dihydrolase, and soft rot of potato (*Solanum tuberosum* L.). These tests are not useful, however, for differentiating *P. syringae* pv. *syringae* from *P. syringae* pv. *tomato* (Okabe) Young, Dye, & Wilkie, with which *P. syringae* pv. *syringae* is most likely to be confused.

The two organisms can be differentiated by their fatty acid profiles. Although the two profiles are nearly identical, *P. syringae* pv. *syringae* contains a significant amount of delta-cis-9,10-methylene hexadecanoic acid, which is absent in *P. syringae* pv. *tomato*. The utilization of erythritol, sucrose, and DL-lactate as sole carbon sources by *P. syringae* pv. *syringae* distinguishes it from both *P. syringae* pv. *tomato* and *P. viridiflava* (Burkholder) Dowson, which do not utilize all compounds. The majority of *P. syringae* pv. *syringae* strains isolated from tomato transplants are ice-nucleation active, which also can differentiate this organism from *P. syringae* pv. *tomato*. Most strains of *P. syringae* pv. *syringae* recovered from tomato produce syringomycin and induce a reaction in the cowpea (*Vigna unguiculata* (L.) Walp.) cultivar California Blackeye 3.

Disease Cycle and Epidemiology

The only known sources of inoculum are other epiphytic hosts. The two most common sources in transplant fields are rye (*Secale* L.), which is used as a winter cover crop, and wild cherry trees (*Prunus serotina* Ehrh.) planted along the edges of fields. Areas with a large number of cherry trees have had a high incidence of syringae leaf spot each year.

Most outbreaks of syringae leaf spot occur during cool temperatures, whereas bacterial speck and bacterial spot typically occur later in the growing season, when average daytime temperatures have increased. *P. syringae* pv. *syringae* is a weak but opportunistic pathogen; it is found frequently in wounds and as

Fig. 113. Symptoms of syringae leaf spot on a field-grown tomato transplant. (Courtesy R. D. Gitaitis)

a secondary organism in mixed infections with other leaf spot pathogens.

Management

Weekly applications of bactericides adequately manage syringae leaf spot. In most cases of the disease, there is no economic damage and no control measures are necessary. To ensure correct diagnosis of syringae leaf spot, an accurate and thorough characterization of the causal agent is required. It is critical to rule out other diseases that would require more thorough management procedures.

Selected References

Bonn, W. G., and Gitaitis, R. D. 1987. Ice nucleation and disease caused by *Pseudomonas syringae* pv. *syringae* in tomato transplants. (Abstr.) Phytopathology 77:1729.

Gitaitis, R. D., Jones, J. B., Jaworski, C. A., and Phatak, S. C. 1985. Incidence and development of *Pseudomonas syringae* pv. *syringae* on tomato transplants in Georgia. Plant Dis. 69:32-35.

Gitaitis, R. D., Sasser, M. J., Beaver, R. W., McInnes, T. B., and Stall, R. E. 1987. Pectolytic xanthomonads in mixed infections with *Pseudomonas syringae* pv. *syringae, P. syringae* pv. *tomato,* and *Xanthomonas campestris* pv. *vesicatoria* in tomato and pepper transplants. Phytopathology 77:611-615.

Jones, J. B., Gitaitis, R. D., and McCarter, S. M. 1983. Evaluation of indirect immunofluorescence and ice nucleation activity as rapid tests for identifying foliar diseases of tomato transplants incited by fluorescent pseudomonads. Plant Dis. 67:684-687.

Jones, J. B., Gitaitis, R. D., and McCarter, S. M. 1986. Fluorescence on single-carbon sources for separation of *Pseudomonas syringae* pv. *syringae, P. syringae* pv. *tomato,* and *P. viridiflava* on tomato transplants. Plant Dis. 70:151-153.

Jones, J. B., McCarter, S. M., and Gitaitis, R. D. 1981. Association of *Pseudomonas syringae* pv. *syringae* with a leaf spot disease of tomato transplants in southern Georgia. Phytopathology 71:1281-1285.

(Prepared by R. D. Gitaitis)

Seedling Blight and Leaf Spot

Disease management of tomato transplant material is a major concern to transplant producers. Given the high plant density and other conditions in transplant production that favor development of disease, the occurrence of bacterial pathogens can easily compromise production. Several bacterial species can cause blight and leaf spot diseases of tomato seedlings. Some of them do not represent a serious threat, however, occurring rarely or under particular conditions as opportunistic pathogens.

Symptoms

Seedlings with blight and leaf spot manifest mixed symptoms. Plants may show discrete lesions along the leaf veins, and isolated leaf spots sometimes occur. The spots are irregular and vary in size from 2 to 4 mm. Affected tissue is discolored and necrotic. Most infected seedlings have apical or marginal leaf necrosis. The diseased tissue collapses, turns brown, and necrotizes.

Blighting of younger plants causes unrecoverable loss of foliage. However, the disease remains localized on individual plants or on plants within a single tray. Symptom development stops within 1 week of the initial appearance. Symptoms are not observed on older plants. Less than 5% of the plants grown in the greenhouse are typically affected.

Causal Organism

Herbaspirillum huttiense (Leifson) Ding & Yokota is reported to be a causal agent of this unusual leaf spot and blighting of tomato seedlings. This organism was originally isolated from distilled water and until recently was not known as a plant pathogen. The strains of this bacterium isolated from tomato seedlings are gram-negative, slightly curved cells with one or two flagella on one or both poles. They form nonfluorescent, creamy-white colonies and grow at 40°C. In addition, they are oxidase positive but levan, pectate hydrolysis, and arginine dihydrolase negative.

The strong evidence for identifying these strains as *H. huttiense* is based on 16S rRNA sequence analysis. Representative strains had greater than 99% sequence identity with *H. huttiense.* Fatty acid analysis and carbon utilization assays were also useful for showing a close relationship between the tomato strains and *H. huttiense.* The fatty acid composition of two representative tomato strains was very close to that of *H. huttiense* and *H. rubrisubalbicans* (Christopher & Edgerton) Baldani et al..

Disease Cycle and Epidemiology

The origin of these bacterial strains is unknown. In the southern Florida location where the transplants were produced, sugarcane (*Saccharum* L.) was used as a windbreak. Since sugarcane is a host for *H. rubrisubalbicans* and the organism isolated from tomato was closely related to the sugarcane pathogen, it is plausible that sugarcane serves as a source of inoculum. It is also plausible that *H. rubrisubalbicans* is heterogeneous and may consist of at least two bacterial species: one including *H. rubrisubalbicans* and the other representing the strains more closely related to *H. huttiense.* The bacterium affects only tomato seedlings, with no further spreading on older plants, which makes it quite apparent that *H. huttiense* is a weak pathogen able to cause disease only under high-humidity conditions.

Management

Management of seedling blight and leaf spot seems unnecessary, since symptoms stop developing as the plant matures. Even so, the likelihood of the disease even occurring can be reduced by controlling temperature and humidity for transplant production and by following transplant-growing practices that enhance plant growth and vigor.

Selected References

Baldani, J. I., Pot, B., Kirchhof, G., Falsen, E., Baldani, V. L. D., Olivares, F. L., Hoste, B., Kersters, K., Hartmann, A., Gillis, M., and Döbereiner, J. 1996. Emended description of *Herbaspirillum;* inclusion of *[Pseudomonas] rubrisubalbicans,* a mild plant pathogen, as *Herbaspirillum rubrisubalbicans* comb. nov.; and classification of a group of clinical isolates (EF group 1) as *Herbaspirillum* species 3. Int. J. Syst. Bacteriol. 46:802-810.

Ding, L., and Yokota, A. 2004. Proposals of *Curvibacter gracilis* gen. nov., sp. nov. and *Herbaspirillum putei* sp. nov. for bacterial strains isolated from well water and reclassification of *[Pseudomonas] huttiensis, [Pseudomonas] lanceolata, [Aquaspirillum] delicatum* and *[Aquaspirillum] autotrophicum* as *Herbaspirillum huttiense* comb. nov., *Curvibacter lanceolatus* comb. nov., *Curvibacter delicatus* comb. nov. and *Herbaspirillum autotrophicum* comb. nov. Int. J. Syst. Evol. Micro. 54:2223-2230.

Leifson, E. 1962. The bacterial flora of distilled and stored water. III. New species of the genera *Corynebacterium, Flavobacterium, Spirillum* and *Pseudomonas.* Int. Bull. Bacteriol. Nomencl. Taxon. 12:161-170.

Obradovic, A., Jones, J. B., Minsavage, G. V., Dickstein, E. R., and Momol, T. M. 2003. *Pseudomonas huttiensis* associated with leaf necrosis and blighting of tomato seedlings in the greenhouse. Pages 627-630 in: *Pseudomonas syringae* and Related Pathogens—Biology and Genetics. N. S. Iacobellis, A. Collmer, S. W. Hutcheson, J. W. Mansfield, C. E. Morris, J. Murillo, N. W. Schaad, D. E. Stead, G. Surico, and M. S. Ullrich, eds. Kluwer Academic, Dordrecht, the Netherlands.

Obradovic, A., Jones, J. B., Minsavage, G. V., Dickstein, E. R., and Momol, T. M. 2007. A leaf spot and blight of greenhouse tomato seedlings incited by a *Herbaspirillum* sp. Plant Dis. 91:886-890.

(Prepared by A. Obradovic and J. B. Jones)

Bacterial Leaf Blight

Symptoms of tomato leaf blight are associated with several bacterial species, and quite often, leaf blight results from a severe outbreak of bacterial spot or bacterial speck. However, the etiology of blighting can vary. Under conditions conducive for development of the disease, bacterial leaf blight can produce extensive damage. When this disease was observed in southwestern Florida in 1983, it affected nearly 100% of the plants in several fields.

Symptoms

The leaf spots that are characteristic of bacterial leaf blight enlarge to the extent of causing extensive tissue necrosis. Blighting may affect as much as 50% of the foliage. Symptoms diminish rapidly, however, with the onset of favorable plant-growing conditions.

Causal Organism

Bacterial leaf blight of tomato is caused by *Pseudomonas viridiflava* (Burkholder) Dowson, a gram-negative, rod-shaped, aerobic and fluorescent bacterium. The strains induce a hypersensitive reaction in tobacco (*Nicotiana* L.) and are oxidase and arginine dihydrolase negative. They produce a soft rot in potato (*Solanum tuberosum* L.) but do not utilize sucrose. Tomato strains of *P. viridiflava* are positive for citrate utilization, aesculin and gelatin hydrolysis, catalase activity, growth in 5% sodium chloride, casein hydrolysis, and peptonization of litmus milk but negative for nitrate reduction, urease activity, starch hydrolysis, and lipolysis.

Disease Cycle and Epidemiology

P. viridiflava is an opportunistic pathogen that incites leaf spot and blight of tomato. It survives as an epiphyte on various crops and weeds, which may serve as sources of primary inoculum. *P. viridiflava* has also been reported to cause stem rot and tomato pith necrosis (see the section on pith necrosis earlier in Part I), and it is a foliage, fruit, stem, or root pathogen of several other crops in the United States and elsewhere.

The 1983 outbreak of bacterial leaf blight of tomato in Florida was associated with excessive rainfall in February and March, when more than three times the average precipitation for this period was recorded. Disease development was favored by water congestion of plant tissues, as well as other stress factors, before and during the epidemic; for example, cool temperatures and high winds resulted in sand injury and contributed to the disease.

In greenhouses during the winter months, prolonged cool and humid conditions with reduced illumination are ideal for the initiation of bacterial leaf blight. Excessive nitrogen fertilization, resulting in increased succulence and plant vigor, also supports initiation of the disease.

Management

In most cases, no control measures are necessary. Applying good crop management practices and avoiding stressful plant-growing conditions will reduce the likelihood of *P. viridiflava* becoming established and initiating disease. Protective applications of copper bactericides can also result in effective management of the disease.

Selected References

Goumans, D. E., and Chatzaki, A. K. 1998. Characterization and host range evaluation of *Pseudomonas viridiflava* from melon, blite, tomato, chrysanthemum and eggplant. Eur. J. Plant Pathol. 104:181-188.

Jones, J. B., Jones, J. P., McCarter, S. M., and Stall, R. E. 1984. *Pseudomonas viridiflava:* Causal agent of bacterial leaf blight of tomato. Plant Dis. 68:341-342.

(Prepared by J. B. Jones)

Psyllid Yellows and Vein Greening

Psyllid yellows and vein-greening disease of tomato and other solanaceous species (including potato (*Solanum tuberosum* L.)) are prevalent throughout the Americas. The range of the two diseases spans from Central America to Canada and includes the western United States. This entire region is located in the dispersal pathway of the tomato (or potato) psyllid, *Paratrioza cockerelli.*

In the United States, psyllid yellows has occurred in commercial tomato and potato crops since the 1920s–1930s, affecting production in states in the Rocky Mountains, the Great Plains, and the Southwest. Vein greening, however, has been reported only since approximately 2000. This disorder also affects crops in most of the western states, extending into Texas and the Northwest. In Mexico, vein greening occurs in the southwestern and south-central regions. It also occurs as far south as Central America (i.e., Guatemala, Honduras) in solanaceous crops and as far north as Canada in both fields and greenhouses.

The vector of psyllid yellows and vein greening, *P. cockerelli,* can now also be found in New Zealand—a location outside the native ranges of both the pathogen and vector, which are endemic to the Americas. The complex was introduced into New Zealand in 2008 when the tomato psyllid, harboring the vein-greening bacterial pathogen '*Candidatus* Liberibacter solanacearum,' was imported on fresh-market tomato fruit.

In fact, '*Ca.* L. solanacearum' had emerged as an important plant bacterial pathogen in approximately 2004, when symptoms were observed in potato fields in Mexico. (In potato, the disease is referred to as zebra chip.) The disease subsequently became more widespread in potato fields in Mexico, Central America, and the United States. During 2004–2006, psyllid infestations associated with vein-greening symptoms were noted in tomato greenhouses in western coastal Mexico and in the southwestern United States (i.e., Arizona, California, and Nevada) and in field-grown tomatoes (i.e., Sinaloa, Mexico; California, U.S.A.). During 2007–2009, several laboratories showed that the psyllid vector transmitted the obligate bacterial pathogen '*Ca.* L. solanacearum' to tomato (and potato) plants.

Feeding on tomato plants by the tomato psyllid has long been known to cause symptoms in tomato in the western United States, and these symptoms have been referred to as "psyllid yellows." Psyllid yellows disorder has been attributed experimentally to a salivary toxin introduced to the tomato plant during psyllid feeding. Once the plant has been freed from psyllid infestation (using insecticides or physical removal), new growth is free of yellows symptoms, suggesting that a transmissible agent is not responsible for the disorder. However, the recent emergence of the vein-greening disease bacterium has created confusion about the symptomology, because psyllid feeding alone can cause yellows symptoms, whereas feeding by psyllids plus infection by '*Ca.* L. solanacearum' results in a distinct symptom phenotype.

Symptoms

Psyllid Yellows. The most obvious symptom of foliar chlorosis is the basis for the name of the disorder. Symptoms of the dull, yellow, interveinal chlorosis begin on middle-aged to older leaves and eventually progress to newer leaves if psyllids are not controlled. Psyllid-infested tomato plants exhibit vein clearing and then vein thickening, along with shortened internodes, curled leaves (new and old), interveinal chlorosis (primarily in older leaves), and stunted growth overall (Figs. 114 and 115). Older leaves may have a leathery appearance and texture, while the newest leaves may have a purplish discoloration on one or both sides, depending on the temperature and relative humidity as well as the tomato variety.

The application of insecticides for rigorous management of psyllids immediately upon their migration to tomato crops appeared to have successfully managed the disease until approximately the year 2000. Then, a distinct symptom phenotype that accompanied the psyllid infestations was not alleviated, even when plants were successfully rid of the psyllids. This new disease of tomato has been named tomato vein greening and was recently shown to be associated with a biotic causal agent (see Causal Organism, later in this section).

Vein Greening. Tomato plants with vein-greening disease exhibit a variety of symptoms, depending on plant age at the time of transmission of the bacterial pathogen, '*Ca.* L. solanacearum.' Tomato plants infected during early stages of growth develop vein clearing followed by vein greening, bright interveinal chlorosis, foliar curling, shortened internodes, and stunting of growth above the point of initial infection (Figs. 116 and 117). Some varieties exhibit purpling on newer leaves. Often, infected plants produce more flowers than usual, but most of the blossoms abort. Few fruit are set after the onset of foliar symptoms. Those set before infestation/infection appear to develop normally, but they are often pale in color and have an acidic flavor and substandard brix. Fruit set at the onset of infection will be deformed and smaller than normal.

Tomato plants infected with '*Ca.* L. solanacearum' during later stages of plant growth exhibit similar symptoms on leaves, which emerge 10–21 days postinoculation. However, the remainder of the plant is normal in size and other characteristics. Long-term infection results in necrosis of leaves, overall wilting, and eventually death of the plants.

Fig. 114. Symptoms of psyllid yellows disease on the lower leaves of a tomato plant 40 days after an infestation of adult psyllids, including leathery appearance, interveinal chlorosis, vein thickening, and foliar curling. (Courtesy J. K. Brown)

Fig. 115. Symptoms of psyllid yellows disease alone in tomato plants exposed to psyllid feeding for approximately 3 weeks. (Courtesy J. K. Brown)

Causal Organism

'*Candidatus* Liberibacter solanacearum' Liefting et al., the pathogen that causes psyllid yellows and vein greening, is a fastidious bacterium assigned to a relatively new genus, Liberibacter (ca. 1997). Its closest relatives are distinct '*Ca.* Liberibacter' species that cause citrus-greening disease (three unique strains or species) and carrot yellows (one unique strain or species). '*Ca.* L. solanacearum' is classified as an Alphaproteobacteria and placed in the order Rhizobiaceae (Rhizobiales). Bacterial cells vary somewhat in length, ranging from 2 to 4 µm, but are consistently reported as 0.15–0.20 µm in width.

The fastidious bacterium multiplies in the plant host and in the psyllid vector, *Paratrioza cockerelli* (Sulc) (Hemiptera: Psyllidae). '*Ca.* L. solanacearum' appears to be transmitted in a circulative, persistent manner. Although the bacterium can be detected in psyllid eggs at a high frequency (80–100%), it is not

Fig. 116. Early symptoms of vein-greening disease on the newest growth of tomato seedlings 3–4 weeks after inoculation. (Courtesy J. K. Brown)

Fig. 117. Symptoms of vein-greening disease in tomato plants, including bright-yellow foliar discoloration together with veinal greening beginning on newer leaves and progressing to older leaves. These symptoms distinguish infection by '*Ca.* Liberibacter solanacearum' from symptoms of psyllid yellows, which lack vein greening and are manifested primarily as a dull interveinal chlorosis, initially on middle and older leaves, following foliar vein clearing and thickening on newer leaves. (Courtesy J. K. Brown)

clear if this presence in the egg results in viable infection of the larval instars to the adult stage.

Disease Cycle and Epidemiology

Psyllid yellows and vein-greening disease of tomato result only subsequent to feeding by *P. cockerelli*. '*Ca.* L. solanacearum' is not borne in true seed nor is it transmissible via sap or mechanical means to any notable degree. It is not entirely clear whether the bacterium is transovarially transmitted. Even though '*Ca.* L. solanacearum' has been detected in eggs by polymerase chain reaction (PCR), it is not known if the bacterium is viable and infects the offspring to any significant extent.

Tomato psyllids migrate from where they overwinter or oversummer on wild hosts to cultivated hosts as the overseasoning hosts senesce (i.e., annuals) or drop their leaves (i.e., perennials). Increasingly warmer temperatures, which occur as a result of seasonal changes, correlate with changes in overseasoning flora. As migrating psyllids seek lush plants for food and reproduction, they disperse into cultivated fields, where they feed and oviposit. Since there is no evidence for reverse migration, it can only be assumed that a low-level base population of *P. cockerelli* remains in these locales. The psyllids may survive on annuals or perennials that become available to sustain them until temperatures become optimal and overseasoning hosts are available once again.

Eggs hatch and produce multiple larval stages, the last of which molts into the teneral and then mature adult stage, which further disperses. The life cycle is temperature dependent; at 24–26°C, development from egg to adult takes approximately 17–20 days. Reproduction is sexual, with males fertilizing eggs prior to oviposition by females. Multiple life cycles per season occur under optimal weather conditions and availability of food. High temperatures are particularly damaging to psyllids, and prolonged exposure to heat leads to mortality and migration to cooler locations.

The host range of *P. cockerelli* spans a variety of uncultivated and cultivated solanaceous species. Cultivated hosts of the vector and pathogen include eggplant (*Solanum melongena* L.), pepper (*Capsicum annuum* L.), potato, and tomato. Among the primary uncultivated plants serving as reproductive hosts are jimsonweed (*Datura stramonium* L.), desert-thorn (*Lycium* L.), and Peruvian groundcherry (*Physalis peruviana* L.). *Solanum* spp. (i.e., various nightshades), *Datura* spp., and *Lycium* spp. that are native to the deserts of southwestern and northwestern Mexico and the southwestern United States (including Texas) are reported as overseasoning hosts. Northward migrations of tomato psyllids to the United States originate in these regions.

Populations of *P. cockerelli* also appear to overseason in the highlands in Central America and perhaps southern Mexico and may be the sources of infestation for these areas. Studies are in progress to confirm these preliminary observations. Although the tomato psyllid has been associated with species in a large number of other plant families, reports do not distinguish between colonization and feeding.

Management

Psyllid yellows disorder is managed entirely by insecticide treatments to kill tomato psyllids at all life stages. Monitoring during the migration season, using yellow sticky traps in fields and greenhouses, has facilitated the selective application of insecticides (based on the timing of psyllid arrival into crops) and aimed at preventing infection and psyllid reproduction. If the psyllids are not harboring '*Ca.* L. solanacearum,' the plants recover from the yellowing phenotype, in that subsequently developing leaves are asymptomatic and plant vigor is restored.

In contrast, the tomato vein-greening disease caused by '*Ca.* L. solanacearum' infection is more difficult to manage. Plants do not recover, even when the psyllids are killed or removed. Insecticide treatment to kill the vector is still the primary means of disease management and helps to prevent secondary spread. Roguing of symptomatic plants in greenhouses is often practiced. However, once tomato plants begin to show symptoms of infection, little can be done to improve vigor or production. The bacterium systemically infects plants, reducing plant vigor as well as fruit quality and yield. In commercial production greenhouses and screenhouses that employ nutrient circulation systems, applying micronutrients may be beneficial in treating vein greening. Plants grown in these settings are generally less stressed and may respond better to treatment than field plants, which are repeatedly exposed to '*Ca.* L. solanacearum' and psyllid feeding, along with other biotic and abiotic stresses.

Selected References

Al-Jabr, A. M. 1999. Integrated pest management of tomato/potato psyllid, *Paratrioza cockerelli* (Sulc) (Homoptera: Psyllidae) with emphasis on its importance in greenhouse grown tomatoes. Ph.D. diss. Colorado State University, Fort Collins.

Binkley, A. M. 1929. Transmission studies with the new psyllid-yellows disease of solanaceous plants. Science 70:615.

Brown, J. K., Rehman, M., Rogan, D., Martin, R. R., and Idris, A. M. 2010. First report of "*Candidatus* Liberibacter psyllaurous" (synonym "*Ca.* L. solanacearum") associated with the 'tomato vein-greening' and 'tomato psyllid yellows' diseases in commercial greenhouses in Arizona. Plant Dis. 94:376.

Cranshaw, W. S. 1993. An annotated bibliography of potato/tomato psyllid, *Paratrioza cockerelli* (Sulc) (Homoptera: Psyllidae). Colo. Agric. Exp. Stn. Bull. TB93-5.

Crosslin, J. M., Munyaneza, J. E., Brown, J. K., and Liefting, L. W. 2010. A history in the making: Potato zebra chip disease associated with a new psyllid-borne bacterium—A tale of striped potatoes. Available online at www.apsnet.org/publications/apsnetfeatures/Pages/PotatoZebraChip.aspx

Daniels, L. B. 1934. The tomato psyllid and the control of psyllid yellows of potatoes. Colo. Agric. Exp. Stn. Bull. 410.

Eyer, J. R., and Crawford, R. F. 1933. Observations on the feeding habits of the potato psyllid (*Paratrioza cockerelli* Sulc.) and the pathological history of the "psyllid yellows," which it produces. J. Econ. Entomol. 26:846-850.

Ferguson, G. 2001. Potato psyllid—A new pest in greenhouse tomatoes and peppers. Ontario Ministry of Food and Agriculture, Canada. Available online at www.omafra.gov.on.ca/english/crops/facts/potato_psyllid.htm

Hansen, A. K., Trumble, J. T., Stouthamer, R., and Paine, T. D. 2008. A new huanglongbing species, '*Candidatus* Liberibacter psyllaurous,' found to infect tomato and potato, is vectored by the psyllid *Bactericera cockerelli* (Sulc). Appl. Environ. Microbiol. 74:5862-5865.

Liefting, L. W., Sutherland, P. W., Ward, L. I., Paice, K. L., Weir, B. S., and Clover, G. R. G. 2009. A new '*Candidatus* Liberibacter' species associated with diseases of Solanaceous crops. Plant Dis. 93:208-214.

Liefting, L. W., Weir, B. S., Pennycook, S. R., and Clover, G. R. G. 2009. '*Candidatus* Liberibacter solanacearum,' associated with plants in the family Solanaceae. Int. J. Syst. Evol. Microbiol. 59:2274-2276.

Lindford, M. R. 1939. The potato and tomato psyllid and its control on tomatoes. Colo. Agric. Exp. Stn. Bull. 454.

Liu, D., and Trumble, J. T. 2004. Tomato psyllid behavioral responses to tomato plant lines and interactions of plant lines with insecticides. J. Econ. Entomol. 97:1078-1085.

Munyaneza, J. E., Sengoda, V. G., Crosslin, J. M., Garzón-Tiznado, J. A., and Cardenas-Valenzuela, O. G. 2009. First report of "*Candidatus* Liberibacter solanacearum" in tomato plants in Mexico. Plant Dis. 93:1076.

Munyaneza, J. E., Sengoda, V. G., Crosslin, J. M., Garzón-Tiznado, J. A., and Cardenas-Valenzuela, O. G. 2009. First report of "*Candidatus* Liberibacter solanacearum" in pepper plants in Mexico. Plant Dis. 93:1076.

Romney, V. E. 1939. Breeding areas of the tomato psyllid, *Paratrioza cockerelli* (Sulc). J. Econ. Entomol. 32:150-151.

(Prepared by J. K. Brown)

Diseases Caused by Phytoplasmas

In 1992, the wall-less, nonhelical prokaryotes known as *mycoplasma-like organisms* (MLOs) were renamed *phytoplasmas*. Phylogenetic studies based on the sequence analysis of the 16S ribosomal RNA (rRNA) and ribosomal protein (rp) gene sequences showed that phytoplasmas form a large, distinct monophylectic cluster within the class Mollicutes, with 15 specific subclades or 16Sr groups. Several distinct taxa have been described as *Candidatus* (*Ca*). Most of the phytoplasmal diseases found in tomato (*Solanum lycopersicum* L.) belong to the aster yellows taxon, '*Ca.* Phytoplasma asteris' (16SrI); the faba bean phyllody taxon, '*Ca.* Phytoplasma aurantifolia' (16SrII); or the clover proliferation taxon, '*Ca.* Phytoplasma trifoli' (16SrVI). In central Italy, phytoplasmas from four taxa were found present in tomato plants with symptoms of tomato big bud.

Tomato Big Bud

Tomato big bud (BB) has been reported in Australia, Italy, Japan, and Jordan and in several regions of the United States. Losses of up to 30% have been reported from Australia.

Symptoms

The phytoplasmas involved in tomato big bud vary depending on the geographic location; thus, it is not surprising that the symptoms vary, as well. In Australia, symptoms include purple veins, thickening of stems, dwarfing of leaves, and proliferation of shoots. In the U.S. states of Arkansas and New York, infected plants have thickened apical stems and a stiff, erect growth habit (Fig. 118). In the state of California, disease symptoms include leaf proliferation of lateral shoots, hypertrophy of calyxes, and greening of flower petals. Similar symptoms occur on infected plants in Jordan.

There are also several common symptoms of tomato big bud, as observed in New York State. Leaves are small, distorted, and yellowish green, and aerial roots may occur. In addition, internodes are shortened and flower buds are enlarged and imperfectly developed (Fig. 119). These abnormal flowers fail to set fruit, and immature fruit become deformed. Normal fruit do not usually develop.

It often takes up to 6 weeks for symptoms of tomato big bud to develop.

Causal Organisms, Disease Cycle, and Epidemiology

Phytoplasmas from several groups cause tomato big bud, and phytoplasmas from different groups have been associated with the disease in different locations (Table 4). In Arkansas, the disease is associated with a phytoplasma belonging to 16SrI-A and has been shown to be transmitted by the aster leafhopper (*Macrosteles quadrilineatus* Forbes). In New York, microscopy and response to antibiotics suggest that a phytoplasma is the

Fig. 118. Plant infected with the tomato big bud phytoplasma, showing stiff and erect growth. (Courtesy J. E. Thomas)

Fig. 119. Greatly enlarged flower buds on a plant infected with the tomato big bud phytoplasma. (Courtesy T. A. Zitter)

TABLE 4. Phytoplasma Diseases of Tomato, by Insect Vector, Geographic Location, 16S RNA Subgroup, and Reference(s)

Disease (Strain)	Insect Vector	Geographic Location	16S RNA Subgroup Classification	Reference(s)
Tomato big bud (BB)	*Macrosteles quadrilineatus*	United States (Arkansas)	16SrI-A	Lee et al., 1993
Tomato big bud (BLTV)	*Circulifer tenellus*	United States (California)	16SrVI-A	Lee et al.,1993; Shaw et al., 1993
Tomato big bud (BB)	Unknown	Eastern Jordan	16SrVI	Anfoka et al., 2003
Tomato big bud (BB)	Unknown	Central Italy	16SrI, 16SrIII, 16SrV, 16SrXII	Del Serrone et al., 2001
Tomato big bud (BB)	Unknown	Bulgaria, southern Italy	16SrXII-A	Albanese et al., 1998; Lee et al., 1998; Vibio et al., 1996
Tomato big bud (BB)	*Orosius argentatus*	Australia	16SrII	Gibb et al., 1995; Greber 1966; Hutton and Grylls 1956
Tomato yellows (TY)	*Macrosteles striifrons*	Japan, Mexico, United States (Arizona, Texas)	16SrI-B	Elateek et al., 2008; Okuda et al., 1997

causal organism, but neither the 16SrRNA subgroup classification nor the insect vector has been determined.

In California, tomato big bud is caused by the beet leafhopper-transmitted virescence agent (BLTVA), a member of the 16SrVI-A group and transmitted by the insect *Circulifer tenellus* (Baker). The host range for the pathogen and vector include celery (*Apium* L.), radish (or daikon) (*Raphanus sativus* L.), plantain (*Plantago major* L.), and Chinese cabbage (*Brassica rapa* L.). The beet leafhopper population is low in California as a result of the state's Curly Top Virus Control Program, and hence, the overall incidence of BLTVA is low.

In central Italy, some intriguing observations were made: Four different phytoplasmas were detected in tomatoes showing symptoms of big bud. Infection by more than one phytoplasma group (16SrI and XII) was determined in some cases. This was also the first report of phytoplasmas from 16SrIII or 16SrV being detected in tomato with big bud symptoms. In southern Italy and Bulgaria, a phytoplasma from 16SrXII (stolbur group) was associated with tomato big bud.

In Jordan, the disease is caused by a phytoplasma belonging to 16SrVI, but the insect vector is unknown. In Australia, the disease is caused by a phytoplasma belonging to 16SrII, and the insect vector has been identified as the common brown leafhopper (*Orosius argentatus* (Evans)).

Management

In North America, tomato big bud is of limited economic importance. However, in Australia, Italy, and Jordan, this disease can have a significant impact on the tomato industry.

Establishing an effective management program requires understanding the epidemiology of the disease. Knowing the taxon to which the phytoplasma belongs allows predicting the potential insect vectors and plant hosts. Controlling weeds along the borders of and within tomato fields and using labeled insecticides usually provides a sufficient level of disease control. In the case of organic production, applying row covers at planting can protect the plants from insects.

Selected References

Albanese, G., Polizzi, G., Grimaldi, V., Collodoro, S., Tessitori, M., and Davis, R. E. 1998. Detection and molecular identification of phytoplasma in diseased tomato plants in Calabria (Italy). Phytopathol. Mediterr. 37:83-87.

Anfoka, G. H., Khalil, A. B., and Fattash, I. 2003. Detection and molecular characterization of a phytoplasma associated with big bud disease of tomato in Jordan. J. Phytopathol. 151:223-227.

Del Serrone, P., Marzachi, C., Bragaloni, M., and Galeffi, P. 2001. Phytoplasma infection of tomato in central Italy. Phytopathol. Mediterr. 40:137-142.

Gibb, K. S., Padovan, A. C., and Mogen, B. D. 1995. Studies on sweet potato little-leaf phytoplasma detected in sweet potato and other plant species growing in northern Australia. Phytopathology 85:169-174.

Granett, A. L., and Provvidenti, R. 1974. Tomato big bud in New York. Plant Dis. Rep. 58:211-214.

Greber, R. S. 1966. Identification of the virus causing papaw yellow crinkle with tomato big bud virus by transmission tests. Queensl. J. Agric. Anim. Sci. 23:147-153.

Hiruki, C., and Wang, K. 2004. Clover proliferation phytoplasmas: "*Candidatus* Phytoplasma trifolii." Int. J. Syst. Evol. Microbiol. 54:1349-1353.

Hutton, E. M., and Grylls, N. E. 1956. Legume "little leaf"; a virus disease of subtropical pasture species. Aust. J. Agric. Res. 7:85-97.

IRPCM Phytoplasma/Spiroplasma Working Team—Phytoplasma Taxonomy Group. 2004. "*Candidatus* Phytoplasma," a taxon for the wall-less, non-helical prokaryotes that colonize plant phloem and insects. Int. J. Syst. Evol. Microbiol. 54:1243-1255.

Lee, I.-M, Gundersen-Rindal, D. E., Davis, R. E., and Bartoszyk, I. M. 1998. Revised classification scheme of phytoplasmas based on RFLP analyses of 16S rRNa and ribosomal protein gene sequences. Int. J. Syst. Bacteriol. 48:1153-1169.

Lee, I.-M., Gundersen-Rindal, D. E., Davis, R. E., Bottner, K. D., Marcone, C., and Seemüller, E. 2004. "*Candidatus* Phytoplasmas asteris," a novel phytoplasma taxon associated with aster yellows and related diseases. Int. J. Syst. Evol. Microbiol. 54:1037-1048.

Lee, I.-M., Hammond, R. W., Davis, R. E., and Gundersen, D. E. 1993. Universal amplification and analysis of pathogen 16S rDNA for classification and identification of mycoplasmalike organisms. Phytopathology 83:834-842.

Shaw, M. E., Kirkpatrick, B. C., and Golino, D. A. 1993. The beet leafhopper-transmitted virescence agent causes tomato big bud disease in California. Plant Dis. 77:290-295.

Zreik, L., Carle, P., Bové, J. M., and Garnier, M. 1995. Characterization of the mycoplasmalike organism associated with witches'-broom disease of lime and proposition of a *Candidatus* taxon for the organism, "*Candidatus* Phytoplasma aurantifolia." Int. J. Syst. Bacteriol. 45:449-453.

(Prepared by H. M. Griffiths and T. A. Zitter)

Other Phytoplasmal Diseases

Tomato yellows (TY) has been detected on tomato in Japan, and the causal organism has been shown to belong to 16SrI-B. This phytoplasma is transmitted by the vector *Macrosteles striifrons,* a species of leafhopper (Table 4). The economic importance of TY is unknown. If management is warranted, a labeled insecticide can be applied. Alternatively, row covers can be used from the beginning of the season.

A new disease of tomato, *Hoja de perejil* (parsley leaf), was recently reported in Bolivia; it is associated with a phytoplasma that may belong to a new subgroup of 16SrI. Symptoms begin with adventitious sprouting of axillary buds and rapid elongation of side shoots, which subsequently break through the crowns of normal leaves. The leaves of the side shoots are small and fernlike and generate a large, bushy plant. In addition, flower size is small and fruit set is minimal. Infection rates as high as 60% were observed in some fields planted with tomato cultivar Rio Grande. Hedgerow vines belonging to the milkweed family (subfamily Asclepiadoiadeae, family Apocynaceae) also tested positive for the new 16SrI phytoplasma. Insect vectors are unknown at this time. Because disease levels are high, control measures aimed at eliminating hedgerow vines are recommended.

Selected References

Jones, P., Arocha, Y., Antesana, O., Montellano, E., and Franco, P. 2005. *Hoja de perejil* (parsley leaf) of tomato and morrenia little leaf, two new diseases associated with a phytoplasma in Bolivia. Plant Pathol. 54:235.

Okuda, S., Prince, J. P., Davis, R. E., Dally, E. L., Lee, I.-M., Mogen, B., and Kato, S. 1997. Two groups of phytoplasmas from Japan distinguished on the basis of amplification and restriction analysis of 16S rDNA. Plant Dis. 81:301-305.

(Prepared by H. M. Griffiths and T. A. Zitter)

Diseases Caused by Viruses

Viruses are an important group of tomato pathogens that occur in most regions of the world. Plant viruses are obligate parasites that can replicate only in living cells, and thus, they share an intimate relationship with their hosts. Viral pathogens are complex molecules composed of a nucleic acid core (either ribonucleic acid (RNA) or deoxyribonucleic acid (DNA)) and a protein coat. They can be seen with an electron microscope but are too small to be seen with a light microscope. The virus particles, which are also called virions, may be short or long rods or polyhedral in shape. Viruses can be transmitted among plants mechanically (in sap), by grafting, and by seed, but they are most commonly transmitted by vectors such as aphids, beetles, fungi, leafhoppers, nematodes, thrips, and whiteflies or by unidentified vectors.

In most cases, the diseases caused by viruses are diagnosed by symptoms that include leaf mosaic (i.e., light- and dark-green or yellow areas), chlorosis, necrosis, malformation (e.g., twisting, puckering), and stunting. Diagnosis is not always straightforward, however, because some of these symptoms are similar to those caused by nutritional deficiencies and herbicide damage. Complicating the situation further is the fact that a single plant can be infected with more than one virus. Thus, in many regions of the world, it is difficult, if not impossible, to visually diagnose the virus or viruses responsible for an infection.

Virus diseases are difficult to control and can only be managed, at best, to minimize losses. The most effective management strategy is to use resistant cultivars, and genes for resistance have been identified and bred into many modern-day tomato cultivars. However, resistance-breaking strains are a common threat to the long-term survival of a given cultivar of tomato. Chemical treatments are not completely successful in treating virus diseases, for several reasons. Insects frequently develop resistance to insecticides (common for whitefly-transmitted viruses), and insects are often able to transmit viruses in a very limited time frame (common for aphid-transmitted viruses in a nonpersistent manner). Also, the incidence and severity of virus diseases varies from season to season because of the complex interrelationships among the pathogen host and vector, the virus source, and the environment. Given this unpredictability, growers often do not apply management strategies on a regular basis.

This section describes viruses known to infect tomatoes worldwide. Since the first edition of this compendium was published in 1991, there has been a virtual explosion of new virus diseases, especially for whitefly-transmitted begomoviruses belonging to the family Geminiviridae. Within this section, viruses are discussed according to their means of transmission. See also Table 5, which identifies accepted tomato viruses in the genus *Begomovirus*, and Table 6, which identifies tomato viruses of other genera.

TABLE 5. Accepted Begomoviruses Reported to Infect Tomato

Begomovirus	Acronym	Natural (N) or Experimental (E) Infection	GenBank Accession Number: Component A	GenBank Accession Number: Component B	Reference(s)
Abutilon mosaic virus	AbMV	E	NC 001928	NC 001929	Unseld et al., 2000
Ageratum yellow vein virus	AYVV	N	AB100305 (tomato) X74516 NC_004626	Beta DNA NC_003403 (Weed)	Tan and Wong, 1993; Tan et al., 1995
Bean calico mosaic virus	BCaMV	E	NC_003504	NC_003505	Brown et al., 1999
Bean dwarf mosaic virus	BDMV	E	NC_001931	NC_001930	Hou et al., 1998
Chino del tomate virus	CdTV	N	NC_003830	NC_003831	Brown and Nelson, 1988; Jiang et al., 1998
Cotton leaf curl Alabad virus	CLCuAV	E	NC_004582	*Not reported*	Harrison et al., 1997; Liu et al., 1998; Zhou et al., 1998
Cotton leaf curl Kokhran virus	CLCuKV	E	NC_004583	*Not reported*	Harrison et al., 1997; Liu et al., 1998; Zhou et al., 1998
Cotton leaf curl Multan virus	CLCuMV	E	NC_004607	*Not reported*	Harrison et al., 1997; Liu et al., 1998; Zhou et al., 1998
Honeysuckle yellow vein mosaic virus	HYVMV	N	NC_003609	*Not reported*	Kitamura et al., 2004
Papaya leaf curl virus	PaLCuV	N	Y15934	AY244706	Chowda et al., 2005
Pepper golden mosaic virus	PepGMV	N	NC_004101	NC_004096	Brown and Poulos, 1990; Holguín-Peña et al., 2004; Paximadis et al., 1999; Torres-Pacheco et el., 1996
Pepper huasteco yellow vein virus	PHYVV	N	NC_001359	NC_001369	Torres-Pacheco et el., 1996
Pepper leaf curl Bangladesh virus	PepLCBV	N	NC_004192	*Not reported*	Chowda et al., 2005
Potato yellow mosaic Panama virus	PYMPV	N	NC_002048	NC_002049	Engel et al., 1998

(continued)

[a] This virus is discussed as *Tomato yellow mosaic virus* later in this section.
[b] This virus is also known as *Sida golden mosaic Honduras virus*.
[c] See the discussion of this virus later in this section.

TABLE 5. *(continued)*

Begomovirus	Acronym	Natural (N) or Experimental (E) Infection	GenBank Accession Number		Reference(s)
			Component A	Component B	
Potato yellow mosaic Trinidad virus	PYMTV-[TT]	N	NC_004638	NC_004644	Umaharan et al., 1998
Potato yellow mosaic virus[a]	PYMV	N	NC_001934	NC_001935	Buragohain et al., 1994; Guzman et al., 1997; Polston et al., 1997; Roberts et al., 1998; Urbino et al., 2004
Sida golden mosaic Costa Rica virus	SiGMCRV	E	NC_004657	NC_004658	Jovel and Ruiz, 2003
Sida golden mosaic virus	SiGMV	N	NC_002046	NC_002047	Al-Aqeel, 2003
Sida yellow vein virus[b]	SiYVV	E	NC_004659	NC_004660	Unseld et al., 2000
Tobacco curly shoot virus	TbCSV	N	NC_003722	*Not reported*	Li et al., 2002, 2004
Tobacco leaf curl Japan virus	TbLCJV	N	NC_004654	*Not reported*	Kitamura et al., 2004
Tobacco leaf curl Kochi virus	TbLCKoV	N	NC_004641	*Not reported*	Kitamura et al., 2004
Tobacco leaf curl Yunnan virus	TbLCYV	N	NC_004356	*Not reported*	Li et al., 2004; Zhou et al., 2001
Tobacco leaf curl Zimbabwe virus	TbLCZV	E	NC_002817	*Not reported*	Paximadis and Rey, 2001
Tomato chlorotic mottle virus	ToCMoV	N	NC_003664	NC_003665	Ambrozevicius et al., 2002; Galvao et al., 2003
Tomato golden mosaic virus	TGMV	N	NC_001507	NC_001508	Bisaro et al., 1982; Hamilton et al., 1984; Matyis et al., 1975
Tomato golden mottle virus	ToGMoV	N	NC_008058	NC_008057	*No published reference found.*
Tomato leaf curl Bangalore virus	ToLCBV	N	NC_003891	*Not reported*	McGrath and Harrison, 1995; Morris et al., 2002; Muniyappa et al., 2000
Tomato leaf curl Bangladesh virus	ToLCBDV	N	NC_004614	*Not reported*	Shih et al., 1998a
Tomato leaf curl Gujarat virus	ToLCGV	N	NC_004558	NC_004559	Chakraborty et al., 2003a, 2003b
Tomato leaf curl Karnataka virus	ToLCKV	N	NC_003897	*Not reported*	Chatchawankanphanich and Maxwell, 2002
Tomato leaf curl Laos virus	ToLCLV	N	NC_004613	*Not reported*	*No published reference found.*
Tomato leaf curl Malaysia virus	ToLCMV	N	NC_004648	*Not reported*	Shih et al., 1998b
Tomato leaf curl New Delhi virus	ToLCNDV	N	NC_004611	NC_004612	Padidam et al., 1995; Shih et al., 2003
Tomato leaf curl Sinaloa virus	ToLCSiV	N	NC_009606	NC_009805	Brown et al., 1993; Idris and Brown, 1998
Tomato leaf curl Sri Lanka virus	ToLCSLV	N	NC_004647	*Not reported*	*No published reference found.*
Tomato leaf curl Taiwan virus	ToLCTWV	N	NC_003898	*Not reported*	*No published reference found.*
Tomato leaf curl Vietnam virus	ToLCVV	N	NC_004153	*Not reported*	*No published reference found.*
Tomato leaf curl virus	ToLCV	N	NC_003896	Satellite DNA NC_002743	Dry et al., 1993
Tomato mosaic Havana virus	ToMHaV	N	NC_003867	NC_003868	Martinez Zubiaur et al., 1998
Tomato mottle Taino virus	ToMoTaV	N	NC_001828	NC_001917	Ramos et al., 1997
Tomato mottle virus[c]	ToMoV	N	NC_001938	NC_001939	Abouzid et al., 1992; Polston et al., 1993
Tomato rugose mosaic virus	ToRMV	N	NC_002555	NC_002556	Ribeiro et al., 2003; Santos et al., 2003
Tomato severe leaf curl virus	ToSLCV	N	NC_004642	*Not reported*	Ala Poikela, 2003; Nakhla et al., 1994
Tomato severe rugose virus	ToSRV	N	AY029750	AY029751	*No published reference found.*
Tomato yellow leaf curl China virus	TYLCCNV	N	NC_004044	Satellite DNA AJ784151	Li et al., 2004; Yin et al., 2001
Tomato yellow leaf curl Gezira virus	TYLCGV	N	AY044137	*Not reported*	Idris and Brown, 2001, 2005
Tomato yellow leaf curl Malaga virus	TYLCMalV	N	NC_004569	*Not reported*	Monci et al., 2002
Tomato yellow leaf curl Sardinia virus	TYLCSV	N	NC_003828	*Not reported*	Kheyr Pour et al., 1991; Noris et al., 1994
Tomato yellow leaf curl Thailand virus	TYLCTHV	N	NC_000869	NC_000870	Rochester et al., 1994; Sawangjit et al., 2005
Tomato yellow leaf curl virus[c]	TYLCV	N	X15656	*Not reported*	Antignus and Cohen, 1994; Navot et al., 1991

TABLE 6. Viruses (Less Begomoviruses) Known to Infect Cultivated Tomato

Aphid-Transmitted Viruses: Nonpersistent Mode of Transmission

Virus[a,b]	Abbreviation	Genus (Family)	Other Means of Transmission[c]	Reference(s)[d]
Alfalfa mosaic virus[e]	AMV	*Alfamovirus* (Bromoviridae)	M, S	Porter, 1935
Cucumber mosaic virus[e]	CMV	*Cucumovirus* (Bromoviridae)	M, S	Johnson, 1926
Datura distortion mosaic virus	DDMV	*Potyvirus* (Potyviridae)	M, G	Mali, 1985
Datura necrosis virus	DNV	*Potyvirus* (Potyviridae)	M	Badami and Kassanis, 1959
Eggplant green mosaic virus	EGMV	*Potyvirus* (Potyviridae)	M	Ladipo, 1976
Eggplant severe mottle virus	ESMV	*Potyvirus* (Unassigned)	M	Ladipo et al., 1988
Henbane mosaic virus	HMV	*Potyvirus* (Potyviridae)	M	Hamilton, 1932
Marigold mottle virus	MaMoV	*Potyvirus* (Potyviridae)	M	Naqvi et al., 1981
Peanut stunt virus	PSV	*Cucumovirus* (Bromoviridae)	M, S	Mink et al., 1969
Pepper mottle virus	PepMoV	*Potyvirus* (Potyviridae)	M, G	Purcifull et al., 1975
Pepper severe mosaic virus	PeSMV	*Potyvirus* (Potyviridae)	M, G	Feldman and Gracia, 1977
Pepper veinal mottle virus	PVMV	*Potyvirus* (Potyviridae)	M, G	Ladipo and Roberts, 1977
Peru tomato mosaic virus	PTV	*Potyvirus* (Potyviridae)	M, G	Raymer et al., 1972; Fribourg, 1979
Potato aucuba mosaic virus	PAMV	*Potexvirus* (Alphaflexiviridae)	M[f]	Dykstra, 1939
Potato virus A	PVA	*Potyvirus* (Potyviridae)	M	MacLachlan et al., 1953
Potato virus M	PVM	*Carlavirus* (Betaflexiviridae)	M, G	Grieco et al., 1997
Potato virus V	PVV	*Potyvirus* (Potyviridae)	M, G	Fribourg and Nakashima, 1984
Potato virus Y[e]	PVY	*Potyvirus* (Potyviridae)	M	Johnson, 1930
Tobacco etch virus[e]	TEV	*Potyvirus* (Potyviridae)	M	Johnson, 1930
Tobacco vein mottling virus	TVMV	*Potyvirus* (Potyviridae)	M, G	Gooding and Sun, 1972
Tomato aspermy virus	TAV	*Cucumovirus* (Bromoviridae)	M, G, S	Brierley et al., 1955
Tomato mild mottle virus	ToMMV	Unassigned (Potyviridae)	M	Walkey et al., 1994
Wild potato mosaic virus	WPMV	*Potyvirus* (Potyviridae)	M	Jones and Fribourg, 1979; Jones et al., 1980

Aphid-Transmitted Viruses: Persistent Mode of Transmission

Virus[a,b]	Abbreviation	Genus (Family)	Other Means of Transmission[c]	Reference(s)[d]
Beet western yellows virus	BWYV	*Polerovirus* (Luteoviridae)	G	Duffus 1981
Potato leafroll virus	PLRV	*Polerovirus* (Luteoviridae)	G	Natti et al., 1953; Dykstra, 1930
(syn. *Solanum yellows virus*)	SYV	*Luteovirus* (Luteoviridae)		Milbrath and Duffus, 1978
(syn. *Tomato yellow top virus*)	TYTV	*Luteovirus* (Luteoviridae)		Braithwaite and Blake, 1961
Tobacco mottle virus	TMoV	*Umbravirus* (Unassigned)	M[f]	Smith, 1945; Smith, 1946a
Tobacco necrotic dwarf virus	TNDV	Unassigned (Luteoviridae)	G	Abeygunawardena et al., 1967; Kubo and Takanami, 1977
Tobacco vein distorting virus	TVDV	*Polerovirus* (Luteoviridae)	G[f]	Smith, 1946c
Tobacco yellow vein (assistor) virus	TYVAV	*Luteovirus* (Unassigned)	M	Adams and Hull, 1972

Beetle-Transmitted Viruses

Virus[a,b]	Abbreviation	Genus (Family)	Other Means of Transmission[c]	Reference(s)[d]
Eggplant mosaic virus	EMV	*Tymovirus* (Tymoviridae)	M, S	Dale, 1954
(str. *Andean potato latent virus*)	(APLV)			Gibbs and Harrison, 1969
(syn. Abelia latent virus)	—			Waterworth et al., 1975
Physalis mottle virus	PhyMV	*Tymovirus* (Tymoviridae)	M	Lee et al., 1979
(syn. *Belladonna mottle virus*)	(BeMV)			
Andean potato mottle virus	APMoV	*Comovirus* (Secoviridae)	M, G	Fribourg et al., 1977

(*continued*)

[a] Viruses for which the original publication could be obtained to verify the susceptibility of mainly *Solanum lycopersicum*.
[b] Viruses include those previously identified as "tentative" by the International Committee on Taxonomy of Viruses. Names of viruses that are italicized are identified as "accepted" in *Virus Taxonomy: Ninth Report of the International Committee on Taxonomy of Viruses* (King et al., 2012).
[c] G = grafting; M = mechanical transmission; P = pollen transmission; S = seed transmission. The manner of transmission identified is for the virus in general and not necessarily for infections of *S. lycopersicum*. If a virus is mechanically transmissible, it is also transmissible by grafting in most cases.
[d] The first reference provides the identification of the virus; any additional references provide supporting information.
[e] See the discussion of this virus later in this section.
[f] Requires a so-called helper virus for aphid transmission.
[g] Not normally transmitted mechanically unless special procedures are used.

TABLE 6. (*continued*)

Fungus-Transmitted Viruses

Virus[a,b]	Abbreviation	Genus (Family)	Other Means of Transmission[c]	Reference(s)[d]
Cucumber necrosis virus (Olpidium bornovanus)	CNV (CuNV)	*Tombusvirus* (Tombusviridae)	M	Obermeier et al., 2001
Potato mop-top virus (Spongospora subterranea)	PMTV	*Pomovirus* (Virgaviridae)	M, G	Harrison and Jones, 1970
Tobacco necrosis virus (Olpidium brassicae)	TNV	*Necrovirus* (Tombusviridae)	M[g]	Price, 1940
Tobacco stunt virus (Olpidium brassicae)	TStV	*Varicosavirus* (Unassigned)	M, G	Hiruki, 1975

Leafhopper-Transmitted Viruses

Virus[a,b]	Abbreviation	Genus (Family)	Other Means of Transmission[c]	Reference(s)[d]
Beet curly top virus[e]	BCTV	*Curtovirus* (Geminiviridae)	M	Severin, 1929
Solanum apical leaf curling virus	SALCV	*Bigeminivirus?* (Geminiviridae)	G	Hooker and Salazar, 1983
Tobacco yellow dwarf virus	TYDV	*Mastrevirus* (Geminiviridae)	G	Thomas and Bowyer, 1984

Nematode-Transmitted Viruses

Virus[a,b]	Abbreviation	Genus (Family)	Other Means of Transmission[c]	Reference(s)[d]
Arabis mosaic virus	ArMV	*Nepovirus* (Secoviridae)	M, G, S	Cadman and Harrison, 1960
(syn. Raspberry yellow dwarf virus)	—			
Pepper ringspot virus	PepRSV	*Tobravirus* (Virgaviridae)	M, G	Silberschmidt, 1963
Potato black ringspot virus	PBRSV	*Nepovirus* (Secoviridae)	M, G	Salazar and Harrison, 1978
Potato virus U	PVU	*Nepovirus* (Secoviridae)	M, G, S	Jones et al., 1983
Strawberry latent ringspot virus	SLRSV	Unassigned (Secoviridae)	M, S	Schmelzer, 1969
Tobacco rattle virus	TRV	*Tobravirus* (Virgaviridae)	M, S	Schmelzer, 1957; Cadman and Harrison, 1959
Tobacco ringspot virus	TRSV	*Nepovirus* (Secoviridae)	M, S	Wingard, 1928; Price, 1940
Tomato black ring virus	TBRV	*Nepovirus* (Secoviridae)	M, G, S	Smith, 1946b
Tomato ringspot virus	ToRSV	*Nepovirus* (Secoviridae)	M, S	Imle and Samson, 1937; Samson and Imle, 1942

Thrips-Transmitted Viruses

Virus[a,b]	Abbreviation	Genus (Family)	Other Means of Transmission[c]	Reference(s)[d]
Capsicum chlorosis virus	CaCV	*Tospovirus* (Bunyaviridae)	M	McMichael et al., 2002
Impatiens necrotic spot virus	INSV	*Tospovirus* (Bunyaviridae)	M	De Ávila et al., 1992
Tobacco streak virus	TSV	*Ilarvirus* (Bromoviridae)	M	Gracia and Feldman, 1974
Tomato chlorotic spot virus	TCSV	*Tospovirus* (Bunyaviridae)	M	De Ávila et al., 1990
Tomato spotted wilt virus[e]	TSWV	*Tospovirus* (Bunyaviridae)	M	Brittlebank,1919; Samuel et al., 1930
Tomato yellow ring virus	TYRV	*Tospovirus* (Bunyaviridae)	M	Hassani-Mehraban et al., 2005
(syn. Tomato Varamin virus)	ToVV			Ghotbi et al., 2005
Tomato zonate spot virus	TZSV	*Tospovirus* (Bunyaviridae)	M	Dong et al., 2008

Treehopper-Transmitted Virus

Virus[a,b]	Abbreviation	Genus (Family)	Other Means of Transmission[c]	Reference(s)[d]
Tomato pseudo-curly top virus	TPCTV	*Topocuvirus* (Geminiviridae)	G	Giddings et al., 1951; Simons and Coe, 1958

Whitefly-Transmitted Viruses (non-Begomoviruses)

Virus[a,b]	Abbreviation	Genus (Family)	Other Means of Transmission[c]	Reference(s)[d]
Cowpea mild mottle virus	CPMMV	*Carlavirus* (Betaflexiviridae)	M, S	Brunt and Kenten, 1973
(str. Tomato pale chlorosis virus)	—			Cohen and Antignus, 1982
Potato yellow vein virus[e]	PYVV	*Crinivirus* (Closteroviridae)	G	Salazar et al., 2000
Tomato chlorosis virus[e]	ToCV	*Crinivirus* (Closteroviridae)	G (poorly)	Wisler et al., 1998
Tomato infectious chlorosis virus[e]	TICV	*Crinivirus* (Closteroviridae)	G (poorly)	Duffus et al., 1994
Tomato necrotic dwarf virus	ToNDV	*Nepovirus* (Secoviridae)	G	Larsen et al., 1984

(*continued*)

TABLE 6. (*continued*)

Picorna-Like Whitefly-Transmitted Viruses (non-Begomoviruses)				
Virus[a,b]	Abbreviation	Genus (Family)	Other Means of Transmission[c]	Reference(s)[d]
Tomato torrado virus[e]	ToTV	*Torradovirus* (Secoviridae)	M	Verbeek et al., 2007
Tomato apex necrosis virus[e]	ToANV	*Torradovirus* (Secoviridae)	M	Turina et al., 2007
Tomato marchitez virus[e]	ToMarV	*Torradovirus* (Secoviridae)	M	Verbeek et al., 2008
Tomato chocolate spot virus[e]	ToCSV	*Torradovirus* (Secoviridae)	M, G	Batuman et al., 2010

Viruses Transmitted by Unknown Means				
Virus[a,b]	Abbreviation	Genus (Family)	Other Means of Transmission[c]	Reference(s)[d]
Cymbidium ringspot virus	CyRSV (CymRSV)	*Tombusvirus* (Tombusviridae)	M	Hollings et al., 1977
Datura yellow vein virus	DYVV	*Nucleorhavdovirus* (Rhabdoviridae)	G	Thomas and Dietzgen, 1991
Eggplant mottled dwarf virus	EMDV	*Nucleorhavdovirus* (Rhabdoviridae)	M, G	El Maataoui et al., 1985
(syn. Tomato vein yellowing virus)	TVYV			Kano et al., 1985
(syn. Tomato vein clearing virus)	TVCV			Castellano and Martelli, 1987
(syn. Pelargonium vein clearing virus)	PVCV			—
(syn. Pittosporum vein yellowing virus)	PVYV			—
Lettuce necrotic stunt virus[e]	LNSV	*Tombusvirus* (Tombusviridae)	M	Obermeier et al., 2001
Moroccan pepper virus	MPV	*Tombusvirus* (Tombusviridae)	M, G	Fischer and Lockhart, 1977
Ourmia melon virus	OuMV	*Ourmiavirus* (Unassigned)	M, S	Lisa et al., 1988
Parietaria mottle virus	PMoV	*Ilarvirus* (Bromoviridae)	M, G	Caciagli et al., 1989
(str. Tomato necrotic spot virus)	(ToNSV)			Batuman et al., 2009
Pelargonium zonate spot virus	PZSV	*Anulavirus* (Bromoviridae)	M, S, P	Martelli and Cirulli, 1969; Gallitelli, 1982
Pepino mosaic virus[e]	PepMV	*Potexvirus* (Alphaflexiviridae)	M, G	Jones et al., 1980
Physalis mosaic virus	PMV	*Tymovirus* (Unassigned)	M	Peter and Derks, 1974
Potato virus X	PVX	*Potexvirus* (Alphaflexiviridae)	M	Ainsworth, 1934
Ribgrass mosaic virus	RMV	*Tobamovirus* (Virgaviridae)	M, G	Holmes, 1950
Sunn-hemp mosaic virus	SHMV	*Tobamovirus* (Virgaviridae)	M	Capoor, 1962
Tobacco mosaic virus[e]	TMV	*Tobamovirus* (Virgaviridae)	M, G, S	Miller and Thornberry, 1958
Tobacco mosaic satellite virus	—	Satellite virus	M, S	Valverde et al., 1991
Tomato bushy stunt virus[e]	TBSV	*Tombusvirus* (Tombusviridae)	M, G, S	Smith, 1935
Tomato double-virus streak	ToMV + PVX	*Tobamovirus* + *Potexvirus* (Unassigned)	M	Ainsworth, 1933
Tomato mosaic virus[e]	ToMV	*Tobamovirus* (Virgaviridae)	M, S	Jarrett, 1930

Selected References

Abeygunawardena, D. V. W., Karandawela, C. B., and Bandaranayake, C. M. 1967. Yellow net virus disease of flue-cured tobacco. Trop. Agric. 123:37-50.

Abouzid, A. M., Polston, J. E., and Hiebert, E. 1992. The nucleotide sequence of tomato mottle virus, a new geminivirus isolated from tomatoes in Florida. J. Gen. Virol. 73:3225-3229.

Adams, A. N., and Hull, R. 1972. Tobacco yellow vein, a virus dependent on assistor viruses for its transmission by aphids. Ann. Appl. Biol. 71:135-140.

Ainsworth, G. C. 1933. An investigation of tomato virus diseases of the mosaic "stripe," streak group. Ann. Appl. Biol. 20:421-428.

Ainsworth, G. C. 1934. A comparison of certain English and Canadian potato viruses found infecting tomatoes. Ann. Appl. Biol. 21:581 587.

Al-Aqeel, H. A. 2003. Characterization of two begomoviruses isolated from *Sida santaremensis* Monteiro and *Sida acuta* Burm. f. M.S. thesis, University of Florida, Gainesville.

Ala Poikela, M. 2003. Detection of begomoviruses in tomato crops in Nicaragua. Minor Field Studies International Office, Swedish University of Agricultural Sciences, Publ. 242:18.

Ambrozevicius, L. P., Calegario, R. F., Fontes, E. P. B., de Carvalho, M. G., and Zerbini, F. M. 2002. Genetic diversity of begomovirus infecting tomato and associated weeds in southeastern Brazil. Fitopatol. Bras. 27:372-377.

Anfoka, G. H., Abhary, M., and Nakhla, M. K. 2005. Molecular identification of species of the tomato yellow leaf curl virus complex in Jordan. J. Plant Pathol. 87:65-70.

Antignus, Y., and Cohen, S. 1994. Complete nucleotide sequence of an infectious clone of a mild isolate of tomato yellow leaf curl virus (TYLCV). Phytopathology 84:707-712.

Attathom, S., Chiemsombat, P., Kositratana, W., and Sae-Ung, N. 1994. Complete nucleotide sequence and genome analysis of bipartite tomato yellow leaf curl virus in Thailand. Kasetsart J. 28:632-639.

Badami, R. S., and Kassanis, B. 1959. Some properties of three viruses isolated from a diseased plant of *Solanum jasminoides* Paxt. from India. Ann. Appl. Biol. 47:90-97.

Batuman, O., Kuo, Y.-W., Palmieri, M., Rojas, M. R., and Gilbertson, R. L. 2010. Tomato chocolate spot virus, a member of a new torradovirus species that causes a necrosis-associated disease of tomato in Guatemala. Arch. Virol. 155:857-869.

Batuman, O., Miyao, G., Kuo, Y.-W., Chen, L.-F., Davis, R. M., and Gilbertson, R. L. 2009. An outbreak of a necrosis disease of tomato in California in 2008 was caused by a new Ilarvirus species related to *Parietaria mottle virus*. Plant Dis. 93:546.

Bisaro, D. M., Hamilton, W. D., Coutts, R. H., and Buck, K. W. 1982. Molecular cloning and characterization of the two DNA components of tomato golden mosaic virus. Nucleic Acids Res. 10:4913-4922.

Braithwaite, B. M., and Blake, C. D. 1961. Tomato yellow top virus: Its distribution, characteristics, and transmission by the aphid *Macrosiphum euphorbiae* (Thom.). Aust. J. Agric. Res. 12:1100-1107.

Brierley, P., Smith, F. F., and Doolittle, S. P. 1955. Some hosts and vectors of tomato aspermy virus. Plant Dis. Rep. 39:152-156.

Brittlebank, C. C. 1919. Tomato diseases. J. Dept. Agric. Victoria. 17:231-235.

Brown, J. K., and Nelson, M. R. 1988. Transmission, host range and virus-vector relationships of chino del tomate virus, a whitefly-transmitted geminivirus from Sinaloa, Mexico. Plant Dis. 72:866-869.

Brown, J. K., and Poulos, B. T. 1990. Serrano golden mosaic virus: A newly identified whitefly-transmitted geminivirus of pepper and tomato in the United States and Mexico. Plant Dis. 74:720.

Brown, J. K., Idris, A. M., and Fletcher, D. C. 1993. Sinaloa tomato leaf curl virus, a newly described geminivirus of tomato and pepper in West Central Mexico. Plant Dis. 77:1262.

Brown, J. K., Ostrow, K. M., Idris, A. M., and Stenger, D. C. 1999. Biotic, molecular, and phylogenetic characterization of bean calico mosaic virus, a distinct *Begomovirus* species with affiliation in the squash leaf curl virus cluster. Phytopathology 89:273-280.

Brunt, A. A., and Kenten, R. H. 1973. Cowpea mild mottle, a newly recognized virus infecting cowpeas (*Vigna unguiculata*) in Ghana. Ann. Appl. Biol. 74:67-74.

Buragohain, A. K., Sung, Y. K., Coffin, R. S., and Coutts, R. H. A. 1994. The infectivity of dimeric potato yellow mosaic geminivirus clones in different hosts. J. Gen. Virol. 75:2875-2861.

Caciagli, P., Boccardo, G., and Lovisolo, O. 1989. Parietaria mottle virus, a possible new Ilarvirus from *Parietaria officinalis* (Urticaceae). Plant Pathol. 38:577-584.

Cadman, C. H., and Harrison, B. D. 1959. Studies on the properties of soil-borne viruses of the tobacco-rattle type occurring in Scotland. Ann. Appl. Biol. 47:542-556.

Cadman, C. H., and Harrison, B. D. 1960. Studies on the behavior in soils of tomato black ring, raspberry ringspot, and arabis mosaic viruses. Virology 10:1-20.

Capoor, S. P. 1962. Southern sunn hemp mosaic virus: A strain of tobacco mosaic virus. Phytopathology 52:393-397.

Castellano, M. A., and Martelli, G. P. 1987. Tomato vein yellowing in Italy, a disease caused by eggplant mottled dwarf virus. Phytopathol. Mediterr. 26:46-50.

Chakraborty, S., Pandey, P. K., Banerjee, M. K., Kalloo, G., and Fauquet, C. M. 2003a. A new begomovirus species causing tomato leaf curl disease in Varanasi, India. Plant Dis. 87:313.

Chakraborty, S., Pandey, P. K., Banerjee, M. K., Kalloo, G., and Fauquet, C. M. 2003b. *Tomato leaf curl Gujarat virus,* a new *Begomovirus* species causing a severe leaf curl disease of tomato in Varanasi, India. Phytopathology 93:1485-1495.

Chatchawankanphanich, O., and Maxwell, D. P. 2002. *Tomato leaf curl Karnataka virus* from Bangalore, India, appears to be a recombinant begomovirus. Phytopathology 92:637-645.

Chowda Reddy, R. V., Colvin, J., Muniyappa, V., and Seal, S. 2005. Diversity and distribution of begomoviruses infecting tomato in India. Arch. Virol. 150:845-867.

Cohen, S., and Antignus, Y. 1982. A noncirculative whitefly-borne virus affecting tomatoes in Israel. Phytoparasitica 10:101-109.

Crescenzi, A., Comes, S., Napoli, C., Fanigliulo, A., Pacella, R., and Accotto, G. P. 2004. Severe outbreaks of tomato yellow leaf curl Sardinia virus in Calabria, Southern Italy. Commun. Agric. Appl. Biol. Sci. 69:575-580.

Dale, W. T. 1954. Sap-transmissible mosaic diseases of solanaceous crops in Trinidad. Ann. Appl. Biol. 41:240-247.

De Ávila, A. C., De Haan, P., Kitajima, E. W., Lormelink, R., de O. Resende, R., Goldback, R. W., and Peters, D. 1992. Characterization of a distinct isolate of tomato spotted wilt virus (TSWV) from impatiens sp. in The Netherlands. J. Phytopathol. 134:133-151.

De Ávila, A. C., Hugnueot, C., de O. Resende, R., Kitajima, E. W., Goldbach, R. W., and Peters, D. 1990. Serological differentiation of 20 isolates of tomato spotted wilt virus. J. Gen. Virol. 71:2801-2807.

Dong, J.-H., Cheng, X.-F, Yin, Y.-Y, Fang, Q., Ding, M. Li, T.-T., Zhang, L.-X., Su, X.-X., McBeath, J. H., and Zhang, Z.-K. 2008. Characterization of tomato zonate spot virus, a new tospovirus in China. Arch. Virol. 153:855-864.

Dry, I. B., Krake, L. R., Rigden, J. E., and Rezaian, M. A. 1997. A novel subviral agent associated with a geminivirus: The first report of a DNA satellite. Proc. Natl. Acad. Sci. USA 94:7088-7093.

Dry, I. B., Rigden, J. E., Krake, L. R., Mullineaux, P. M., and Rezaian, M. A. 1993. Nucleotide sequence and genome organization of tomato leaf curl Geminivirus. J. Gen. Virol. 74:147-151.

Duffus, J. E. 1981. Beet western yellows virus—A major component of some potato leaf roll-affected plants. Phytopathology 71:193.

Duffus, J. E., Liu, H.-Y., and Wisler, G. C. 1996. Tomato infectious chlorosis virus—A new clostero-like virus transmitted by *Trialeurodes vaporariorum*. Eur. J. Plant Pathol. 102:219-226.

Dykstra, T. P. 1930. Leaf roll transmission from potato to other solanaceous plants by means of *Myzus persicae*. (Abstr.) Phytopathology 20:853.

Dykstra, T. P. 1939. A study of viruses infecting European and American varieties of the potato, *Solanum tuberosum*. Phytopathology 29:40-67.

Edwardson, J. R., and Christie, R. G. 1997. Viruses Infecting Peppers and Other Solanaceous Crops, vol. 1 and 2. University of Florida Agricultural Experiment Station, Institute of Food and Agricultural Sciences, Gainesville.

El Maataoui, M., Lockhart, B. E. L., and Lesemann, D. E. 1985. Biological, serological, and cytopathological properties of tomato vein-yellowing virus, a rhabdovirus occurring in tomato in Morocco. Phytopathology 75:109-115.

Engel, M., Fernandez, O., Jeske, H., and Frischmuth, T. 1998. Molecular characterization of a new whitefly-transmissible bipartite geminivirus infecting tomato in Panama. J. Gen. Virol.7 9:2313-2317.

Fauquet, C. M., Mayo, M. A., Maniloff, J., Desselberger, U., and Ball, L. A., eds. 2005. Virus Taxonomy: Eighth Report of the International Committee on Taxonomy of Viruses. Elsevier Academic, London.

Feldman, J. M., and Gracia, O. 1977. Pepper severe mosaic virus: A new potyvirus from pepper in Argentina. Phytopathol. Z. 89:146-160.

Fischer, H. U., and Lockhart, B. E. L. 1977. Identification and comparison of two isolates of tomato bushy stunt virus from pepper and tomato in Morocco. Phytopathology 67:1352-1355.

Fribourg, C. E. 1979. Host plant reactions, some properties, and serology of Peru tomato virus. Phytopathology 69:441-445.

Fribourg, C. E., and Nakashima, J. 1984. Characterization of a new potyvirus from potato. Phytopathology 74:1363-1369.

Fribourg, C. E., Jones, R. A. C., and Koenig, R. 1977. Andean potato mottle, a new member of the cowpea mosaic virus group. Phytopathology 67:969-974.

Gallitelli, D. 1982. Properties of a tomato isolate of Pelargonium zonate spot virus. Ann. Appl. Biol. 100:457-466.

Ghotbi, T., Shahraeen, N., and Winter, S. 2005. Occurrence of tospoviruses in ornamental and weed species in Markazi and Tehran provinces in Iran. Plant Dis. 89:425-429.

Gibbs, A. J., and Harrison, B. D. 1969. Eggplant mosaic virus, and its relationship to Andean potato latent virus. Ann. Appl. Biol. 64:225-231.

Giddings, N. J., Bennett, C. W., and Harrison, A. L. 1951. A tomato disease resembling curly top. Phytopathology 41:415-417.

Gooding, G. V., Jr., and Sun, M. 1972. A newly recognized virus disease of burley tobacco in North Carolina. (Abstr.) Phytopathology 62:803.

Gracia, O., and Feldman, J. M. 1974. Tobacco streak virus in pepper. J. Phytopathol. 80:313-323.

Grieco, F., Di Franco, A., and Gallitelli, D. 1997. Potato virus M in tomato crops in Southern Italy (Apulia). J. Plant Pathol. 78:45-49.

Guzman, P., Arredondo, C. R., Emmatty, D., Portillo, R. J., and Gilbertson, R. L. 1997. Partial characterization of two whitefly-transmitted geminiviruses infecting tomatoes in Venezuela. Plant Dis. 81:312.

Hamilton, M. A. 1932. On three new virus diseases of *Hyoscyamus niger*. Ann. Appl. Biol. 19:550-567.

Hamilton, W. D. O., Stein, V. E., Coutts, R. H. A., and Buck, K. W. 1984. Complete nucleotide sequence of the infectious cloned DNA components of tomato golden mosaic virus: Potential coding regions and regulatory sequences EMBO J. 3:2197-2205.

Hanssen, I. M., Lapidot, M. and Thomma, B. P. H. J. 2010. Emerging viral diseases of tomato crops. Mol. Plant-Microbe Interact. 23:539-548.

Harrison, B. D. 1958. Raspberry yellow dwarf, a soil-borne virus. Ann. Appl. Biol. 46:221-229.

Harrison, B. D., and Jones, R. A. C. 1970. Host range and some properties of potato mop-top virus. Ann. Appl. Biol. 65:393-402.

Harrison, B. D., Liu, Y. L., Khalid, S., Hameed, S., Otim-Nape, G. W., and Robinson, D. J. 1997. Detection and relationships of cotton leaf curl virus and allied whitefly-transmitted geminiviruses occurring in Pakistan. Ann. Appl. Biol.1 30:61-75.

Hassani-Mehraban, A., Saaijer, J., Peters, D., Goldbach, R., and Kormelink, R. 2005. A new tomato-infecting tospovirus from Iran. Phytopathology 95:852-858.

Hiruki, C. 1975. Host range and properties of tobacco stunt virus. Can. J. Bot. 53:2425-2434.

Holguín-Peña, R. J., Vásquez-Juárez, R., and Rivera-Bustamante, R. F. 2004. *Pepper golden mosaic virus* affecting tomato crops in the Baja California Peninsula, Mexico. Plant Dis. 88:221.

Hollings, M., Stone, O. M., and Barton, R. J. 1977. Pathology, soil transmission and characterization of cymbidium ringspot, a virus from cymbidium orchids and white clover (*Trifolium repens*). Ann. Appl. Biol. 85:233-248.

Holmes, F. O. 1950. Internal-browning disease of tomato caused by strains of tobacco-mosaic virus from Plantago. Phytopathology 40:487-492.

Hooker, W. J., and Salazar, L. F. 1983. A new plant virus from the high jungle of the Eastern Andes: Solanum apical leaf curling virus (SALC V). Ann. Appl. Biol. 103:449-454.

Hou, Y. M., Paplomatas, E. J., and Gilbertson, R. L. 1998. Host adaptation and replication properties of two bipartite geminiviruses and their pseudorecombinants. Mol. Plant-Microbe Interact. 11:208-217.

Idris, A. M., and Brown, J. K. 1998. Sinaloa tomato leaf curl geminivirus: Biological and molecular evidence for a new subgroup III virus. Phytopathology 88:648-657.

Idris, A. M., and Brown, J. K. 2001. Three previously unidentified begomoviral genotypes from tomato exhibiting leaf curl disease symptoms from central Sudan. Plant Dis. 85:1209.

Idris, A. M., and Brown, J. K. 2005. Evidence for interspecific-recombination for three monopartite begomoviral genomes associated with the tomato leaf curl disease from central Sudan. Arch. Virol. 150:1003-1013.

Imle, E. P., and Samson, R. W. 1937. Studies on a ring-spot type of virus disease of tomato. (Abstr.) Phytopathology 27:132.

Jarrett, P. H. 1930. Streak—A virus disease of tomatoes. Ann. Appl. Biol. 17:248-259.

Jeffrey, J. L., Pooma, W., and Petty, I. T. 1996. Genetic requirements for local and systemic movement of tomato golden mosaic virus in infected plants. Virology 223:208-218.

Jiang, H., Hou, Y.-M., Guzman, P., and Gilbertson, R. L. 1998. Complete nucleotide sequence of the infectious clone of tomato leaf crumple geminivirus. Phytopathology 88(Suppl.):S44.

Johnson, J. 1926. Mosaic diseases on differential hosts. Phytopathology 16:141-149.

Johnson, E. M. 1930. Virus diseases of tobacco in Kentucky. Ky. Agric. Exp. Stn. Res. Bull. 306.

Jones, R. A. C., and Fribourg, C. E. 1979. Host plant reactions, some properties, and serology of wild potato mosaic virus. Phytopathology 69:446-449.

Jones, R. A. C., Fribourg, C. E., and Koenig, R. 1983. A previously undescribed nepovirus isolated from potato in Peru. Phytopathology 73:195-198.

Jones, R. A. C., Koenig, R., and Lesemann, D. E. 1980. Pepino mosaic virus, a new potexvirus from pepino (*Solanum muricatum*). Ann. Appl. Biol. 94:61-68.

Jovel, J., and Ruiz, E. 2003. Pathogenicity of Sida golden mosaic Costa Rica virus (SiGMCRV) on several crop plants. Manejo Integrado de Plagas y Agroecologia 70:19-29.

Kano, T., Namba, S., Yamashita, S., Doi, Y., and Yora, K. 1985. Tomato vein clearing virus, a sap-transmissible rhabdovirus in tomato. Ann. Phytopathol. Soc. Jpn. 51:606-612.

Kheyr Pour, A., Bendahmane, M., Matzeit, V., Accotto, G. P., Crespi, S., and Gronenborn, B. 1991. Tomato yellow leaf curl virus from Sardinia is a whitefly transmitted monopartite geminivirus. Nucleic Acids Res. 19:6763-6769.

King, A. M. Q., Lefkowitz, E. J., Adams M. J., and Carstens, E. B. 2012. Virus Taxonomy: Ninth Report of the International Committee on Taxonomy of Viruses. Elsevier, London.

Kirthi, N., Maiya, S. P., Murthy, M. R. N., and Savithri, H. S. 2002. Evidence for recombination among the tomato leaf curl virus strains/species from Bangalore, India. Arch. Virol. 147:255-272.

Kitamura, K., Murayama A., and Ikegami, M. 2004. Evidence for recombination among isolates of *Tobacco leaf curl Japan virus* and *Honeysuckle yellow vein mosaic virus*. Arch. Virol. 149:1221-1229.

Kubo, S., and Takanami, Y. 1977. Tobacco necrotic dwarf virus. (In Japanese.) Ann. Phytopathol. Soc. Jpn. 43:76-77.

Ladipo, J. L. 1976. A green mosaic disease of eggplant (*Solanum melongena*) in Nigeria. Plant Dis. Rep. 60:1068-1072.

Ladipo, J. L., and Roberts, I. M. 1977. Pepper veinal mottle virus associated with a streak disease of tomato in Nigeria. Ann. Appl. Biol. 87:133-138.

Ladipo, J. L., Lesemann, D. E., and Koenig, R. 1988. Host ranges serology and cytopathology of eggplant and tomato strains of eggplant severe mottle virus, a new potyvirus from Nigeria. J. Phytopathol. 122:359-371.

Larsen, R. C., Duffus, J. E., and Liu, H.-Y. 1984. Tomato necrotic dwarf—A new type of whitefly-transmitted virus. (Abstr.) Phytopathology 74:795.

Lee, R. F., Niblett, C. L., Hubbard, J. D., and Johnson, L. B. 1979. Characterization of belladonna mottle virus isolates from Kansas and Iowa. Phytopathology 69:985-989.

Li, Z. H., Li, G. X., Xie, Y., Zhang, Z. K., and Zhou, X. P. 2002. Tomato leaf curl disease in Yunnan is caused by tobacco curly shoot virus. Chin. J. Virol. 18:355-361.

Li, Z. H., Zhou, X. P., Zhang, X., and Xie, Y. 2004. Molecular characterization of tomato-infecting begomoviruses in Yunnan, China. Arch. Virol. 149:1721-1732.

Lisa, V., Milne, R. G., Accotto, G. P., Boccardo, G., Caciagli, P., and Parvizy, R. 1988. Ourmia melon virus, a virus from Iran with novel properties. Ann. Appl. Biol. 112:291-302.

Liu, Y., Robinson, D. J., and Harrison, B. D. 1998. Defective forms of cotton leaf curl virus DNA-A that have different combinations of sequence detection, duplication, inversion and rearrangement. J. Gen. Virol. 79:1501-1508.

MacLachlan, D. S., Larson, R. H., and Walker, J. C. 1953. Strain Interrelationships in Potato Virus A. Univ. Wis. Agric. Exp. Stn. Res. Bull. 180.

Mali, V. R., Nirmal, D. D., Patel, K. V., and Vyanjane, N. T. 1985. Studies on a virus causing distortion mosaic on *Datura fastuosa*. Indian Phytopathol. 38:413-417.

Martelli, G. P., and Cirulli, M. 1969. Le virosi delle piante ortensi in Puglia. III. Una maculatura gialla del pomodoro causata dal virus della necrosis perinervale del tabacco (Tobacco streak virus). (In Italian.) Phytopathol. Mediterr. 8:154-156.

Martinez Zubiaur, Y., de Blas, C., Quinones, M., Castellanos, C., Peralta, E. L., and Romero, J. 1998. Havana tomato virus, a new bipartite geminivirus infecting tomatoes in Cuba. Arch. Virol. 143:1757-1772.

Matyis, J. C., Silva, D. M., Oliveira, A. R., and Costa, A. S. 1975. Purification and morphology of tomato golden mosaic virus. (In Portuguese.) Summa Phytopathol. 1:267-275.

McGrath, P. F., and Harrison, B. D. 1995. Transmission of tomato leaf curl geminiviruses by *Bemisia tabaci:* Effects of virus isolate and vector biotype. Ann. Appl. Biol. 126:307-316.

McMichael, L. A., Persley, D. M., and Thomas, J. E. 2002. A new tospovirus serogroup IV species infecting capsicum and tomato in Queensland, Australia. Australas. Plant Pathol. 31:231-239.

Milbrath, G. M., and Duffus, J. E. 1978. Solanum yellows virus, an apparently distinct member of the luteovirus group. (Abstr.) Phytopathol. News 12:170.

Miller, P. M., and Thornberry, H. H. 1958. A new viral disease of tomato and pepper. Phytopathology 48:665-670.

Mink, G. I., Silbernagel, M. J., and Saksena, K. N. 1969. Host range, purification, and properties of the western strain of peanut stunt virus. Phytopathology 59:1625-1631.

Monci, F., Sanchez-Campos, S., Navas-Castillo, J., and Moriones, E. 2002. A natural recombinant between the geminiviruses *Tomato yellow leaf curl Sardinia virus* and *Tomato yellow leaf curl virus* exhibits a novel pathogenic phenotype and is becoming prevalent in Spanish populations. Virology 303:317-326.

Morris, J., Harju, V., Cambridge, J., Boonham, N., and Henry, C. 2002. Virus transmission studies and diagnostic detection of four begomovirus isolates in selected crop hosts and in *Bemisia tabaci* biotype B. Bull. OEPP 32(1):41-45.

Muniyappa, V., Venkatesh, H. M., Ramappa, H. K., Kulkarni, R. S., Zeidan, M., Tarba, C. Y., Ghanim, and M., Czosnek, H. 2000. Tomato leaf curl virus from Bangalore (ToLCV-Ban4): Sequence comparison with Indian ToLCV isolates, detection in plants and insects, and vector relationships. Arch. Virol. 145:1583-1598.

Nakhla, M. K., Maxwell, M. D., Hidayat, S. H., Lange, D. R., Loniello, A. O., Rojas, M. R., Maxwell, D. P., Kitajima, E. W., Rojas, A.,

Anderson, P., and Gilbertson, R. L. 1994. Two geminiviruses associated with tomatoes in Central America. (Abstr.) Phytopathology 84:1155.

Naqvi, Q. A., Hadi, S., and Mahmood, K. 1981. Marigold mottle virus in Aligarh, India. Plant Dis. 65:271-275.

Natti, J. J., Kirkpatrick, H. C., and Ross, A. F. 1953. Host range of potato leaf roll virus. Am. Potato J. 30:55-64.

Navot, N., Pichersky, E., Zeidan, M., Zamir, D., and Czosnek, H. 1991. Tomato yellow leaf curl virus: A whitefly-transmitted geminivirus with a single genomic component. Virology 185:151-161.

Noris, E., Hidalgo, E., Accotto, G. P., and Moriones, E. 1994. High similarity among the tomato yellow leaf curl virus isolates from the west Mediterranean basin: The nucleotide sequence of an infectious clone from Spain. Arch. Virol.1 35:165-170.

Obermeier, C., Sears, J. L., Liu, H.-Y., Schlueter, K. O., Ryder, E. J., Duffus, J. E., Koike, S. T., and Wisler, G. C. 2001. Characterization of distinct tombusviruses that cause diseases of lettuce and tomato in the western United States. Phytopathology 91:797-806.

Padidam, M., Beachy, R. N., and Fauquet, C. M. 1995. Tomato leaf curl geminivirus from India has a bipartite genome and coat protein is not essential for infectivity. J. Gen. Virol. 76:25-35.

Paximadis, M., and Rey, M. E. C. 2001. Genome organization of *Tobacco leaf curl Zimbabwe virus,* a new, distinct monopartite begomovirus associated with subgenomic defective DNA molecules. J. Gen. Virol. 82:3091-3097.

Paximadis, M., Idris, A. M., Torres-Jerez, I., Villarreal, A., Rey, M. E. C., and Brown, J. K. 1999. Characterization of tobacco geminiviruses in the Old and New World. Arch. Virol. 144:703-717.

Peter, D., and Derks, A. F. L. M. 1974. Host range and some properties of Physalis mosaic virus, a new virus of the turnip yellow mosaic virus group. Neth. J. Plant Pathol. 80:124-132.

Polston, J. E., Bois, D., Ano, G., Poliakoff, F., and Ubrino, C. 1997. Occurrence of a strain of potato yellow mosaic geminivirus infecting tomato in the Eastern Caribbean. Plant Dis. 82:126.

Polston, J. E., Hiebert, E., McGovern, R. J., Stansly, P. A., and Schuster, D. J. 1993. Host range of tomato mottle virus, a new geminivirus infecting tomato in Florida. Plant Dis. 77:1181-1184

Pospieszny, H., Borodynko, N., Obrępalska-Stęplowska, A., and Hasiów, B. 2007. The first report of Tomato torrado virus in Poland. Plant Dis. 91:1364.

Porter, D. R. 1935. Insect transmission, host range, and field spread of potato calico. Hilgardia 9:383-394.

Price, W. C. 1940. Comparative host range of six plant viruses. Am. J. Bot. 27:530-541.

Purcifull, D. E., Zitter, T. A., and Hiebert, E. 1975. Morphology, host range, and serological relationships of pepper mottle virus. Phytopathology 65:559-562.

Qian Y. J., and Zhou, X. P. 2005. Pathogenicity and stability of a truncated DNA beta associated with *Tomato yellow leaf curl China virus.* Virus Res. 109:159-163.

Ramos, P. L., Guerra, O., Peral, R., Oramas, P., Guevara, R. G., and Rivera-Bustamante, R. 1997. Taino tomato mottle virus, a new bipartite Geminivirus from Cuba. Plant Dis. 81:1095.3.

Rampersad, S. N. 2003. Proposed strategies for begomovirus disease management in tomato in Trinidad. Plant Health Progress. doi:10.1094/PHP-2003-1006-01-HM.

Raymer, W. B., Kahn, R. P., Hikida, H. R., and Waterworth, H. E. 1972. A new tomato virus from Peru. (Abstr.) Phytopathology 62:784.

Ribeiro, S. G., Ambrozevicius, L. P., Avila, A. C., Bezerra, I. C., Calegario, R. F., Fernandes, J. J., Lima, M. F., de Mello, R. N., Rocha, H., and Zerbini, F. M. 2003. Distribution and genetic diversity of tomato-infecting begomoviruses in Brazil. Arch. Virol. 148:281-295.

Roberts, E. J. F., Buck, K. W., and Coutts, R. H. A. 1998. Characterization of potato yellow mosaic virus as a geminivirus with a bipartite genome. Intervirology 29:162-169.

Rochester, D. E., DePaulo, J .J., Fauquet, C. M., and Beachy, R. N. 1994. Complete nucleotide sequence of the geminivirus tomato yellow leaf curl virus, Thailand isolate. J. Gen. Virol. 75:477-485.

Rojas, A., Kvarnheden, A., Marcenaro, D., and Valkonen, J. P. T. 2005. Sequence characterization of *Tomato leaf curl Sinaloa virus* and *Tomato severe leaf curl virus:* Phylogeny of New World begomoviruses and detection of recombination. Arch. Virol. 150:1281-1299.

Salazar, L. F., and Harrison, B. D. 1978. Host range and properties of potato black ringspot virus. Ann. Appl. Biol. 90:375-386.

Salazar, L. F., Müller, G., Querci, M., Zapata, J. L., and Owens, R. A. 2000. Potato yellow vein virus: Its host range, distribution in South America and identification as a crinivirus transmitted by *Trialeurodes vaporariorum*. Ann. Appl. Biol. 137:7-19.

Samson, R. W., and Imle, E. P. 1942. A ring-spot type of virus disease of tomato. Phytopathology 32:1037-1047.

Samuel, G., Bald, J. G., and Pittman, H. A. 1930. Investigations on "spotted wilt" of tomatoes in Australia. Aust. Commonw. Advis. Counc. Sci. Ind. Bull. 44: 8-11.

Sánchez-Campos, S., Díaz, J. A., Monci, F., Bejarano, E. R., Reina, J., Navas-Castillo, J., Aranda, M. A., and Moriones, E. 2002. High genetic stability of the begomovirus *Tomato yellow leaf curl Sardinia virus* in southern Spain over an 8-year period. Phytopathology 92:842-849.

Santos, C. D. G., de Avila, A. C., and de O Resende, R. 2003. Studies of the interaction of a new begomovirus isolate from tomato with the whitefly. Fitopatol. Bras. 28:664-673.

Sawangjit, S., Chatchawankanphanich, O., Chiemsombat, P., Attathom, T., Dale, J., and Attathom, S. 2005. Molecular characterization of tomato-infecting begomoviruses in Thailand. Virus Res. 109:1-8.

Schmelzer, K., 1957. Untersuchung uber den Wirtspflanzenkreis des Tabakmauchen-virus. (In Dutch.) Phytopathol Z. 30:281-314.

Schmelzer, K. 1969. Das latent erdbeerringflecken-virus aus *Evonymus, Robinia* und *Aesculus.* (In Dutch.) Phytopathol. Z. 66:1-24.

Severin, H. H. P. 1929. Additional host plants of curly top. Hilgardia 3:595-629.

Shih, S. L., Tsai, W. S., Nakhla, M. K., Maxwell, D. P., Rashid, M. H., and Green, S. K. 1998a. Molecular comparison of two tomato leaf curl viruses from Bangladesh. Zhiwu Baohuxue Hui Huikan 40:436-437.

Shih, S. L., Roff, M. M. N., Nakhla, M. K., Maxwell, D. P., and Green, S. K. 1998b. A new geminivirus associated with a leaf curl disease of tomato in Malaysia. Zhiwu Baohuxue Hui Huikan 40:434-435.

Shih, S. L., Tsai, W. S., Green, S. K., Khalid, S., Ahmad, I., Rezaian, M. A., and Smith, J. 2003. Molecular characterization of tomato and chili leaf curl begomoviruses from Pakistan. Plant Dis. 87:200.1.

Silberschmidt, K. 1963. Studies on a Brazilian tomato ringspot disease. Phytopathol Z. 46:209-229.

Simons, J. N., and Coe, D. M. 1958. Transmission of pseudo-curly top virus in Florida by a treehopper. Virology 6:43-48.

Smith, K. M. 1935. A new virus disease of the tomato. Ann. Appl. Biol. 22:731-741.

Smith, K. M. 1945. Transmission by insects of a plant virus complex. Nature 155:174.

Smith, K. M. 1946a. Tobacco rosette: A complex virus disease. Parasitology 37:21-24.

Smith, K. M. 1946b. Tomato black-ring: A new virus disease. Parasitology 37:126-130.

Smith, K. M. 1946c. The transmission of a plant virus complex by aphides. Parasitology 37:131-134.

Sylvester, E. S. 1954. Yellow-net virus disease of tomato. Phytopathology 44:219-220.

Tan, H. P. N., and Wong, S. M. 1993. Some properties of Singapore Ageratum yellow vein virus. J. Phytopathol. 139:165-176.

Tan, P. H., Wong, S. M., Wu, M., Bedford, I. D., Saunders, K., and Stanley, J. 1995. Genome organization of ageratum yellow vein virus, a monopartite whitefly-transmitted geminivirus isolated from a common weed. J. Gen. Virol. 76:2915-2922.

Thomas, J. E., and Bowyer, J. W. 1984. Tobacco yellow dwarf virus. Descriptions of Plant Viruses, no. 278. C.A.B. International Mycological Institute, Kew, England.

Thomas, J. E., and Dietzgen, R. G. 1991. Characterisation of datura yellow vein virus, a newly described rhabdovirus from Australia. Ann. Appl. Biol. 118:339-349.

Torres-Pacheco, I., Garzón-Tiznado, J. A., Brown, J. K., Becerra-Flora, A., and Rivera-Bustamante, R. F. 1996. Detection and distribution of geminiviruses in Mexico and the southern United States. Phytopathology 86:1186-1192.

Turina, M., Ricker, M. D., Lenzi, R., Masenga, V., and Ciuffo, M. 2007. A severe disease of tomato in the Culiacan area (Sinaloa, Mexico) is caused by a new picorna-like viral species. Plant Dis. 91:932-941.

Umaharan, P., Padidam, M., Phelps, R. H., Beachy, R. N., and Fauquet, C. M. 1998. Distribution and diversity of geminiviruses in Trinidad and Tobago. Phytopathology 88:1262-1268.
Unseld, S., Ringel, M., Hofer, P., Hohnle, M., Jeske, H., Bedford, I. D., Markham, P. G., and Frischmuth, T. 2000. Host range and symptom variation of pseudorecombinant virus produced by two distinct bipartite geminiviruses. Arch. Virol. 145:1449-1454.
Urbino, C., Polston, J. E., Patte, C. P., and Caruana, M.-L. 2004. Characterization and genetic diversity of *Potato yellow mosaic virus* from the Caribbean. Arch. Virol. 149:417-424.
Valverde, R. A., Heick, J. A., and Dodds, J. A. 1991. Interactions between satellite tobacco mosaic virus, helper tobamoviruses, and their hosts. Phytopathology 81:99-104.
Verbeek, M., Dullemans, A. M., van den Heuvel, J. F. J. M., Maris, P. C., and van der Vlugt, R. A. A. 2007. Identification and characterisation of tomato torrado virus, a new plant picorna-like virus from tomato. Arch. Virol. 152:881-890.
Verbeek, M., Dullemans, A. M., van den Heuvel, J. F. J. M., Maris, P. C., and van der Vlugt, R. A. A. 2008. Tomato marchitez virus, a new plant picorna-like virus from tomato related to tomato torrado virus. Arch. Virol. 153:127-134.
Walkey, D. G. A., Spence, N. J., Clay, C. M., and Miller, A. 1994. A potyvirus isolated from solanaceous hosts. Plant Pathol. 43:931-937.
Waterworth, H. E., Kaper, J. M., and Koenig, R. 1975. Purification and properties of a tymovirus from Abelia. Phytopathology 65:891-896.
Wingard, S. A. 1928. Host and symptoms of ring spot, a virus disease of plants. J. Agric. Res. 37:127.
Wisler, G. C., Li, R. H., Liu, H.-Y., and Duffus, J. E. 1998. Tomato chlorosis virus: A new whitefly-transmitted, phloem-limited, bipartite closterovirus of tomato. Phytopathology 88:402-409.
Yin, Q., Yang, H., Gong, Q., Wang, H., Liu, Y., Hong, Y., and Tien, P. 2001. Tomato yellow leaf curl China virus: Monopartite genome organization and agroinfection of plants. Virus Res. 81:69-76.
Zhou, X., Liu, G. W., Robinson, D. J., and Harrison, B. D. 1998. Four DNA-A variants among Pakistani isolates of cotton leaf curl virus and their affinities to DNA-A of geminivirus isolates from okra. J. Gen. Virol. 79:915-923.
Zhou, X. P., Xie, Y., and Zhang, Z. K. 2001. Molecular characterization of a distinct Begomovirus infecting tobacco in Yunnan, China. Arch. Virol. 146:1599-1606.

(Prepared by T. A. Zitter)

Begomoviruses

Tomato-Infecting Begomoviruses

Begomoviruses are whitefly-transmitted (*Bemisia tabaci* complex), single-stranded (ss) DNA viruses. They are a major constraint to tomato production in many tropical and subtropical regions of the world, and their importance has increased since the 1990s for three primary reasons: (1) the movement of known begomoviruses throughout the world; (2) the appearance of new begomoviruses resulting from the movement of biotypes or their vectors into new regions; and (3) changes in cultural practices associated with more intense production schedules.

Begomovirus is now the largest genus of plant viruses. Approximately 90 begomoviruses (accepted and tentative) have been reported to infect tomato. However, many of these reports include only a partial genome sequence, and a complete genome sequence is required for speciation; thus, only 50 begomoviruses have been approved as species. Table 5 (pages 71–72) lists the approved *Begomovirus* spp. known to infect tomato either experimentally (i.e., only in greenhouse transmission studies) or naturally (i.e., sequence came from infected field plants). Many of these viruses have been identified since 1995, and for most of them, there is a lack of sequence data along with a lag in the reporting of biological data (e.g., response to resistance genes, host range, ecology, recognition of strains). Researchers expect even more new begomoviruses to be added to the list of pathogens of tomato, based in part on the long list of tentative species.

All begomoviruses have unique geminate particle morphology and are able to be transmitted by members of *Bemisia tabaci* (Gennadius) (or were once able, in the case of a few viruses). Begomoviruses can be divided into two groups: those with a monopartite genome (approximately 2,600 nucleotides) and those with bipartite genomes (approximately 2,500 nucleotides per component). Begomoviruses with monopartite genomes appear to have originated in the Old World, and at least one—*Tomato yellow leaf curl virus* (TYLCV)—occurs in the New World because of its spread across continents during the 1990s. The bipartite begomoviruses occur in both the Old and New Worlds. (The TYLCV complex has been reviewed in a book edited by Czosnek.)

Begomoviruses cause two general types of foliar symptoms: yellow edges and upward curling of leaf margins, as illustrated by TYLCV (Fig. 155, page 100), and yellow mottling and puckering of the leaf tissues, as illustrated by *Tomato mottle virus* (ToMoV) (Fig. 147, page 95). Both viruses are known for causing the first set of symptoms: yellow margins and upward curling of leaves. The second set of symptoms are caused by the bipartite viruses of Mexico, Central America, and South America and are quite different when compared with those caused by TYLCV. All begomoviruses can cause severe stunting and plant malformation if plants are infected at a young age. In general, it is not possible to identify specific species of geminiviruses based only on symptoms; molecular methods are required (e.g., DNA hybridization, polymerase chain reaction (PCR)).

Begomoviruses and DNA-β satellites

Not only have new begomoviruses been discovered, but new diseases have emerged from mixtures of monopartite begomoviruses and subviral entities known as DNA-β (beta) satellites. DNA-β satellites are single-stranded, circular DNAs that depend on a monopartite begomovirus for replication, movement, and transmission to new hosts. DNA-β satellites are approximately 1,360 nucleotides in size and share no sequence homology with begomoviruses. Many of the diseases caused by monopartite begomoviruses in parts of the Old World involve DNA-β satellites.

More than 100 DNA-β satellites have been reported and all were found in the Old World, with a center of origin presumed to be in Asia. The genome encodes one gene, *βC1*, which suppresses the plant host's ability to resist the helper begomovirus. DNA-β satellites increase symptom expression of monopartite begomoviruses, which, in the absence of a satellite, replicate to low titers and cause no symptoms or mild or atypical symptoms in the host. When present, the DNA-β satellite is the main determinant of symptom expression in the infected host.

Most of the DNA-β satellites are associated with diseases of cotton (*Gossypium* L.), but the number of diseases in tomato that can be attributed to DNA-β satellites is increasing. Reports of the involvement of DNA-β satellites in tomato diseases have come from China, India, Mali, and Pakistan. Moreover, it is likely that the number of begomoviruses associated with DNA-β satellites will increase and that DNA-β satellites will have a greater role in producing new diseases in tomato, as well as in other crops.

In general, all tomato diseases caused by begomoviruses can be managed by the application of similar strategies:
1. using a vector/host-free period for 3 months
2. selecting planting sites that are not adjacent to old tomato fields
3. using virus-free seedlings (i.e., seedlings produced in vector-proof houses)
4. protecting transplant seedlings during the first 3 weeks by using special insect-restricting cloth
5. applying appropriate insecticides to kill the vector
6. using resistant cultivars

Selected References

Anderson, P. K., and Morales, F. J., eds. 2005. Whitefly and whitefly-borne virus in the tropics: Building a knowledge base for global action. Publ. 341. Centro Internacional de Agricultura Tropical (CIAT), Cali, Colombia.

Briddon, R. W., Brown, J. K., Moriones, E., Stanley, J., Zerbini, M., Zhou, X., and Fauquet, C. M. 2008. Recommendations for the classification and nomenclature of the DNA-β satellites of begomoviruses. Arch. Virol. 153:763-781.

Czosnek, H., ed. 2007. Tomato Yellow Leaf Curl Virus Disease: Management, Molecular Biology, Breeding for Resistance. Springer, Dordrecht, the Netherlands.

Jones, D. R. 2003. Plant viruses transmitted by whiteflies. Eur. J. Plant Pathol. 109:195-219.

Mansoor, S., Zafar, Y., and Briddon, R. W. 2006. Geminivirus disease complexes: The threat is spreading. Trends Plant Sci. 11:209-212.

Rojas, M. R., Gilbertson, R. L., Russell, D. R., and Maxwell, D. P. 1993. Use of degenerate primers in the polymerase chain reaction to detect whitefly-transmitted geminiviruses. Plant Dis. 77:340-347.

Stanley, J., Bisaro, D. M., Briddon, R. W., Brown, J. K., Fauquet, C. M., Harrison, B. D., Rybicki, E. P., and Stenger, D. C. 2005. Family Geminiviridae. In: Virus Taxonomy: Eighth Report of the International Committee on Taxonomy of Viruses. C. M. Fauquet, M. A. Mayo, J. Maniloff, U. Desselberger, and L. A. Ball, eds. Elsevier Academic, London.

Wyatt, S. D., and Brown, J. K. 1996. Detection of subgroup III geminivirus isolates in leaf extracts by degenerate primers and polymerase chain reaction. Phytopathology 86:1288-1293.

(Prepared by J. E. Polston)

Alfalfa mosaic virus

Alfalfa mosaic virus (AMV) occurs worldwide but is generally not considered of major economic importance in tomato. Exceptions are noted when tomato crops are located near older alfalfa fields or permanent pastures. Perhaps the heaviest losses to tomato have occurred in the United States in California's Imperial Valley, where losses of 10–15% have been reported and damage has often been spread uniformly throughout tomato fields. In other parts of California, damage from AMV infection has been limited to those areas of tomato fields that are closest to alfalfa fields. Secondary virus spread rarely occurs. In the late 1990s, the virus became problematic for tomato production in Italy and France.

Symptoms

Symptoms of alfalfa mosaic depend on the particular virus strain, host cultivar, stage of growth, and environmental conditions. Early foliar symptoms consist of yellowing (Fig. 120) or bronzing of the terminals and necrosis of the leaflet veins. In advanced stages of the disease, plants stop growing and the leaflets curl downward. Reddish-brown discoloration of the phloem on the main stem near the soil line can be observed by scraping the epidermis of the stem. This brown discoloration often extends into the terminal shoots and roots. The fungi responsible for Fusarium wilt and Verticillium wilt can also cause internal discoloration, but in these cases, the discoloration is generally limited to the crown and lower stems.

Fruit symptoms are usually present, as well, with severity depending on the age of the plant at the time of infection. Typical symptoms include varying degrees of necrotic spots, both externally and internally, and distortion (Fig. 121).

Causal Agent

AMV is the type species of the genus *Alfamovirus* in the family Bromoviridae. The AMV genome consists of three distinct single-stranded RNA components, plus a fourth RNA component: a subgenomic messenger RNA, which encodes for viral coat protein. The three genomic RNAs and either the fourth RNA or the coat protein are needed to initiate infection. The coat protein of various Ilarviruses (e.g., *Tobacco streak virus*) can be substituted to activate the AMV genome. AMV is serologically unrelated to any other virus.

In a 2000 study, strains from four countries were divided into two subgroups. Subgroup I contained strains from Italy, the United States, and England, while subgroup II contained all of the French strains and two English strains. The existence of the two subgroups did not correlate with differences in host

Fig. 120. Symptoms of alfalfa mosaic on tomato leaves. (Courtesy J. E. Thomas)

Fig. 121. Symptoms of alfalfa mosaic on tomato fruit. (Courtesy D. G. A. Walkey)

range or symptoms (i.e., necrotic or nonnecrotic) induced in tomato; rather, it reflected mainly a geographic structure based on variations in the amino acid sequence of the coat protein. Vacuolate inclusions are commonly found in the cytoplasm (Fig. 122).

Disease Cycle and Epidemiology

AMV occurs naturally in 150 species in 21 plant families and has been transmitted experimentally to more than 430 species in 51 dicotyledonous families. The virus is readily sap transmitted. In addition to tomato, AMV affects a variety of other crops, including potato (*Solanum tuberosum* L.), celery (*Apium* L.), lettuce (*Lactuca sativa* L.), bean (*Phaseolus* L.), pepper (*Capsicum annuum* L.), pea (*Pisum sativum* L.), chickpea (*Cicer arietinum* L.), alfalfa (*Medicago sativa* L.), red clover (*Trifolium pratense* L.) and white clover (*T. repens* L.), and tobacco (*Nicotiana* L.). The virus occurs naturally and often symptomlessly in many herbaceous and woody hosts.

AMV is transmitted in nature by at least 22 aphid species in a nonpersistent manner. Vectors include the pea aphid (*Acyrthosiphon pisum* Harris), the green peach aphid (*Myzus persicae* (Sulzer)), and the blue alfalfa aphid (*A. kondoi* Shinji).

There are no reports of seed transmission of AMV in tomato. Even so, seed transmission does occur commonly in alfalfa cultivars and has been reported in chili pepper and in apple of Peru (*Nicandra physalodes* (L.) Scop.).

Management

Because of the nonpersistent manner of virus spread, spraying to control aphids will not control primary spread of AMV. It is possible to avoid infection by not planting tomato near fields of alfalfa or red clover. The minimum distance necessary to avoid infection is not known, but a few hundred yards of separation is usually sufficient.

No resistant or tolerant tomato cultivars are available. Sources of resistance to AMV have been identified in the wild tomato *Lycopersicon hirsutum* Humb. & Bonpl., and this resistance is effective against most strains. The single dominant gene *Am* has been mapped to the short arm of tomato chromosome 6, which includes resistance to other tomato pathogens.

Selected References

Finetti-Sialer, M., Di Franco, A., Papanice, M. A., and Gallitelli, D. 1997. Tomato necrotic yellows induced by a novel strain of *Alfalfa mosaic virus*. J. Plant Pathol. 79:115-120.

Jaspars, E. M. J., and Bos, L. 1980. Alfalfa mosaic virus. Descriptions of Plant Viruses, no. 229. C.A.B. International Mycological Institute, Kew, England.

Parrella, G., Lanave, C., Marchoux, G., Finetti-Sialer, M. M., Di Franco, A., and Gallitelli, D. 2000. Evidence for two distinct subgroups of *Alfalfa mosaic virus* (AMV) from France and Italy and their relationships with other AMV strains. Arch. Virol. 145:2659-2667.

Parrella, G., Laterrot, H., Gebre Selassie, K., and Marchoux, G. 1998. Inheritance of resistance to *Alfalfa mosaic virus* in *Lycopersicon hirsutum* f. *glabratum* PI 134417. J. Plant Pathol. 80:241-243.

Parrella, G., Laterrot, H., Marchoux, G., and Gebre-Selassie, K. 1997. Screening *Lycopersicon* accessions for resistance to *Alfalfa mosaic virus*. J. Genet. Breed. 51:75-78.

Parrella, G., Moretti, A., Gognalons, P., Lesage, M.-L., Marchoux, G., Gebre-Selassie, K., and Caranta, C. 2004. The *Am* gene controlling resistance to *Alfalfa mosaic virus* in tomato is located in the cluster of dominant resistance genes on chromosome 6. Phytopathology 94:345-350.

Statewide Integrated Pest Management (IPM) Program. 1998. Integrated Pest Management for Tomatoes. 4th ed. Publ. 3274. Statewide IPM Program, Agriculture and Natural Resources, University of California.

(Prepared by T. A. Zitter)

Beet curly top virus

Beet curly top virus (BCTV) (or curly top disease of tomato) infects a wide range of crops and weeds. The virus occurs in the arid and semi-arid regions of western North America from Mexico to Canada and in the eastern Mediterranean Basin, where the leafhopper vector is common. The disease, which is also called western blight, can cause heavy losses in staked tomatoes, for which plants are more widely spaced. BCTV can also be a serious problem for processing tomatoes in California's San Joaquin Valley during years with early and high populations of beet leafhopper (Fig. 123).

Symptoms

Plants infected with BCTV at an early stage usually die, while those infected later may survive. Surviving plants remain yellowed, however, and have erect terminals and a stunted habit. Leaves become thickened and crisp and roll upward, as the petioles curve downward (Fig. 124). Leaves turn a dull yellow and have purplish veins.

Fruit ripen prematurely (Fig. 125). In fact, immature, dull, and wrinkled fruit are characteristic of curly top. When plants are infected after fruit set, both normal and affected fruit can appear on the same stem.

Causal Agent

BCTV is the type species of the genus *Curtovirus* within the family Geminiviridae. The virus has isometric particles

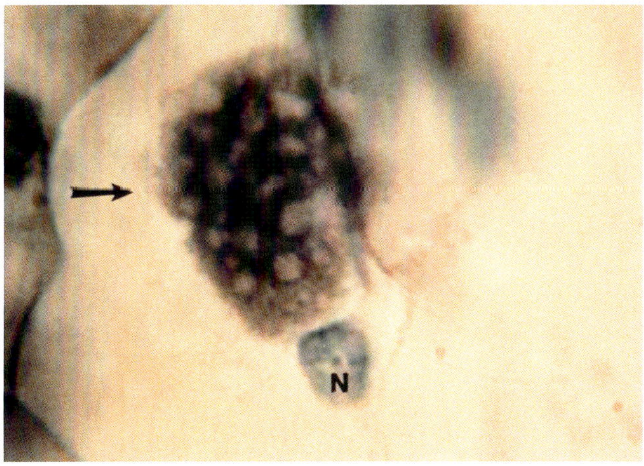

Fig. 122. Vacuolate inclusions of *Alfalfa mosaic virus* (AMV) (N = nucleus). (Courtesy R. G. Christie)

Fig. 123. Tomato field severely affected by *Beet curly top virus* (BCTV). (Courtesy W. M. Wintermantel)

approximately 20 nm in diameter and occurring in pairs. The curtoviruses are defined as circular, single-stranded DNA genomes (monopartite) and are leafhopper transmitted; they can be distinguished from members of the genus *Mastrevirus* on the basis of differences in genome organization.

The beet leafhopper (*Circulifer tenellus* Baker) is the only natural vector of BCTV in North America, whereas two leafhopper species (*C. tenellus* and *C. opacipennis* Lethierry) transmit the virus in the Mediterranean area. The virus, which is restricted to the phloem, is not transmissible mechanically, but it can be transmitted inefficiently by a pin-prick method using a high concentration of inoculum and somewhat more effectively by a high-pressure injector method.

All curtoviruses were previously defined as strains of BCTV. Currently, the complete nucleotide sequence of six curtovirus isolates is known. All of the isolates are vectored by *C. tenellus*, but they are separated into five distinct virus species on the basis of host range, symptom severity in sugar beet (*Beta vulgaris* L.), and nucleotide sequence. Three *Curtovirus* spp. are known to infect tomato: BCTV, *Beet severe curly top virus* (BSCTV), and *Beet mild curly top virus* (BMCTV).

Although BCTV was once prevalent, it is rarely found today in North America. The two most prevalent curtoviruses in U.S. tomatoes are BSCTV and BMCTV.

Disease Cycle and Epidemiology

BCTV is transmitted by leafhoppers to a wide range of hosts, including more than 300 plant species in 45 families of dicotyledonous plants, especially Asteraceae, Brassicaceae, Chenopodiaceae, Fabaceae, and Solanaceae. No monocotyledonous hosts have been reported. Besides tomato, other important crops affected are sugar beet, Swiss chard (*B. vulgaris* L. subsp. *cicla* (L.) W. D. J. Koch), spinach (*Spinacia oleracea* L.), green pepper (*Capsicum annuum* L.), beans (*Phaseolus* L.), and cucurbits (Cucurbitaceae). Infected crops play an important role in the epidemiology of the virus. Many hosts to which the leafhopper transmits the virus, including tomato, will not sustain the vector. There are no reports of seed transmission.

The virus–vector relationship between BCTV and the leafhopper is circulative, persistent, and nonpropagative, with no transovarial passage of the virus through the egg. Virus transmission can occur after as brief an acquisition time as 1 min, but the frequency of transmission is maximized after a 2- or 3-day acquisition access period. The minimum latent period in the vector before transmission can occur is rarely less than 4 h. Infective vectors have been known to transmit the virus for life, but the proportion of test plants infected decreases when the leafhoppers are not exposed to infected source plants. Transmission efficiency depends on the amount of virus in the vector, which in turn depends on the virus titer in the source plant.

Because curly top is a major disease of sugar beet, its occurrence, survival, and spread have been well documented. Much of the spread of BCTV into tomato fields depends on the seasonal cycle of the leafhopper vector. Beet leafhoppers are well adapted to desert conditions (e.g., intermountain U.S. West) and are able to overwinter in uncultivated areas on limited vegetation. The long-distance spread of BCTV from wild weeds has been clearly described. In spring, when wild hosts begin to dry up, the leafhoppers migrate into valleys to feed on spring crops. Sugar beets are a common host for the insects and can serve as a virus reservoir, as can wild mustard (*Brassica* L.), Russian thistle (*Salsola* L.), and an array of wild species common to agricultural regions of the western United States. Because tomato is not a desired host for leafhoppers, the insects remain in tomato fields for a limited time but long enough to transmit the virus. Little or no secondary spread occurs in the field, and the pattern of infection often suggests an influx of viruliferous leafhoppers occurring during migratory flights.

Management

A range of management strategies can be used to reduce the occurrence of BCTV in tomato. Insecticides should be sprayed on weeds when necessary to prevent leafhoppers from migrating into tomato crops. However, application of insecticides directly to the tomato crop does not prevent leafhoppers from transmitting the virus and does not reduce the incidence of disease.

Weeds such as Russian thistle and wild mustard should be eliminated to reduce some of the most significant primary summer hosts of the beet leafhopper, and tomatoes should not be planted near overwintered sugar beets—an important winter virus reservoir. Plowing down reservoir crops prior to planting tomato avoids movement of the vector from the reservoir crop into the tomato field at plow-down.

Double-row planting of processing tomato fields seems to reduce losses from BCTV. It is not known if this practice discourages leafhoppers from visiting tomato plants or if it allows healthy plants to compensate for the loss of infected plants.

High levels of resistance to BCTV are known to exist in several wild species of *Solanum*. Some released cultivars have increased levels of tolerance to establishment of infection and leafhopper preference.

Fig. 124. Tomato plant exhibiting classic symptoms of beet curly top: upward rolling of leaves, along with yellowing and stunting. (Courtesy W. M. Wintermantel)

Fig. 125. Tomato plant infected with *Beet curly top virus* (BCTV), showing premature ripening, leaf curling, and vein purpling. (Courtesy W. M. Wintermantel)

Selected References

Bennett, C. W. 1971. The Curly Top Disease of Sugarbeet and Other Plants. Monogr. 7. American Phytopathological Society, St. Paul, MN.

Chen, L.-F., and Gilbertson, R. L. 2008. Beet mild curly top virus. Pages 195-215 in: Characterization, Diagnosis and Management of Plant Viruses. Vol. 1, Industrial Crops. G. P. Rao, S. M. Paul Khurana, and S. Lenardon, eds. Studium Press, Houston, TX.

Creamer, R., Luque-Williams, M., and Howo, M. 1996. Epidemiology and incidence of beet curly top geminivirus in naturally infected weed hosts. Plant Dis. 80:533-535.

Martin, M. W., and Thomas, P. E. 1986. Levels, dependability, and usefulness of resistance to tomato curly top disease. Plant Dis. 70:136-141.

Stenger, D. C., and McMahon, C. L. 1997. Genotypic diversity of beet curly top virus population in the western United States. Phytopathology 87:737-744.

Strausbaugh, C. A., Wintermantel, W. M., Gillen, A. M., and Eujayl, I. A. 2008. Curly top survey in the western United States. Phytopathology 98:1212-1217.

Thomas, P. E., and Martin, M. W. 1971. Vector preference, a factor of resistance to curly top virus in certain tomato cultivars. Phytopathology 61:1257-1260.

Thomas, P. E., and Martin, M. W. 1972. Characterization of a factor of resistance in curly top virus-resistant tomatoes. Phytopathology 62:954-958.

Thomas, P. E., and Mink, G. I. 1979. Beet curly top virus. Descriptions of Plant Viruses, no. 210. C.A.B. International Mycological Institute, Kew, England.

(Prepared by T. A. Zitter and W. M. Wintermantel)

Cucumber mosaic virus

Cucumber mosaic virus (CMV) is considered an important disease in the temperate regions of the world but can occur wherever tomatoes are grown. This virus is distributed worldwide and can infect nearly 1,200 plant species. In addition, numerous isolates of CMV have been reported as strains, some of which are specific for tomato.

Symptoms

In the early stages of infection with CMV, tomato plants are yellow, bushy, and considerably stunted. The leaves may show a mottle similar to that caused by *Tomato mosaic virus* (ToMV). The most characteristic symptom of CMV is filiformity, or shoestring-like leaf blades (Fig. 126). This can be confused with a symptom of ToMV known as "fern-leaf," in which the blade of the leaflet is not as completely suppressed as in a shoestring leaflet but is abnormally long and narrow. The symptoms caused by CMV can be transitory, with the bottom leaves or newly developed top leaves showing severe symptoms and the middle leaves appearing almost normal. Severely affected plants produce few fruit, and the fruit that are produced tend to be small, are often mottled or necrotic, and have delayed maturity.

CMV symptoms in tomato can be classified into two distinct groups, depending on the growing conditions. In greenhouse-grown tomato, CMV-infected plants may express a general chlorosis followed by an easily observed mosaic symptom on young, systemically infected leaves. These symptoms tend to be somewhat transient, as newly emerging leaves may be smaller than normal and show varying degrees of deformation, such as shoestringing. Internode elongation is reduced, leading to overall stunting, but a bushy growth habit may not occur.

A different set of symptoms can be observed in field-grown tomato plants. The occurrence of mosaic symptoms on young, systemically infected leaves typically includes initial chlorosis and mosaic. If plants are young at the time of infection, the mosaic symptoms are rather transient, given severe leaf deformation and, in particular, an upward cupping or rolling of leaves. If plants are somewhat more mature at the time of infection, mosaic symptoms may be more obvious, in part because they are not masked by severe leaf deformation and cupped or rolled leaves. Tomato plants that are young at the time of infection remain small (sometimes severely stunted) and have a bushy appearance.

Causal Agent

CMV is the type species of the genus *Cucumovirus* in the family Bromoviridae. To cause infection, cucumoviruses require three functional pieces of single-stranded RNA (RNAs 1, 2, and 3). The genome is encapsidated in three icosahedral particles approximately 29 nm in diameter. Two subgenomic RNAs (RNA4 and RNA4A) also occur. RNA4A translates the 2b protein, which is encapsidated with the RNA3 particle, and occurs in CMV subgroup II strains. The coat protein is also encoded on RNA3 and is expressed from subgenomic RNA4. A fifth RNA (RNA5) is also present in some CMV strains.

In addition to genomic and subgenomic RNAs, some strains of CMV encapsidate a satellite RNA (satRNA), which is a small RNA molecule that is completely dependent on the viral genome for its replication and spread but does not supply the helper virus with any essential function. Some satRNAs intensify symptoms, inducing systemic necrosis or chlorosis in tomato, whereas the majority of satRNAs attenuate the virus-induced symptoms. In 1988, a severe outbreak of lethal necrosis, affecting thousands of acres of processing tomatoes in southern Italy, was associated with the occurrence of CMV and satRNA (Fig. 127).

CMV strains have been divided into two subgroups (subgroups I and II) on the basis of serological relationships and sequence analysis. Subgroup I strains have been further divided into IA and IB, differentiated in cowpea (*Vigna unguiculata* (L.) Walp.) by inducing a systemic mosaic or necrotic local lesions on inoculated leaves, respectively. Some strains are host specific, infecting certain hosts in the same family, such as the legume strain of CMV. CMV is serologically related to *Tomato aspermy virus* (TAV) and *Peanut stunt virus* (PSV).

Fig. 126. Tomato plant infected with *Cucumber mosaic virus* (CMV), with leaves displaying "shoestringing." (Courtesy S. K. Green)

So-called empty virus crystals located in a vacuole may appear as a neat hexagonal outline (Fig. 128) but more often appear as empty circles. Such empty crystals are diagnostic of CMV but not cucumoviruses in general.

Disease Cycle and Epidemiology

Because of CMV's wide host range, numerous weeds can serve as reservoirs for the virus and contribute to its spread to crops. In addition to tomato, the host range includes many important vegetable crops: carrot (*Daucus carota* L.), celery (*Apium* L.), cucurbits (Cucurbitaceae), legumes (Fabaceae (Leguminosae)), lettuce (*Lactuca sativa* L.), pepper (*Capsicum annuum* L.), and spinach (*Spinacia oleracea* L.). Many ornamental plants are also suitable hosts, such as anemone (*Anemone* L.), candytuft (*Iberis* L.), columbine (*Aquilegia* L.), dahlia (*Dahlia* Cav.), delphinium (*Delphinium* L.), geranium (*Geranium* L.), petunia (*Petunia* Juss.), phlox (*Phlox* L.), viola (*Viola* L.), and zinnia (*Zinnia* L.), and various woody and semiwoody plants can serve as hosts, such as banana (*Musa* L.), ixora (*Ixora* L.), and passion fruit (*Passiflora edulis* Sims).

More than 75 aphid species, including the green peach aphid (*Myzus persicae* (Sulzer)) and the cotton aphid (*Aphis gossypii* Glover), are capable of transmitting CMV in a nonpersistent, stylet-borne manner. Generally, the virus is acquired by all instars within 1 min, but the instars' ability to transmit the virus declines and is lost within several hours. Transmission efficiency varies with the aphid species, virus strain, host plant species, environmental conditions, and time of the year. Experimentally, mechanical transmission of CMV occurs easily, but efficiency is increased by the addition of a reducing agent (e.g., 10 mM sodium sulfite) to the buffer when preparing inoculum.

Since CMV is not a stable virus, like *Tobacco mosaic virus* (TMV), it is not likely transmitted by workers touching infected plants. The virus is not carried in tomato seeds but can be carried in the seeds of 19 other plant species.

Fig. 127. Field infected by *Cucumber mosaic virus* (CMV) and satellite RNA. (Courtesy D. Gallitelli)

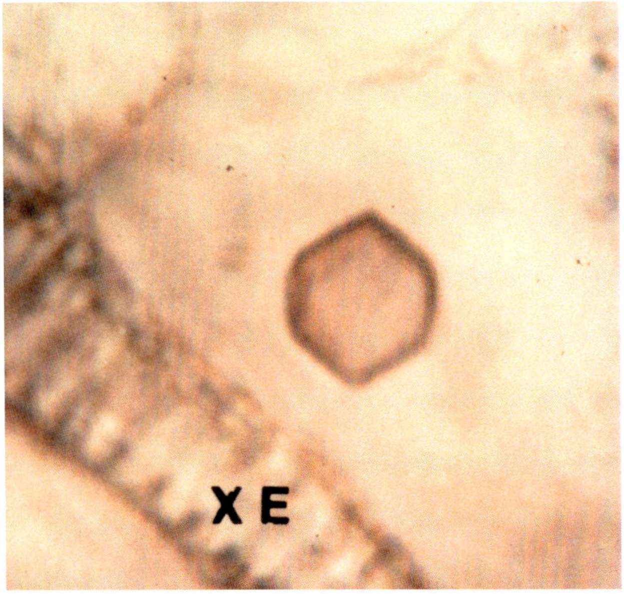

Fig. 128. Empty virus crystal of *Cucumber mosaic virus* (CMV) located in the vacuole (XE = xylem element). (Courtesy R. G. Christie)

Management

Sources of resistance and tolerance to CMV have been reported in wild tomato relatives but not in *Solanum lycopersicum* L. No cultivars with CMV resistance have been released.

Eradicating weed hosts of CMV is often difficult because of the extensive host range of this virus. However, elimination of several key perennial or biennial weeds located near the tomato crop may reduce severe virus pressure. In addition, isolation of tomato fields by growing taller, nonsusceptible barrier crops, such as corn (*Zea mays* L.), may delay initial infection. Applications of insecticides and mineral oil sprays have been used to control this virus, as well as other nonpersistently transmitted viruses.

Alternative methods of reducing CMV losses have also been reported. Cross-protection (i.e., inoculation of plants with mild, protective strains of CMV) and the use of satRNAs have been described in China. The use of transgenic tomato genotypes, expressing either the CMV coat protein or a satRNA, has protected plants from CMV infection or the severe effects of some satellite RNA-associated CMV infections. The application of plant growth-promoting rhizobacteria, which induced systemic resistance or enhanced plant growth, has also provided protection against CMV and may be integrated into other pest management schemes.

A key factor in reducing yield losses caused by CMV in tomato involves delaying infection and allowing the plant to mature. The plant thereby develops a resistance to infection or a tolerance that does not affect yield.

Selected References

Francki, R. I. B., Mossop, D. W., and Hatta, T. 1979. Cucumber mosaic virus. Descriptions of Plant Viruses, no. 213. C.A.B. International Mycological Institute, Kew, England.

Gallitelli, D. 2000. The ecology of *Cucumber mosaic virus* and sustainable agriculture. Virus Res. 71:9-21.

Gonsalves, D., Provvidenti, R., and Edwards, M. C. 1982. Tomato white leaf: The relation of an apparent satellite RNA and *Cucumber mosaic virus*. Phytopathology 72:1533-1538.

Murphy, J. F., Reddy, M. S., Ryu, C.-M., Kloepper, J. W., and Li, R. 2003. Rhizobacteria-mediated growth promotion of tomato leads to protection against *Cucumber mosaic virus*. Phytopathology 93:1301-1307.

Murphy, J. F., Sikora, E. J., Sammons, B., and Kaniewski, W. K. 1998. Performance of transgenic tomatoes expressing cucumber mosaic virus *cp* gene under epidemic conditions. HortScience 33:1032-1035.

Palukaitis, P., and Garcia-Arenal, F. 2003. Cucumoviruses. Adv. Virus Res. 62:241-323.

Palukaitis, P., Roossinck, M. J., Dietzgen, R. G., and Francki, R. I. B. 1992. Cucumber mosaic virus. Adv. Virus Res. 41:281-348.

Zitter, T. A., and Murphy, J. F. 2009. Cucumber mosaic. The Plant Health Instructor. doi: 10.1094/PHI-I-2009-0518-01.

(Prepared by T. A. Zitter and J. F. Murphy)

Pepino mosaic virus

Pepino mosaic virus (PepMV) is an emerging disease in greenhouse tomato production. The virus was first described in 1980 on pepino (*Solanum muricatum* Aiton) in Peru. It was a relatively minor problem for almost 20 years, until it was first observed infecting greenhouse tomato in Europe in 1999. The disease has since been reported as widespread across much of Europe, North America, and South America.

Symptoms

Tomato plants infected with PepMV develop a variety of symptoms depending on cultivar, age of infection, and environmental conditions. In general, yellow mosaic and spots develop on leaves in younger plants (Fig. 129), which may produce bright-yellow patches, severe leaf distortion, or bubbling on older leaves (Fig. 130). In some cases, the infected leaflets become narrower and more pointed to the growth point, resembling nettleheads. Some strains of PepMV cause severe necrosis and necrotic lesions on stems and petioles of infected tomato plants (Fig. 131).

Mature fruit show uneven ripening with symptoms that resemble marbling (flaming) (Fig. 132). Symptoms may also be observed on a few fruit in a single cluster while other clusters have no symptoms, even though the fruit may contain high titers of PepMV. Other tomato-infecting viruses (e.g., *Potato virus X*) may cause similar symptoms. Therefore, molecular or serological tests (i.e., enzyme-linked immunosorbent assay (ELISA)) are necessary to make a positive diagnosis.

In most cases, the yield loss from plants infected with PepMV is not significant. Even so, the marketable fruit quality may be considerably reduced.

Fig. 129. Yellow mosaic and spots on younger tomato plants infected with *Pepino mosaic virus* (PepMV). (Courtesy K. S. Ling)

Fig. 130. Bright-yellow spots and patches on plants infected with *Pepino mosaic virus* (PepMV). (Courtesy K. S. Ling)

Causal Agent

Pepino mosaic virus is a member of the genus *Potexvirus* in the family Alphaflexiviridae. PepMV has a typical potexvirus genome with a single-stranded RNA. The first open reading frame (ORF) encodes the genes associated with virus replication, followed by a triple gene block coding for virus cell-to-cell movement. The last ORF contains the coat protein gene.

Interestingly, tomato isolates from different geographic regions have a very high genome sequence similarity (99–100%), which suggests a close evolutionary relationship or a common origin. Even the pepino isolate of PepMV shares a high nucleotide sequence identity (96–97%) with the tomato isolates. However, recently gathered sequence data indicate greater variability in additional tomato isolates from Chile and the United States. These emerging strains have only 82% nucleotide sequence identity to the European tomato strain.

PepMV is serologically related to two other potexviruses: *Narcissus mosaic virus* (NMV) and *Cactus virus X* (CVX).

Fig. 131. Some strains of *Pepino mosaic virus* (PepMV) cause necrosis or necrotic lesions on stems and petioles of infected plants. (Courtesy K. S. Ling)

Fig. 132. Symptoms of *Pepino mosaic virus* (PepMV) on mature fruit. Note the discoloration and marbling (flaming). (Courtesy M. Bledsoe)

Disease Cycle and Epidemiology

PepMV has a relatively narrow host range, including plants in three families: Aizoaceae (e.g., New Zealand spinach (*Tetragonia expansa* Murray)), Cucurbitaceae (e.g., cucumber (*Cucumis sativus* L.)), and many species of Solanaceae. Even though PepMV has not been reported as naturally occurring in potato (*S. tuberosum* L.), it produces severe symptoms on several potato cultivars following mechanical inoculation. A survey in central and southern Peru for the natural occurrence of PepMV in *Solanum* species showed that besides the natural hosts pepino and tomato, several wild *Solanum* species are also natural hosts of the virus: *S. chilense* (Dunal) Reiche; *S. chmielewskii* (C. M. Rick, Kesicki, Fobes, & M. Holle) D. M. Spooner, G. J. A (C. M. Rick, Kesicki, Fobes, & M. Holle) D. M. Spooner, G. J. A.; *S. neorickii* D. M. Spooner, G. J. Anderson, & R. K. Jansen (syn. *Lycopersicon parviflorum* C. M. Rick); and *S. peruvianum* L.

It appears that PepMV-contaminated tomato seeds may be responsible for the worldwide emergence of this disease. The virus has been shown to be seedborne even though it is not likely seed transmitted. Easy mechanical transmission from PepMV-contaminated seeds to seedlings can initiate virus infection. Secondary spread of the disease occurs mainly through mechanical contact, including contaminated tools, hands, and clothing. Even though aphids, such as the green peach aphid (*Myzus persicae* (Sulzer)), have been shown not to be vectors of the virus, there has been speculation that pollinating bumble bees (*Bombus* spp.) may play a role in virus spread.

Management

Commercial tomato cultivars have no effective disease resistance against PepMV, even though wild *Solanum* species have some resistance or tolerance. Disease management must include strict greenhouse sanitation and hygiene practices for avoidance and prevention, including planting of certified virus-free seeds, frequent handwashing with soap, and dipping pruning tools in 20% bleach or skim milk between cuttings of plants. In some European countries, workers are advised not to bring supermarket tomatoes to greenhouse tomato production facilities.

Even though roguing of early infected plants is recommended, for late-infected plants, yield loss may be minimized without removing PepMV-infected plants through careful containment practices and thorough decontamination of the entire greenhouse structure between plantings. After harvesting, infected plant tissue needs to be disposed of properly through burning or removal to a landfill to eliminate virus infectivity.

Selected References

Gomez, R., Sempere, R. N., and Aranda, M. A. 2012. Pepino mosaic virus and tomato torrado virus: Two emerging viruses affecting tomato crops in the Mediterranean Basin. Adv. Virus Res. 84:505-532.

Jones, R. A. C., Koenig, R., and Lesemann, D. E. 1980. Pepino mosaic virus, a new potexvirus from pepino (*Solanum muricatum*). Ann. Appl. Biol. 94:61-68.

Ling, K.-S. 2007. Molecular characterization of two *Pepino mosaic virus* variants from imported tomato seed reveals high levels of sequence identity between Chilean and US isolates. Virus Genes 34:1-8.

Ling, K.-S., and Carpenter, L. 2005. *Pepino mosaic virus,* an emerging disease in greenhouse tomato production worldwide: Is seed responsible? Acta Hortic. 695:43-50.

Ling, K.-S., and Scott, J. W. 2007. Sources of resistance to *Pepino mosaic virus* in tomato accessions. Plant Dis. 91:749-753.

Maroon-Lango, C. J., Guaragna, M. A., Jordan, R. L., Hammond, J., Bandla, M., and Marquardt, S. K. 2005. Two unique US isolates of *Pepino mosaic virus* from a limited source of pooled tomato tissue are distinct from a third (European-like) US isolate. Arch. Virol. 150:1187-1201.

Pico, B., Herraiz, J., Ruiz, J. J., and Nuez, F. 2002. Widening the genetic basis of virus resistance in tomato. Sci. Hortic. 94:73-89.

Salomone, A., and Roggero, P. 2002. Host range, seed transmission and detection by ELISA and lateral flow of an Italian isolate of *Pepino mosaic virus*. J. Plant Pathol. 84:65-68.

Soler-Aleixandre, S., López, C., Cebolla-Cornejo, J., and Nuez, F. 2007. Sources of resistance to *Pepino mosaic virus* (PepMV) in tomato. HortScience 42:40-45.

Spence, N. J., Basham, J., Mumford, R. A., Hayman, G., Edmondson, R., and Jones, D. R. 2006. Effect of *Pepino mosaic virus* on the yield and quality of glasshouse-grown tomatoes in the UK. Plant Pathol. 55:595-606.

van der Vlugt, R. A. A., Stijger, C. C. M. M., Verhoeven, J. Th. J., and Lesemann, D.-E. 2000. First report of *Pepino mosaic virus* on tomato. Plant Dis. 84:103.

Verhoeven, J. Th. J., van der Vlugt, R. A. A., and Roenhorst, J. W. 2003. High similarity between tomato isolates of *Pepino mosaic virus* suggests a common origin. Eur. J. Plant Pathol. 109:419-425.

(Prepared by K.-S. Ling)

Potato virus Y

Potato virus Y (PVY) is the type member of the genus *Potyvirus* in the family Potyviridae—the largest group of plant viruses and one of the most economically important. PVY occurs worldwide but has a narrow host range, causing important diseases in solanaceous crops, including tomato. Scattered reports of PVY infection of tomato occurred in the United States during the 1930s and 1940s. Since the 1990s, PVY has affected tomato crops in Argentina, Australia (Queensland), Canada (Ontario), Guadeloupe, Taiwan, and the United States (Florida), as well as several Mediterranean countries (Israel, Morocco, and Tunisia). Serious outbreaks have occurred in France since the 1980s and in central and southern Italy since 2002.

Symptoms

Potato virus Y produces a wide range of symptoms in tomato based on virus strain, cultivar infected, plant age, and environmental factors. Originally called vein-banding mosaic because of the prominent dark-green bands associated with the main veins in tomato, tobacco (*Nicotiana* L.), and pepper (*Capsicum annuum* L.), the general symptoms in tomato consist of faint mottling and slight distortion.

However, field symptoms of PVY can be decidedly more severe, consisting of dark-brown, dead areas in the laminae of nearly mature leaflets. The terminal leaflets may show severe necrosis, but in many cases, all leaflets are affected (Fig. 133). Subsequently formed leaves exhibit mild rugosity, slight distortion, and mild mottling. Leaflets of plants infected for some time roll downward and have curved petioles, giving plants a drooping appearance. The necrotic strains of PVY reported from Italy cause mosaic and leaf distortion, flower necrosis, and irregular fruit ripening.

Causal Agent

PVY particles appear as long, flexuous rods (11 × 730 nm) and contain single-stranded RNA, which is infectious. In addition to solanaceous crops, members of Amaranthaceae, Chenopodiaceae, Asteraceae, and Fabaceae are included in the experimental host range.

PVY is readily transmitted mechanically. It is also transmitted by many aphid species in a nonpersistent manner. The most efficient vector is the green peach aphid (*Myzus persicae* (Sulzer)); it can acquire the virus in less than 60 sec and transmit it in less than 60 sec. Most feeding aphids fail to transmit after 1 h; however, aphids can retain the virus for longer than 24 h if they are not allowed to probe. Plants infected with PVY contain a helper factor: a labile, proteinaceous material that enables the transmission of the virus and other potyviruses.

Most strains of PVY are closely related serologically, as determined by enzyme-linked immunosorbent assay (ELISA) and using polyclonal and monoclonal antibodies. Serologically, PVY is distantly related to *Tobacco etch virus* (TEV), *Henbane mosaic virus* (HMV), and *Potato virus A* (PVA). PVY isolates have traditionally been classified into PVY^{O}, PVY^{N}, and PVY^{C} groups. Recently, the use of a host-independent classification based on a restriction fragment length polymorphism (RFLP) pattern of the coat protein gene after immunocapture reverse transcription polymerase chain reaction (IC-RT-PCR) has allowed the regrouping of PVY isolates into three main clusters: potato PVY^{N}, potato PVY^{O}, and nonpotato PVY^{NP} isolates (e.g., tomato, pepper, tobacco, jimsonweed (*Datura stramonium* L.), and black nightshade (*Solanum nigrum* L.)).

Cytoplasmic cylindrical (pinwheel) inclusions occur mainly in epidermal tissue (Fig. 134). Nuclear inclusions present in plants infected with TEV are lacking in hosts of PVY.

Disease Cycle and Epidemiology

In temperate areas, perennial plants rarely act as virus reservoirs. However, potato (*S. tuberosum* L.) is an important reservoir host in temperate zones for tomato and other solanaceous crops, such as pepper. Even in subtropical areas of southern Florida, potato has served as the originating source of PVY for pepper and tomato plantings. Near-total crop failure has occurred when primary inoculum has been detected early in the crop and a high aphid population has been recorded. In tropical and subtropical areas, weeds such as potato vine (*S. atropurpureum* Schrank), black nightshade, and groundcherry (*Physalis* L.) can serve as oversummering reservoirs for subsequent crop infection in the fall and spring. Black nightshade is the most important overwintering source for PVY in Italy.

Tomatoes may be infected only with PVY, or they may be coinfected with PVY and other aphid-transmitted viruses (e.g., TEV, *Cucumber mosaic virus* (CMV)), *Tomato mosaic virus* (ToMV). Plants doubly infected at an early age rarely yield acceptable fruit.

In southern Florida, mixed and overlapping plantings of tomato and pepper undoubtedly contribute to the development of a large aphid population and may contribute to the spread of PVY. The virus does not appear to be transmitted through seed.

Fig. 133. Necrosis of leaflets on a tomato plant infected by *Potato virus Y* (PVY). (Courtesy H. Laterrot)

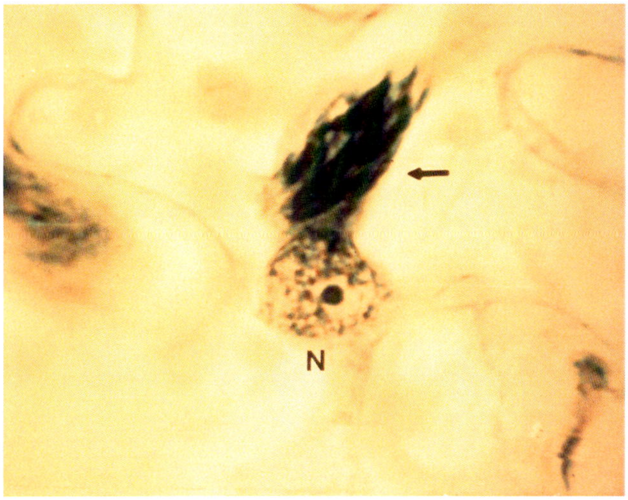

Fig. 134. Cylindrical (pinwheel) inclusions of *Potato virus Y* (PVY) (N = nucleus). (Courtesy R. G. Christie)

Management

Controlling PVY spread through vector control has proven unsuccessful, and destroying oversummering/overwintering virus reservoirs is not practical because of these plants' widespread occurrence in major tomato production areas. Thus, tomato growers have resorted to management methods that delay the entry of primary inoculum into the crop. The main control methods are (1) isolating tomato plantings from areas used to produce earlier tomato and pepper crops, (2) planting in areas that are less likely to support aphid populations and virus reservoirs, and (3) spraying with mineral oil to reduce the frequency of transmission when primary inoculum is kept low. To reduce the risk of spreading PVY in glasshouse operations, pruning tools should be dipped in a 10% solution of sodium hypochlorite; dipping tools for as little as 2 sec can eliminate PVY transmission.

Resistance to PVY and to TEV has been found in *Lycopersicon hirsutum* Humb. & Bonpl. PI 247087. Resistance is conferred by a single recessive gene or two very tightly linked recessive loci. The gene, *pot-1*, was mapped to the short arm of tomato chromosome 3 and is in the vicinity of the recessive *py-1* locus for resistance to tomato corky root rot. A resistance-breaking isolate has been characterized in France.

Selected References

Blanco-Urgoiti, B., Sanchez, F., Dopazo, J., and Ponz, F. 1996. A strain-type clustering of potato virus Y based on the genetic distance between isolates calculated by RFLP analysis of the amplified coat protein gene. Arch. Virol. 141:1-18.

Chagas, C. M., Vicente, M., Alba, A. P. C., and July, J. R. 1977. *Solanum atropurpureum* Schrank, a natural reservoir of potato virus Y (PVY). Phytopathol. Z. 90:147-151.

Conover, R. A., and Fulton, R. W. 1953. Occurrence of potato Y virus on tomatoes in Florida. Plant Dis. Rep. 37:460-462.

de Bokx, J. A., and Huttinga, H. 1981. Potato virus Y. Descriptions of Plant Viruses, no. 242. C.A.B. International Mycological Institute, Kew, England.

Legnani, R., Gebre-Selassie, K., Nono Wondim, R., Gognalons, P., Moretti, A., Laterrot, H., and Marchoux, G. 1995. Evaluation and inheritance of the *Lycopersicon hirsutum* resistance against Potato virus Y. Euphytica. 86:219-226.

Morel, C., Gognalons, P., Guilbaud, L., Caranta, C., Gebre-Selassie, K., Marchoux, G., and Jacquemond, M. 2000. Biological and molecular characterization of two tomato strains of potato virus Y (PVY). Acta Physiol. Plant. 22:336-343.

Parrella, G., Ruffel, S., Moretti, A., Morel, C., Palloix, A., and Caranta, C. 2002. Recessive resistance genes against potyviruses are localized in colinear genomic regions of the tomato (*Lycopersicon* spp.) and pepper (*Capsicum* spp.) genomes. Theor. Appl. Genet. 105:855-861.

Simons, J. N., Conover, R. A., and Walter, J. M. 1956. Correlation of occurrence of potato virus Y with areas of potato production in Florida. Plant Dis. Rep. 40:531-533.
Stobbs, L. W., Poysa, V., and VanSchagen, J. G. 1994. Susceptibility of cultivars of tomato and pepper to a necrotic strain of potato virus Y. Can. J. Plant Pathol. 16:43-48.
Zitter, T. A., and Tsai, J. H. 1981. Viruses infecting tomato in Southern Florida. Plant Dis. 65:787-791.

(Prepared by T. A. Zitter)

Potato yellow vein virus

Potato yellow vein probably emerged as a disease of potato in the region of southern Colombia and northern Ecuador in the early 1940s, but its etiology remained undetermined until 2000. The disease has since disseminated throughout the highlands of Colombia, Ecuador, and northern and central Peru, mainly in potato-producing regions located 2,000–3,400 m above sea level.

In 2005, a serious outbreak of a new disease was observed in tomato fields located in the central highlands of Colombia (Caqueza, Cundinamarca) at an altitude of 1,800 m. This outbreak affected tomato planted in one of the nation's most important regions for growing potato (*Solanum tuberosum* L.). Disease incidence was also associated with the presence of a high population of the greenhouse whitefly on diseased tomato plants.

Symptoms

Tomato plants that are susceptible to potato yellow vein may show intense foliar chlorosis and malformation, followed by foliar necrosis (Fig. 135). Fruit production is reduced because of the lower quantity and size of fruit produced, and the fruit tend to remain green and show some variegation.

Causal Agent

Potato yellow vein is caused by *Potato yellow vein virus* (PYVV), a member of the genus *Crinivirus,* family Closteroviridae. The PYVV isolate from tomato has filamentous particles of two modal lengths (approximately 650 and 900 nm) and contains an apparent tripartite genome of positive-sense, single-stranded RNA. The tomato isolate of PYVV from Colombia is closely related to *Beet pseudoyellows virus* (BPYV) and *Cucurbit yellow stunting disorder virus* (CYSDV), which are both criniviruses. Little pathogenic variability seems to exist among the known isolates of PYVV from potato or tomato characterized thus far in Latin America.

Fig. 135. Severe foliar symptoms on a tomato plant infected with *Potato yellow vein virus* (PYVV). (Courtesy F. J. Morales)

PYVV is transmitted by the greenhouse whitefly (*Trialeurodes vaporariorum* Westwood) in a semipersistent manner. This virus is also transmitted by grafting but not sexually through seed. Transmission and dissemination of PYVV in potato occurs mostly through infected tubers used as planting materials. In addition to potato and tomato, the following plant species have been reported as hosts of PYVV in Colombia: Madagascar periwinkle (*Catharanthus roseus* (L.) G. Don), knotweed (*Polygonum* L.), bitter dock (*Rumex obtusifolius* L.), and marigold (*Tagetes* L.).

Disease Cycle and Epidemiology

Potato yellow vein is currently distributed in the equatorial Andean region of South America, but potato and tomato production co-exist in the Andean region spanning from Venezuela to Bolivia. This is also the main area affected by the greenhouse whitefly vector—a common pest of many crops, including tomato, in the Andean Highlands. Thus, this disease is expected to become widespread in tomato plantings throughout the Andean region of South America.

Movement of PYVV-infected potato tubers from the Andean region to other parts of the world where the greenhouse whitefly exists may result in further dissemination of this emerging disease of tomato in temperate regions of the world. This whitefly species thrives at altitudes above 1,000 m, especially in dry environments and during dry seasons of the year.

Management

Insecticide abuse has been identified as a major cause of problems associated with a high population of the greenhouse whitefly in tomato, potato, and common bean (*Phaseolus* L.) in the Andean region. The use of new systemic insecticides at planting time is recommended, together with cultural practices designed to avoid simultaneous and sequential plantings of susceptible crops. Use of traditional insecticides should be discouraged, because the greenhouse whitefly has become resistant to most old insecticides. Moreover, these products are still toxic to the beneficial fauna that control whiteflies under natural conditions. Crop residues are a major source of viruliferous whiteflies and thus should be destroyed.

Tomato seedlings should be physically protected before transplanting, and young tomato plants should be chemically protected with new systemic insecticides until they form fruit. Planting during rainy seasons usually reduces the incidence of whitefly-borne diseases, but it increases the risk of high incidences of fungal and bacterial pathogens in humid, tropical environments.

Selected References

Livieratos, I. C., Eliasco, E., Müller, G., Olsthoorn, R. C. L., Salazar, L. F., Pleij, C. W. A., and Coutts, R. H. A. 2004. Analysis of the RNA of *Potato yellow vein virus:* Evidence for a tripartite genome and conserved 3'-terminal structures among members of the genus *Crinivirus.* J. Gen. Virol. 85:2065-2075.
Morales, F. J., Martinez, A. K., Olaya, C., and Velasco, A. C. 2004. Detección en tomate (*Lycopersicon esculentum*) del virus del amarillamiento de las nervaduras de la papa (*Potato yellow vein virus*) en Cundinamarca, Colombia. (In Spanish.) Fitopatol. Colomb. 28:40-44.
Moreno, J. D. 1966. Incidencia del virus del amarillamiento de venas en papa en el Ecuador y su transmisión a través de los tubérculos. (In Spanish.) Turrialba 16:15-24.
Salazar, L. F., Müller, G., Querci, M., Zapata, J. L., and Owens, R. A. 2000. Potato yellow vein virus: Its host range, distribution in South America and identification as a crinivirus transmitted by *Trialeurodes vaporariorum.* Ann. Appl. Biol. 137:7-19.

(Prepared by F. J. Morales)

Tobacco etch virus

Tobacco etch virus (TEV) occurs in North and South America, as well as the Philippines, Taiwan, Thailand, and Turkey. The disease has occurred sporadically over widely separated areas since the 1930s, causing the heaviest losses for tomato in two regions of Venezuela and in the United States along the Atlantic coast of southern Florida. In nature, the virus causes diseases of solanaceous crops, including tomato, pepper (*Capsicum annuum* L.), and tobacco (*Nicotiana* L.).

Symptoms

The symptoms of tomato plants infected with *Tobacco etch virus* are far more severe than those infected with *Potato virus Y* (PVY). Characteristic symptoms of TEV include intense mottling, puckering, and rugosity of leaves (Fig. 136). Plants infected at an early age are severely stunted.

The fruit of TEV-infected plants are mottled and never achieve marketable size, which means they must be discarded as culls (Fig. 137). There is a linear relationship between the length of time a plant is infected and the number and size of fruit it produces: The earlier the infection, the lower the yield.

Causal Agent

Tobacco etch virus is in the genus *Potyvirus* and the family Potyviridae. TEV particles are flexuous filaments, have a diameter of 12–13 × 730 nm, and contain single-stranded RNA that is translated into a single polyprotein. The virus is readily sap transmitted to a wide experimental host range of more than 149 species in 19 dicotyledonous families.

Variants or strains of TEV have been identified on the basis of symptom development in tobacco cultivars, tomato and pepper genotypes, and other hosts, such as jimsonweed (*Datura stramonium* L.). Three strains of TEV were differentiated in Florida and five strains were differentiated in California on the basis of symptoms in pepper. Variation in symptom response of tomato genotypes suggests that the response may be strain specific. TEV is serologically related to but distinguishable from other potyviruses, including PVY.

Members of *Potyvirus* that have been cytologically investigated induce the formation of intracellular cytoplasmic inclusions (pinwheels) in host cells. TEV causes crystalline nuclear inclusions of distinct shapes, as well as cytoplasmic inclusions. The flat-plate inclusion located in the nucleus (Fig. 138, right arrow), plus the type II cylindrical inclusions (Fig. 138, left arrow), are diagnostic for a common isolate of TEV. An isolate of TEV (TEV-5) taken in Oxnard, California, induces back-to-back, dome-shaped nuclear inclusions and has a slightly different host range.

Fig. 136. Symptoms of *Tobacco etch virus* (TEV) on the foliage of infected tomato plants. (Courtesy T. A. Zitter)

Disease Cycle and Epidemiology

The occurrence of TEV in commercial tomato fields is closely associated with infection of other solanaceous crops (especially pepper) and natural weed hosts that serve as virus reservoirs. Weeds that are virus reservoirs in both temperate and subtropical areas of North and South America include horsenettle (*Solanum carolinense* L.), thistle (*Cirsium vulgare* (Savi) Ten.), lambsquarters (*Chenopodium album* L.), and sicklepod (*Senna obtusifolia* (L.) Irwin & Barneby). Other weed hosts include jimsonweed, black nightshade (*Solanum nigrum* L.), Brazilian nightshade (*Solanum seaforthianum* L.), groundcherry (*Physalis* L.), and toadflax (*Nuttallanthus canadensis* (L.) D. A. Sutton).

There are no reports of seed transmission of TEV in any host plant. The virus can be transmitted by 10 aphid species in a nonpersistent manner. The green peach aphid (*Myzus persicae* (Sulzer)) is an efficient vector. Like other potyviruses, TEV can be acquired by an aphid in less than 60 sec and then transmitted by the insect in less than 60 sec. Third and fourth instars and adults can transmit TEV.

Management

No commercial tomato cultivars have resistance to TEV, but two tomato accessions have been identified as tolerant to the virus: *Solanum pennellii* Correll LA 716 and *S. pimpinellifolium* L. LA 1478. No symptoms appeared on either accession, but the virus was detected by enzyme-linked immunosorbent

Fig. 137. Symptoms of *Tobacco etch virus* (TEV) on tomato fruit. (Courtesy T. A. Zitter)

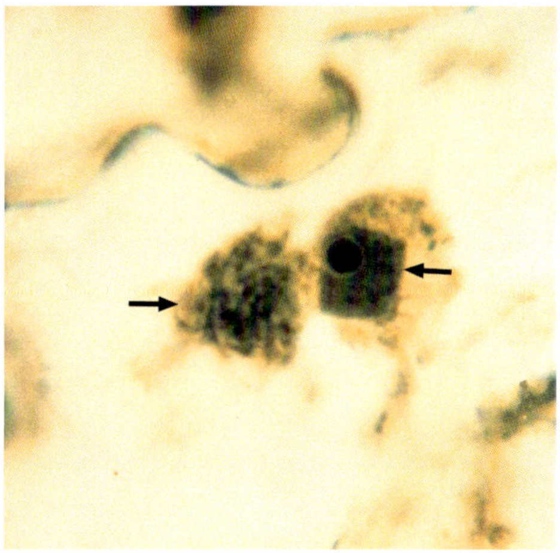

Fig. 138. Cytoplasmic inclusion body (left arrow) and nuclear inclusion body (right arrow) of *Tobacco etch virus* (TEV). (Courtesy R. G. Christie)

assay (ELISA) in noninoculated apical leaves. *Lycopersicon hirsutum* Humb. & Bonpl. PI 247087, previously reported to be resistant to PVY, was also found resistant to TEV. Although TEV can infect PI 247087, systemic spread does not occur. A single recessive gene or two very tightly linked recessive loci are involved in resistance to both TEV and PVY. The locus has been named *pot-1* and has been mapped to the short arm of tomato chromosome 3 and in the vicinity of the recessive *py-1* locus for resistance to corky root rot.

Given the lack of resistant tomato cultivars, other practices must be used to manage TEV. Entry of the virus into the tomato crop can be delayed by isolating the crop from areas in which peppers are grown, since the pepper crop is a major source of inoculum and aphids. Applying mineral oils weekly can also greatly reduce the rate of virus spread, thereby delaying infection and increasing yield. Efforts to reduce virus spread through vector control have been unsuccessful, however.

To reduce the risk of spreading TEV in glasshouse operations, pruning tools should be dipped in a 0.6% solution of sodium hypochlorite. Treating tools for as briefly as 2 sec can eliminate TEV transmission.

Selected References

Abdalla, O. A., Desjardins, P. R., and Dodds, J. A. 1991. Identification, disease incidence, and distribution of viruses infecting peppers in California. Plant Dis. 75:1019-1023.

Anderson, C. W. 1954. *Cassia tora,* a leguminous host of tobacco etch virus. Plant Dis. Rep. 38:736-738.

Demski, J. W. 1979. The epidemiology of tobacco etch virus-infected *Cassia obtusifolia* in relation to pepper. Plant Dis. Rep. 63:647-650.

Herold, F. 1970. Tobacco etch virus in Venezuela. Plant Dis. Rep. 54:344-345.

Lastra, J. R., and de Uzcategui, R. C. 1975. Viruses affecting tomatoes in Venezuela. Phytopathol. Z. 84:253-258.

Legnani, R., Gognalons, P., Gebre-Selassie, K., Marchoux, G., Moretti, A., and Laterrot, H. 1996. Identification and characterization of resistance to tobacco etch virus in *Lycopersicon* species. Plant Dis. 80:306-309.

Parrella, G., Ruffel, S., Moretti, A., Morel, C., Palloix, A., and Caranta, C. 2002. Recessive resistance genes against potyviruses are localized in colinear genomic regions of the tomato (*Lycopersicon* spp.) and pepper (*Capsicum* spp.) genomes. Theor. Appl. Genet. 105:855-861.

Purcifull, D. E., and Hiebert, E. 1982. Tobacco etch virus. Descriptions of Plant Viruses, no. 258. C.A.B. International Mycological Institute, Kew, England.

Zitter, T. A., and Tsai, J. H. 1981. Viruses infecting tomato in southern Florida. Plant Dis. 65:787-791.

(Prepared by T. A. Zitter)

Tobacco mosaic virus and *Tomato mosaic virus*

Tomato mosaic virus (ToMV) and *Tobacco mosaic virus* (TMV) are discussed together because they differ only slightly in host, serological, and cross-protection reactions. In the field, these viruses tend to be restricted to their respective crops. ToMV will be discussed more in this section, however, since its strains are most important for tomato. Although ToMV is not usually a severe problem in greenhouse tomatoes because of the wide use of resistant cultivars, it is still a serious threat worldwide in areas where resistant cultivars are not available and the cultivation of plants includes frequent handling.

Symptoms

The most characteristic symptoms of tomato mosaic caused by common strains of ToMV are mottled areas of light and dark green on the leaves (Fig. 139), although some strains may cause a striking yellow mottling (i.e., *Potato aucuba mosaic virus* (PAMV)). Plants infected in an early stage of growth are usually stunted and have a yellowish cast. Leaves may also be curled, reduced in size, and malformed (fernleaf), although not to the same extent as leaves of plants infected with *Cucumber mosaic virus* (CMV). In the early stages of plant growth, infected cultivars that possess the wilty gene (*wt*) display a prominent upward leaf roll. Stem streaking (i.e., single-virus streak as opposed to double-virus streak) has been caused by at least one strain of ToMV.

In general, ToMV symptoms vary greatly in intensity according to viral strain, cultivar, time of infection (both calendar year and stage of plant growth), light intensity, and temperature. High temperatures usually cause masking of foliar symptoms. However, if tomatoes carrying the *Tm-2* or $Tm2^2$ genotype are exposed to high temperatures, a necrotic reaction can occur in response to infection by common strains of the virus.

The fruit of plants infected with ToMV are not ordinarily markedly disfigured, but they may show uneven ripening and be reduced in size and number. Yellow rings around fruit may occur if fruit ripen at high temperatures (Fig. 140). Various strains of ToMV have been described and named largely on the basis of fruit symptoms (e.g., corky ring, crusty fruit, yellow, Aucuba).

At times, tomato fruit may develop a symptom referred to as internal browning or brownwall. This symptom is evident in mature but unripened fruits and usually involves the fleshy parenchyma, the cells of which often collapse. The symptom is

Fig. 139. Foliage of a plant infected with *Tomato mosaic virus* (ToMV). (Courtesy J. P. Jones)

Fig. 140. Fruit of a plant infected with *Tomato mosaic virus* (ToMV). (Courtesy J. P. Jones)

attributed to a shock reaction following ToMV infection. Additional factors—including high soil moisture, low nitrogen and boron levels, and cultivar sensitivity—also seem to contribute to the severity of this symptom. Growers usually see brownwall on the fruit of the first two clusters but may not attribute it to ToMV infection, since it may appear several days ahead of foliage symptoms.

Causal Agents

TMV is the type virus of the genus *Tobamovirus,* family Virgaviridae, of which ToMV and *Tobacco mild green mosaic virus* (TMGMV) are member species capable of infecting tomato. ToMV is an RNA-containing virus with straight, rod-shaped particles approximately 18 × 300 nm. Virions contain one coat protein that is not required for cell-to-cell movement but has a role in vascular-dependent virus accumulation. The movement protein is involved in cell-to-cell movement; it is associated with plasmodesmata and has a single-stranded nucleic acid binding activity in vitro.

ToMV is readily sap transmissible and is one of the most infectious and persistent viral agents known. It is readily transferred from plant to plant by workers' hands, tools, and clothing, and in rare cases, it is transmitted by insects. Strains of ToMV have been classified by their ability to induce symptoms in plants of certain *Solanum* spp. and to overcome resistance genes. Attenuated strains have been used for cross-protection.

The virions are strong immunogens. Virus-specific antisera for TMV and ToMV are commercially available for use in double-antibody sandwich enzyme-linked immunosorbent assay (DAS-ELISA).

Various members of Solanaceae can be used for identification of tobamoviruses. The host range for ToMV is fairly wide, including many solanaceous species and most species tested in the families Aizoaceae, Amaranthaceae, Chenopodiaceae, and Scrophulariaceae. The host range of TMV is even wider.

Hexagonally shaped virus crystals are commonly found in epidermal cells and trichomes. The inclusion bodies can be seen under a light microscope when epidermal strips are stained in Azure A and heated. The use of heat differentiates *Tobamovirus* inclusions from other plant virus inclusions that will stain in Azure A without heat.

Disease Cycle and Epidemiology

Because of the persistent and infectious nature of ToMV, the virus is readily transmitted by human activities. Infected seed and leaf and root debris are common sources of inoculum. ToMV is present in the external mucilage and testa of tomato seeds (and sometimes the endosperm) but not within the embryo, apparently. The percentage of contaminated seeds can vary greatly. Infection of seedlings from seed inoculum occurs during transplanting.

The handling of plants in standard production practices can effectively spread ToMV. Practices such as plant tying, pruning, removing of plant suckers, and harvesting fruit can all disseminate the virus.

ToMV can also survive in leaf and root debris for varying amounts of time, depending on whether the soil is dry or moist. In dry soil, powdered leaf debris remains infective for 2 years, but in moist soil, leaf debris loses infectivity within 1 month. In root debris, the virus can persist for long periods. It survives in fallow soil for 22 months in root debris at a depth of 120 cm and for longer than 2 years when carried under black plastic covered with a layer of clean compost.

Management

ToMV can be managed using a number of methods, although with varying degrees of success. Both before and during the handling of plants, workers should be encouraged to wash their hands with soap and water to inactivate the virus. Workers' use of tobacco should be discouraged, although tobacco products are not as important a source of ToMV as once thought. Seed treatment with a 10% solution of trisodium phosphate (Na_3PO_4) for at least 15 min or heat treatment of dry seeds for 2–4 days at 70°C eliminates all external and usually all internal virus, apparently without affecting seed germination.

Only healthy transplants should be used, and seedlings should not be transplanted into fields where root debris is present. Inoculation of tomato seedlings with attenuated strains of ToMV is practiced for some greenhouse crops, but the risk of selecting virulent strains is a concern. Cross-protection with the use of attenuated strains has been practiced in Japan for at least 30 years.

Three resistance genes—*Tm-1, Tm-2,* and *Tm-2*^{2}—have been identified in crosses between wild and cultivated tomato. The *Tm-2*^{2} gene is currently used for protection in most commercial cultivars. However, strains capable of overcoming resistance in tomato carrying the *Tm-2*^{2} gene have been reported in Italy, Japan, and Spain. Results suggest that the resistance conferred by this gene against ToMV depends on the occurrence of specific recognition events in the host–pathogen interaction, rather than interference with the fundamental functions of the 30-kDa movement protein.

Selected References

Aramburu, J., and Galipienso, L. 2005. First report in Spain of a variant of *Tomato mosaic virus* (ToMV) overcoming the *Tm-2*^{2} resistance gene in tomato (*Lycopersicon esculentum*). New Dis. Rep. 10:40.

Broadbent, L. 1976. Epidemiology and control of tomato mosaic virus. Annu. Rev. Phytopathol. 14:76-96.

Fletcher, J. T. 1984. Diseases of Greenhouse Plants. Longman, London.

Hollings, M., and Huttinga, H. 1976. Tomato mosaic virus. Descriptions of Plant Viruses, no. 156. C.A.B. International Mycological Institute, Kew, England.

Parrella, G., and Vovlas, C. 2002. Characterization of a ToMV strain isolated from a resistant tomato line. J. Plant Pathol. 84:189.

Pelham, J. 1966. Resistance in tomato to tobacco mosaic virus. Euphytica 15:258-267.

Pelham, J. 1972. Strain-genotype interaction of tobacco mosaic virus in tomato. Ann. Appl. Biol. 71:219-228.

Provvidenti, R., and Hoch, H. C. 1977. Tomato leaf roll caused by the interaction of the wilty gene and tobacco mosaic virus infection. Plant Dis. Rep. 61:500-502.

Rast, A. T. B. 1975. Variability of tobacco mosaic virus in relation to control of tomato mosaic in glasshouse tomato crops by resistance breeding and cross protection. Agric. Res. Rep. 834, Centre for Agricultural Publication and Documentation, Wageningen, the Netherlands.

Schroeder, W. T., Provvidenti, R., and Robinson, R. W. 1967. Incubation temperature and virus strains important in evaluating tomato genotypes for tobacco mosaic virus reaction. Tomato Genet. Coop. 17:47-49.

(Prepared by T. A. Zitter)

Tomato bushy stunt virus and Lettuce necrotic stunt virus

Tomato bushy stunt virus (TBSV) is the type species of the genus *Tombusvirus* in the family Tombusviridae. Until the 1990s, the disease was viewed as a scientific curiosity, rather than an economic threat, because it had rarely occurred in the field. Only three reports (from Argentina, Italy, and Mexico) had been made in tomato since 1935, when the virus was isolated from greenhouse-grown tomatoes in Ireland. Beginning in 2000, the disease has been detected in Europe and the U.S. state of California, and it has become a serious economic threat to tomato production in Morocco and Tunisia.

A virus serologically distinct from *Tombusvirus* species and strains was isolated in the late 1990s from greenhouse-grown tomatoes in the states of Colorado and New Mexico. That virus is called Lettuce necrotic stunt virus (LNSV).

Symptoms

When tomato seedlings are inoculated with TBSV, the first symptoms appear as circular, necrotic, local lesions on leaves within 5 days after infection. These leaves become chlorotic and frequently drop off. Disease progress in the older tissue is slow and may actually stop. If young plants are given excess nitrogen fertilizer, resulting in soft stems, necrotic lesions may develop at the soil line and cause the plants to collapse.

Systemic symptoms consist of twisting of the youngest leaves and tip necrosis, which cause proliferation of lateral shoots and thus lead to a bushy appearance (Fig. 141). Lower leaves on such plants show general chlorosis and purple discoloration. On ripe fruit, symptoms include chlorotic blotching and ring and line patterns (Fig. 142).

The symptoms associated with LNSV appear quite different from those described for TBSV. Foliar symptoms include chlorosis and necrotic speckling of leaves, leaf curling and necrosis, and necrosis of the inflorescence accompanied by flower abortion (i.e., truss necrosis) (Fig. 143). Fruit are not greatly deformed or discolored, as occurs with TBSV, but they can be deeply lobed and exhibit internal browning and necrosis (Fig. 144).

Causal Agent

TBSV is a virus containing single-stranded RNA, with isometric particles approximately 30 nm in diameter. The virus is very stable in crude plant sap and readily mechanically transmitted. It reaches high concentrations in host tissue and thus is easy to extract and purify. Infections are characterized by the formation of multivesicular inclusion bodies. They do not contain virions.

TBSV is soilborne and disseminated through water, but there is no evidence of a natural vector. Seed transmission of the virus is rare, if it occurs at all. Although TBSV has a wide experimental host range, it infects a few dicotyledonous species in widely separated families. Most experimental hosts are not invaded systemically.

In addition to the type genus, 15 other species are recognized, but only *Cymbidium ringspot virus* (CyRSV), *Moroccan pepper virus* (MPV), and LNSV are reported to infect tomato. Most species are serologically interrelated, although to a variable extent. TBSV is serologically related to *Artichoke mottled crinkle virus* (AMCV), *Pelargonium leaf curl virus* (PLCV), and *Petunia asteroid mosaic virus* (PeAMV) and distantly related to MPV.

Disease Cycle and Epidemiology

Tombusvirus-induced diseases usually occur in temperate climates, but there are exceptions with TBSV. Since no natural vectors have been identified for TBSV, its mechanism of dissemination and survival in the soil is unknown. The stability of this virus undoubtedly contributes to its introduction and dissemination. A virus can gain entry into a host when watered

Fig. 141. Characteristic bushy foliage produced by *Tomato bushy stunt virus* (TBSV). (Courtesy C. E. Fribourg)

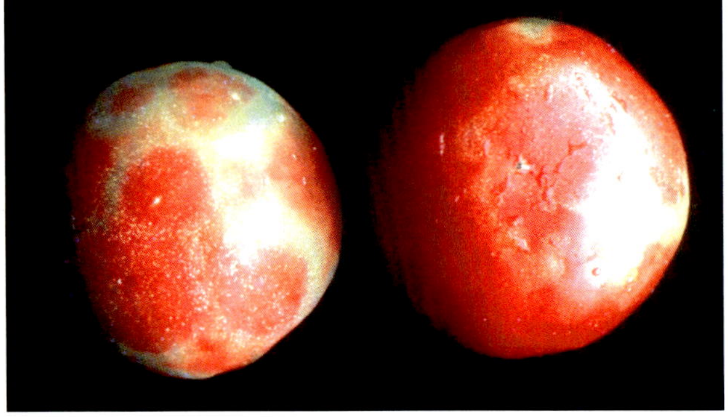

Fig. 142. Symptoms of *Tomato bushy stunt virus* (TBSV) on fruit. (Courtesy E. M. Saadaoui)

Fig. 143. Flower abortion on plant infected with Lettuce necrotic stunt virus (LNSV). (Courtesy G. C. Wisler)

Fig. 144. Internal browning in sectioned fruit, caused by Lettuce necrotic stunt virus (LNSV). (Courtesy G. C. Wisler)

onto the soil. The method of infection is not known, but infection likely occurs through damaged root cells, as has been suggested for *Tomato mosaic virus* (ToMV).

TBSV and related tombusviruses are much more problematic in greenhouse tomato production than production in the field. In the greenhouse, the rapid spread of TBSV is facilitated by the stability of virions in water, their movement in hydroponic culture, and the ease of mechanical transmission via pruning operations and physical contact.

TBSV is not digested nor is its infectivity impaired after passage through the human alimentary tract. This suggests that humans and other animals may act as carriers of TBSV, contributing to its wide distribution in the environment. The discovery of TBSV in river water (presumably resulting from sewage contamination) strengthens this hypothesis. The use of sewage sludge as a fertilizer may contribute to the spread of TBSV and related viruses as a soilborne disease of tomato and other crops.

Management

No strategies have been developed for management of TBSV under field conditions. Effective control under greenhouse conditions requires the removal of plants and thorough cleaning of surfaces in contact with infected plant material or runoff. Recent epidemiological studies indicate that it may be difficult to develop effective and economical control methods.

Selected References

Fisher, H. U., and Lockhart, B. E. L. 1977. Identification and comparison of two isolates of tomato bushy stunt virus from pepper and tomato in Morocco. Phytopathology 67:1352-1355.

Gerik, J. S., Duffus, J. E., Perry, R., Stenger, D. C., and Van Maren, A. F. 1990. Etiology of tomato plant decline in the California desert. Phytopathology 80:1352-1356.

Luis-Artega, M., Rodriquez-Cerezo, E., Fraile, A., Sáez, E., and García-Arenal, F. 1996. Different tomato bushy stunt virus strains that cause disease outbreaks in solanaceous crops in Spain. Phytopathology 86:535-542.

Martelli, G. P. 1981. Tombusviruses. Pages 61-90 in: Handbook of Plant Virus Infections and Comparative Diagnosis. E. Kurstak, ed. Elsevier/North-Holland Biomedical Press, Amsterdam.

Martelli, G. P., Quacquarelli, A., and Russo, M. 1971. Tomato bushy stunt virus. Descriptions of Plant Viruses, no. 69. C.A.B. International Mycological Institute, Kew, England.

Obermeier, C., Sears, J. L., Liu, H. Y., Schlueter, K. O., Ryder, E. J., Duffus, J. E., Koike, S. T., and Wisler, G. C. 2001. Characterization of distinct tombusviruses that cause diseases of lettuce and tomato in the western United States. Phytopathology 91:797-806.

Ohki, T., Uematsu, S., Lesemann, D.-E., Honda, Y., Tsuda, S., and Fujisawa, I. 2005. Characterization of *Tomato bushy stunt virus* newly isolated from nipplefruit (*Solanum mammosum*) in Japan. J. Gen. Plant Pathol. 71:74-79.

Pontis, R. E., Gracia, O., and Feldman, J. M. 1968. Tomato bushy stunt virus on tomato crops in Argentina. Plant Dis. Rep. 52:676-677.

Tomlinson, J. A., and Faithfull, E. M. 1984. Studies on the occurrence of tomato bushy stunt virus in English rivers. Ann. Appl. Biol. 104:485-495.

Tomlinson, J. A., Faithfull, E. M., Flewett, T. H., and Beards, G. 1982. Isolation of infective tomato bushy stunt virus after passage through the alimentary tract of man. Nature 300:637-638.

(Prepared by T. A. Zitter and W. M. Wintermantel)

Tomato chlorosis virus and *Tomato infectious chlorosis virus*

A disease of tomato known as yellow leaf disorder was first seen in the U.S. states of Florida (1989) and California (1993). The disease was originally thought to have abiotic causes, but in the two locations, it was later determined to be caused by a crinivirus: *Tomato chlorosis virus* (ToCV) and *Tomato infectious chlorosis virus* (TICV), respectively.

Since 2002, criniviruses have been considered an emerging group of viruses, and their geographic distribution appears to be increasing. TICV has been reported from Mexico and the United States (California and North Carolina), four countries in the Mediterranean Basin, and the nations of Indonesia, Japan, and Taiwan. ToCV has a wider geographic distribution, having been reported from the United States (Florida, Louisiana, and Colorado), the Caribbean (Puerto Rico and Cuba), eight countries in the Mediterranean Basin, the Canary Islands (off Spain), areas in and around Africa (Morocco, Reunion Island, and South Africa), and Taiwan. In some cases, these viruses have been found together in the same fields and even in the same plants. The symptoms caused by these viruses are often not recognized as having a viral etiology; thus, it is likely that other viruses like these have yet to be described.

As with other viruses, the incidences of ToCV and TICV vary with the growing season and geographic location. Although these criniviruses have the potential to cause severe yield losses in both field and greenhouse tomatoes, in most years and locations, they cause only minor losses. In the United States, TICV and ToCV are primarily a concern in the production of field tomatoes, but the viruses are problematic for both field and greenhouse production in many parts of the world.

Symptoms

TICV and ToCV cause symptoms that can be easily mistaken for those of physiological or nutritional disorders or pesticide phytotoxicity. Typical symptoms are suggestive of magnesium or nitrogen deficiency.

Symptoms of both viruses are identical on tomato and include interveinal yellowing of leaves (Fig. 145 for TICV; Fig. 146 for ToCV), associated loss of photosynthetic capability, leaf brittleness, reduced plant vigor, decreased size and number of fruit, and early plant senescence. Symptoms of interveinal chlorosis or yellowing generally begin to appear approximately 3–4 weeks after initial infection, and the veins remain green. These symptoms typically develop most dramatically on the middle to lower parts of plants, while new growth appears normal.

Criniviruses remain confined to cells associated with the host plant phloem, and foliar symptoms of interveinal yellowing and leaf brittleness are thought to result from plugging of the phloem with large viral inclusion bodies, which probably interferes with normal vascular transport in infected plants. As symptoms develop, plant vigor declines, leading to the production of small fruit, early plant senescence, and reduced plant productivity.

Causal Agents

Yellow leaf disorder in tomato can be caused by two viruses: *Tomato chlorosis virus* (ToCV) and *Tomato infectious chlorosis virus* (TICV). Both viruses are in the genus *Crinivirus* and the family Closteroviridae. Although both produce identical symptoms in tomato, ToCV and TICV can be distinguished from one another and from abiotic disorders through nucleic acid-based assays, some serological assays, host range, and differential whitefly transmission. These viruses are restricted to the phloem tissues of their plant hosts, where vacuolated and pluglike viral inclusions are produced.

Crinivirus genomes consist of approximately 17,000 nucleotides, packaged as bipartite, flexuous, filamentous virus particles. RNA1 encodes proteins predominantly associated with virus replication, whereas RNA2 encodes proteins associated with virion structure, virus movement in the plant, and possibly vector transmission.

Disease Cycle and Epidemiology

As with other plant viruses, the incidences of TICV and ToCV vary from year to year. TICV and ToCV are transmitted

Fig. 145. A, Greenhouse-grown plant exhibiting interveinal chlorosis, caused by *Tomato infectious chlorosis virus* (TICV). **B,** Detail of interveinal chlorosis on tomato leaflets, caused by TICV. (Courtesy W. M. Wintermantel)

Fig. 146. Leaf showing interveinal chlorosis and slight upward curling, caused by *Tomato chlorosis virus* (ToCV). (Courtesy W. M. Wintermantel)

in a semipersistent manner by whiteflies, although each virus is transmitted by a unique set of whitefly species. TICV is transmitted only by the greenhouse whitefly (*Trialeurodes vaporariorum* Westwood). In contrast, ToCV is transmitted by four species of whitefly: most efficiently by the banded-wing whitefly (*T. abutilonea* (Haldeman) and the silverleaf whitefly (*Bemisia argentifolii* Bellows & Perring (formerly *B. tabaci* (Gennadius)) and less efficiently by the sweetpotato whitefly (*B. tabaci*) and the greenhouse whitefly. Both TICV and ToCV can be acquired within a few hours of feeding, although longer feeding periods result in higher rates of transmission. Crinivirus persistence varies by vector species and declines over time after removal of the insects from the source plants. TICV can be transmitted for up to 3 days by its vector, the greenhouse whitefly. ToCV can be transmitted for up to 5 days by the banded-wing whitefly, 2 days by the silverleaf whitefly, and 1 day by the sweetpotato and greenhouse whiteflies.

Both TICV and ToCV can infect plant species from a moderately wide array of taxonomic families. TICV is known to infect 26 species of plants in eight taxonomic families. These hosts include both food and ornamental crop species, such as artichoke (*Cynara scolymus* L.), lettuce (*Lactuca sativa* L.), petunia (*Petunia* Juss.), potato (*Solanum tuberosum* L.), and zinnia (*Zinnia* L.), along with various wild plant species, such as bristly oxtongue (*Picris echioides* L.), groundsel (*Senecio vulgaris* L.), shepherd's purse (*Capsella bursa-pastoris* (L.) Medik.), Wright or sharp-leaf groundcherry (*Physalis acutifolia* (Miers) Sandw.), and sowthistle (*Sonchus oleraceus* L.) ToCV is known to infect a similar number of species: 24 species in seven taxonomic families. Like TICV, ToCV-susceptible hosts can be wild plants as well as ornamental and vegetable crop species. Susceptible hosts include crop plants such as spinach (*Spinacia oleracea* L.) and pepper (*Capsicum annuum* L.); ornamental species such as periwinkle (*Vinca minor* L.), marigold (*Calendula officinalis* L.), and petunia (*Petunia* Juss.); and wild plant species such as Wright groundcherry and black nightshade (*Solanum nigrum* L.).

Although both viruses infect tomato, other hosts can distinguish the viruses. Hosts for differentiation of TICV and ToCV include lettuce, which is a host of TICV but not ToCV; cultivated tobacco (*Nicotiana glutinosa* L.), which is a host of ToCV but not TICV; and Cleveland's tobacco (*N. clevelandii* A. Gray), which exhibits necrotic flecking and interveinal yellowing when infected by TICV but only mild interveinal chlorosis when infected by ToCV.

Management

Management of criniviruses can be very difficult for several reasons, including the potentially high populations of vectors, the ease of transmission by vectors, and the wide host ranges of both vectors and viruses. No crinivirus-resistant tomato cultivars are commercially available. Management is achieved primarily through the control of vector populations with chemical and cultural practices.

Criniviruses cannot spread without whitefly vectors, and if whitefly populations remain suppressed, the spread of these viruses will be kept to a minimum. Although the application of insecticides can effectively reduce whitefly populations, this method is inefficient for the control of criniviruses, since whiteflies can transmit viruses before being killed by an insecticide. In addition, symptoms of TICV and ToCV do not appear until 3–4 weeks after infection occurs; thus, infection can be widespread before symptoms are observed on the first infected plants. By that time, it is often too late to implement vector control measures. Growers should not rely exclusively on a single chemical for vector control, as doing so can lead to developing insecticide-resistant vector populations.

Another effective management strategy is to remove crop, weed, and ornamental hosts from tomato fields and areas adjacent to tomato production. These hosts can serve as source plants for transmission of TICV or ToCV to tomato by whitefly vectors.

Testing of nursery stock and ornamental host plants for the presence of TICV and ToCV can reduce movement of these viruses to new areas. Since the minimum latent period between infection and development of symptoms is 3 weeks, symptomless plants must be tested to identify seedlings in the early stage of infection.

Selected References

Duffus, J. E., Liu, H.-Y., and Wisler, G. C. 1996. Tomato infectious chlorosis virus—A new clostero-like virus transmitted by *Trialeurodes vaporariorum*. Eur. J. Plant Pathol. 102:219-226.

Wintermantel, W. M., and Wisler, G. C. 2006. Vector specificity, host range, and genetic diversity of *Tomato chlorosis virus*. Plant Dis. 90:814-819.

Wintermantel, W. M., Wisler, G. C., Anchieta, A. G., Liu, H.-Y., Karasev, A. V., and Tzanetakis, I. E. 2005. The complete nucleotide sequence and genome organization of *Tomato chlorosis virus*. Arch. Virol. 150:2287-2298.

Wisler, G. C., Li, R. H., Liu, H.-Y., Lowry, D. S., and Duffus, J. E. 1998. Tomato chlorosis virus: A new whitefly-transmitted, phloem-limited, bipartite closterovirus of tomato. Phytopathology 88:402-409.

(Prepared by J. E. Polston and W. M. Wintermantel)

Tomato mottle virus

Tomato mottle virus (ToMoV) is typical, in many ways, of other tomato-infecting New World begomoviruses. It is bipartite, has a relatively narrow host range, and is transmitted readily by the whitefly vector. Also, like other bipartite begomoviruses, ToMoV produces moderate symptoms in tomato. Yield losses from ToMoV result from a reduction in fruit size, not number. Losses can be severe when infection pressure is high and plants are in an early growth stage.

Symptoms

Symptoms of infection by *Tomato mottle virus* are upward leaf curling, foliar mottling and moderate distortion, and overall stunting of the plant (Fig. 147). Symptoms in tomato usually appear 2–3 weeks after infection, depending on the rate of growth of the plant. Symptoms vary with cultivar and are usually more severe in plants infected at early growth stages. Symptoms may also vary depending on whether the tomato has been bred with genetic resistance to other unrelated viruses (Fig. 148).

Causal Agent

ToMoV is a member of the genus *Begomovirus,* the family Geminiviridae. It first appeared in Florida in the late 1980s and was identified in the early 1990s. ToMoV is believed to have emerged as a variant of *Begomovirus* sequences infecting a weed, possibly a species of the genus *Sida* L. Like many other begomoviruses, ToMoV has a narrow host range, and most of the susceptible hosts belong to the Solanaceae. Also like other begomoviruses, ToMoV is restricted to the phloem of the infected host, where large, spherical, nuclear inclusions are produced.

Disease Cycle and Epidemiology

ToMoV is transmitted in a persistent manner by adult whiteflies (*Bemisia tabaci* (Gennadius)). Approximately 15–30 min of feeding by an adult whitefly is required for acquisition of ToMoV. This is followed by a latent period of approximately 6 h. Transmission can then occur following a single 15- to 30-min feeding period. Adult whiteflies can retain the virus for several weeks, spreading it as far as they are able to move, which can be 16 km or more over land and water. Unlike the begomovirus *Tomato yellow leaf curl virus* (TYLCV), ToMoV is rarely acquired by immature whiteflies that develop on infected plants;

Fig. 147. Field-grown plant infected with *Tomato mottle virus* (ToMoV). (Courtesy D. J. Schuster)

Fig. 148. Comparison of symptoms of *Tomato mottle virus* (ToMoV), a bipartite begomovirus **(A),** with those of *Tomato yellow leaf curl virus* (TYLCV), a monopartite begomovirus **(B).** (Courtesy D. J. Schuster)

the adult stage is the primary stage for acquisition and transmission of ToMoV.

The most important reservoir for ToMoV is older tomato plants. Weeds are not known to play a role in the epidemiology of this virus, although ToMoV is known to replicate in some species. A study of the spatial distribution of infected plants in commercial tomato fields revealed that most infections were caused by primary infection, rather than secondary spread.

Management

In general, management of ToMoV and several other bipartite begomoviruses is more often successful than management of TYLCV. Moreover, losses resulting from ToMoV are generally not as great as those resulting from TYLCV.

Management of ToMoV is not as difficult as that of TYLCV. ToMoV can be managed by the use of neonicotinoid insecticides that are applied in the transplant house and again in the field at transplanting using an amount that provides approximately 8 weeks of protection. Once the neonicotinoid has been exhausted, a rotation of insecticides from different classes and growth regulators should be applied. The tomato crop should be removed promptly at the end of the production season.

Selected References

Abouzid, A. M., Polston, J. E., and Hiebert, E. 1992. The nucleotide sequence of *Tomato mottle virus,* a new geminivirus isolated from tomato in Florida. J. Gen. Virol. 73:3225-3229.

McGovern, R. J., Polston, J. E., Danyluk, G. M., Hiebert, E., Abouzid, A. M., and Stansly, P. A. 1994. Identification of a natural weed host of tomato mottle geminivirus in Florida. Plant Dis. 78:1102-1106.

Polston, J. E., Chellemi, D. O., Schuster, D. J., McGovern, R. J., and Stansly, P. A. 1996. Spatial and temporal dynamics of tomato mottle geminivirus and *Bemisia tabaci* (Genn.) in Florida tomato fields. Plant Dis. 80:1022-1028.

Polston, J. E., Hiebert, E., McGovern, R. J., Stansly, P. A., and Schuster, D. J. 1993. Host range of tomato mottle virus, a new geminivirus infecting tomato in Florida. Plant Dis. 77:1181-1184.

(Prepared by J. E. Polston)

Tomato spotted wilt virus

Tomato spotted wilt virus (TSWV) commonly occurs in the tropics and subtropics, but since the 1990s, it has become a serious disease of tomato and other crops in more temperate zones. The disease was first described in 1915 in South Australia and later determined to be caused by a virus. TSWV caused serious losses in tomato crops in Australia for many years and is still a serious problem in New South Wales. Heavy losses have also been reported from Argentina and Spain, and in the United States, the virus has caused losses in Hawaii and in selected counties in Alabama, Arkansas, Florida, Georgia, Louisiana, Mississippi, North Carolina, South Carolina, and Tennessee.

Symptoms

Symptoms of tomato spotted wilt vary, but in most cases, young leaves turn bronze and later develop numerous small, dark spots. Growing tips may die back, and stems of terminals may become streaked (Fig. 149). An affected plant may develop a one-sided growth habit, or the entire plant may be stunted and have drooping leaves, suggesting a wilt disease.

Tomato plants infected early in the season may produce no fruit, while those infected after fruit set produce fruit with chlorotic ringspots. Green fruit have slightly raised areas with faint, concentric rings; on ripe fruit, these rings become more obvious, becoming red and white or red and yellow (Fig. 150). The chlorotic lesions are difficult to observe at the "breaker" stage of picking but are highly visible at full color.

Causal Agent

TSWV is the type species of the genus *Tospovirus* in the family Bunyaviridae. Virions are 80–110 nm in diameter and membrane bound, with four viral encoded structural proteins that are surrounded by a lipid bilayer with surface glycoprotein (GP1 and GP2). The TSWV genome consists of three linear, single-stranded (ss) RNA molecules, referred to as large, medium, and small segments. The viral nucleocapsid (N) protein encapsidates each viral RNA into a pseudocircular nucleocapsid structure. TSWV is extremely variable biologically and serologically. Strains from different areas of the world may differ in their reactions to antibodies against viral structural proteins. TSWV is sap transmissible but quite unstable in vitro, especially at a pH lower than 5.5.

TSWV has one of the widest host ranges of any virus, infecting an estimated 800 plant species in more than 80 plant families (both dicots and monocots). The host range includes many vegetable crops in addition to tomato, such as potato (*Solanum tuberosum* L.), pepper (*Capsicum annuum* L.), celery (*Apium* L.), eggplant (*S. melongena* L.), lettuce (*Lactuca sativa* L.), and many legumes (Fabaceae (Leguminosae)), along with both annual and perennial ornamentals and weed species.

The inclusion bodies of TSWV appear as virus particles with swollen endoplasmic reticulum, which usually maintain a loose network but may be quite compact structures (Fig. 151). The structure and staining features appear to hold across a broad

Fig. 149. Symptoms of *Tomato spotted wilt virus* (TSWV) on foliage. (Courtesy L. L. Black)

Fig. 150. Symptoms of *Tomato spotted wilt virus* (TSWV) on fruit. (Courtesy J. E. Thomas)

host range (including ornamentals), which adds to their diagnostic value.

Disease Cycle and Epidemiology

Important reservoirs of TSWV include species of weeds and perennial ornamentals. The virus and its vectors are frequently spread with the transport of ornamentals and transplants, which may explain, in part, *Tospovirus* diversity in the world.

Natural spread in the field is caused by adult thrips transmitting TSWV in a propagative manner. Seven species are reported as vectors. The two most important species are the western flower thrips (*Frankliniella occidentalis* (Pergande)) and the tobacco thrips (*F. fusca* (Hinds)). Other species that vector TSWV are the tomato thrips (*F. schultzei* (Trybom)), the European flower thrips (*F. intonsa* (Trybom)), the Florida flower thrips (*F. bispinosa* (Morgan)), and the light-brown soybean thrips (*Thrips setosus* Moulton), along with some strains of the onion thrips (*T. tabaci* Lindeman).

Thrips can acquire the virus only during the larval stage, but the virus persists through insect molts to the adult stage and replicates within the insect. Thereafter, adults can transmit the virus during their entire lives. The first site of TSWV infection in larval thrips is the midgut, but the virus must circulate to the salivary gland for transmission to plants by adults to occur. (Interestingly, TSWV ingested by the adult western flower thrips is not retained.) The minimum acquisition period by larvae is 30 min or shorter, and the efficiency of transmission increases with feeding time. A latent period of approximately 3–7 days is required before transmission can occur.

The transmission threshold is 5 min or less, and infection is achieved by shallow feeding on leaf epidermal cells. The western flower thrips prefers to feed on flowers. Tomato plant age can affect feeding on foliage, with younger foliage being more attractive to the tobacco thrips than the western flower thrips. Thrips ovipositional preference varies among hosts, with the tobacco thrips preferring chickweed (*Stellaria media* (L.) Vill.) over tomato. Seed transmission is not known to occur.

Management

Because of the wide host range of TSWV and its vectors and the presence of TSWV in perennial ornamentals and weeds, managing tomato spotted wilt is very difficult. Given this, ornamental stock plants should never be grown in the same greenhouse as vegetable transplants started from seed.

TSWV overwinters in a relatively few abundant winter annual weeds, and dispersal of infectious thrips from these sources to susceptible crops and weeds occurs during a brief period in the spring. To minimize the level of infection, this information should be used in decision making about cultivar selection and planting date of the crop. An integrated management approach has been used successfully in several crops and locations.

Insecticides have been used to reduce larval development and thus limit secondary virus spread. The use of ultraviolet (UV) reflective mulches, a plant activator (acibenzolar-S-methyl), and insecticides has provided excellent management of TSWV in commercial tomato fields.

Resistant cultivars with single dominant gene resistance (*Sw-5*) are currently being used to reduce losses to TSWV. Symptoms appear more visible on fruit (immature and ripening) than on foliage, thus limiting the commercial usefulness of these cultivars. Unfortunately, resistance-breaking isolates of TSWV can break both conventional and transgenic resistance.

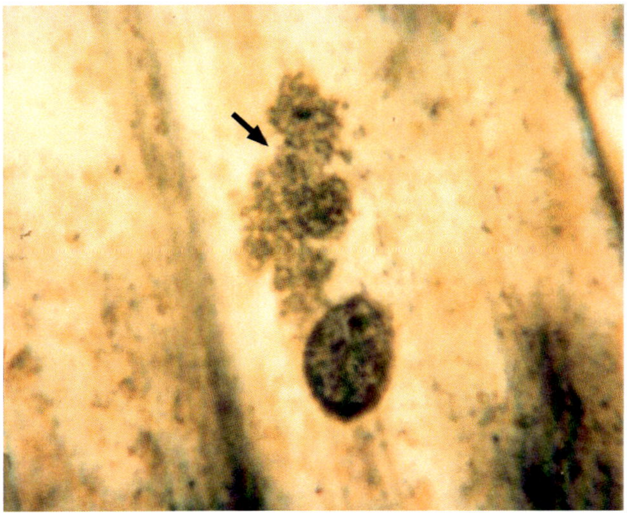

Fig. 151. Inclusion bodies of *Tomato spotted wilt virus* (TSWV). (Courtesy R. G. Christie)

Selected References

Aramburu, J., and Marti, M. 2003. The occurrence in north-east Spain of a variant of *Tomato spotted wilt virus* (TSWV) that breaks resistance in tomato (*Lycopersicon esculentum*) containing the *Sw-5* gene. Plant Pathol. 52:407.

Cho, J. J., Mau, R. F. L., German, T. L., Hartmann, R. W., Yudin, L. S., Gonsalves, D., and Provvidenti, R. 1989. A multidisciplinary approach to management of tomato spotted wilt virus in Hawaii. Plant Dis. 73:375-383.

Finlay, K. W. 1953. Inheritance of spotted wilt resistance in the tomato. II. Five genes controlling spotted wilt resistance in four tomato types. Aust. J. Biol. Sci. 6:153-163.

Groves, R. L., Walgenbach, J. F., Moyer, J. W., and Kennedy, G. G. 2002. The role of weed hosts and tobacco thrips, *Frankliniella fusca*, in the epidemiology of *Tomato spotted wilt virus*. Plant Dis. 86:573-582.

Ie, T. S. 1970. Tomato spotted wilt virus. Descriptions of Plant Viruses, no. 39. C.A.B. International Mycological Institute, Kew, England.

Momol, T. M., Olson, S. M., Funderburk, J. E., Stavisky, J., and Marois J. J. 2004. Integrated management of tomato spotted wilt on field-grown tomatoes. Plant Dis. 88:882-890.

Parrella, G., Gognalons, P., Gebre-Selassie, K., Vovlas, C., and Marchoux, G. 2003. An update on the host range of *Tomato spotted wilt virus*. J. Plant Pathol. 85:227-264.

Sherwood, J. L., German, T. L., Whitfield, A. E., Moyer, J. W., and Ullman, D. E. 2000. Tospoviruses. Pages 1034-1040 in: Encyclopedia of Plant Pathology. O. C. Maloy and T. D. Murray, eds. John Wiley & Sons, New York.

(Prepared by T. A. Zitter and T. M. Momol)

Tomato torrado virus, Tomato apex necrosis virus, *Tomato marchitez virus*, and Tomato chocolate spot virus

During approximately 2003–2004, descriptions of unusual necrosis symptoms in tomato were reported from diverse geographic regions, including the Murcia region of southeastern Spain, the Sinaloa state of northwestern Mexico, and the central area of Guatemala. The disease was referred to as tomato *torrado* (meaning "burned") in Spain, tomato apex necrosis or *marchitez* (meaning "withered") in Mexico, and *mancha de chocolate* (meaning "chocolate spot") in Guatemala.

The symptoms reported from these locations were generally similar and included necrotic spots or lesions in the basal areas of leaflets, which developed into necrotic areas on leaves, stems, and petioles. Eventually, the necrosis spread to and caused a dieback of the shoot apex (hence, the term *tomato apex necrosis*). In some cases, fruit developed necrotic or corky rings and patches, although this symptom tended to be less frequent, particularly in Guatemala.

Also in Guatemala, high incidences of chocolate spot disease were associated with high populations of the greenhouse whitefly. These symptoms were initially thought to be caused by *Tomato spotted wilt virus* (TSWV). However, tests failed to detect TSWV or other previously characterized tomato-infecting viruses, and symptoms of chocolate spot disease developed in TSWV-resistant cultivars (i.e., those carrying the *Sw-5* gene).

Researchers determined that these diseases were caused by a new type of picorna-like virus. Furthermore, the disease in each geographical area (Spain, Mexico, and Guatemala) was caused by a distinct species of this new virus. Thus, in Spain, the virus has been named *Tomato torrado virus* (ToTV); in Mexico, it is called Tomato apex necrosis virus (ToANV) or *Tomato marchitez virus* (ToMarV); and in Guatemala, it is referred to as Tomato chocolate spot virus (ToCSV).

Symptoms

Susceptible tomato plants initially show a mild mottling or crumpling, which develops into necrotic spots that are most pronounced at the bases of infected leaflets (Fig. 152). The necrosis spreads rapidly to shoot apexes, where dieback may occur, and to the stems and petioles (Fig. 153). Green fruit may develop necrotic rings or patches (Fig. 154). Severely infected plants are stunted, and economic loss can result because of reduced fruit yield and quality.

Fig. 152. Initial symptom of Tomato chocolate spot virus (ToCSV): the development of necrotic spots at the bases of infected leaflets. (Courtesy R. L. Gilbertson)

Causal Agent

All three viral species—ToTV, ToANV/ToMarV, and ToCSV—share the following properties: spherical/isometric virions approximately 28 nm in diameter; two genomic single-stranded, plus-sense polyadenylated RNA molecules of approximately 7.0 (RNA1) and 5.0–5.4 (RNA2) kilobases; and multiple (at least three) structural (capsid) proteins. Analysis of the genome sequences of the three viral species has revealed a polyprotein genome strategy. Based on identification of conserved motifs in the predicted polyproteins, RNA1 encodes a helicase, a protease, and an RNA-dependent RNA polymerase, whereas RNA2 encodes a putative movement protein and the capsid proteins. Together, these results indicate that the viruses are picorna-like (i.e., small, spherical RNA viruses), and phylogenetic analyses reveal only distant relationships with previously identified plant picorna-like viruses. Thus, a new genus, *Torradovirus*, has been proposed for these viruses.

The means of plant-to-plant spread of these viruses remains unclear, but disease incidence has been associated with the presence of whiteflies, including the greenhouse whitefly (*Trialeurodes vaporariorum* Westwood) in Guatemala and *Bemisia tabaci* (Gennadius) in Mexico. A report describing ToTV from Poland indicates that the virus is efficiently transmitted by *T. vaporariorum*. However, there is presently no information on the mode of transmission (i.e., nonpersistent versus persistent) of ToTV or the means of transmission of ToANV/ToMarV or ToCSV. These viruses are sap (mechanically) transmissible, although the efficiency of transmission varies depending on the host, with high efficiency to *Nicotiana benthamiana* Domin (a close relative of tobacco) and lower efficiency to tomato.

The host range of the viruses is primarily members of Solanaceae, including tomato; several species of tobacco (*N. benthamiana*, *N. glutinosa* L., and others); husk tomato (*Physalis pubescens* var. *pubescens* L.); and jimsonweed (*Datura stramonium* L.). For ToTV, hosts also include pepper (*Capsicum annuum* L.) and potato (*Solanum tuberosum* L.). ToTV also infects a wide range of weed species, based on a report from Spain, with some species showing symptoms and others remaining asymptomatic.

Disease Cycle and Epidemiology

The almost simultaneous appearance of these diseases in distinct geographical locations raised numerous questions regarding their epidemiology. To date, there is no evidence of seed transmission of these viruses. Moreover, the fact that the

Fig. 153. Extensive necrosis of leaves, petioles, and shoot apices in a plant infected with Tomato chocolate spot virus (ToCSV). (Courtesy R. L. Gilbertson)

Fig. 154. Fruit with symptoms of necrosis or corky ring spot, associated with *Tomato marchitez virus* (ToMarV) infection. (Courtesy C. Sandlin)

sequences of ToTV, ToMarV/ToANV, and ToCSV are approximately 80% identical suggests local emergence, rather than global spread of a single isolate via seed. (RNA1 sequences of ToMarV and ToANV are approximately 95% identical; thus, they are probably isolates of the same species.) These viruses likely exist in local reservoirs (e.g., weeds or other crops) and are introduced into tomato fields by whiteflies.

Additional evidence shows that certain cultivars of tomato are highly susceptible to these viruses. Thus, the emergence of the viruses may have been enhanced by a combination of the widespread planting of susceptible tomato cultivars and outbreaks of large populations of whiteflies.

Further spread of the viruses would be facilitated by the movement of infected tomato transplants or plants, propagative materials of other solanaceous hosts (e.g., potato tubers, in the case of ToTV in Europe), and even viruliferous whiteflies. This could explain the detection of ToTV in Hungary, Poland, and the Canary Islands (off Spain). Fortunately, a number of hybrid tomato cultivars with high levels of resistance to ToTV and ToMarV/ToANV have been identified, which may help limit the spread and economic importance of these viruses.

Management

Hybrid cultivars of tomato with resistance to ToTV and ToMarV/ToANV have been identified and are commercially available. These cultivars should be planted in areas known to have high virus pressure, and cultivars with high known susceptibility should be avoided.

Production of pathogen-free transplants is critical. This can be accomplished by keeping transplants free of whiteflies through production in greenhouses or screenhouses and use of new systemic insecticides (e.g., neonicotinoids) to eliminate or prevent whitefly infestations. In areas with high whitefly populations, a second application of systemic insecticides during transplanting is advisable, or if economically feasible, young plants can be protected with row covers (e.g., Agrobon). During the growing season, whitefly populations should be monitored and managed according to an integrated pest management (IPM) strategy. In addition, weeds should be controlled in and or around transplant greenhouses and screenhouses.

Cultural practices should be followed, as well, such as avoiding sequential plantings of tomatoes (especially plantings of new fields adjacent to old fields known to have infected plants) and controlling weeds in fields. In addition, diseased plants should be rogued, although doing so is not often practical or effective. Following harvest, sanitation is essential; plants should be promptly destroyed by deep plowing or other means. In tropical and subtropical regions, maintaining a tomato-free period of 2–3 months may be effective in reducing inoculum levels.

Selected References

Alfaro-Fernández, A., Córdoba-Sellés, C., Cebrián, M. C., Herrera-Vásquez, J. A., Sánchez-Navarro, J. A., Juárez, M., Espino, A., Martín, R., and Jordá, C. 2008. First report of Tomato torrado virus on weed hosts in Spain. Plant Dis. 92:831.

Batuman, O., Kuo, Y.-W., Palmieri, M., Rojas, M. R., and Gilbertson, R. L. 2009. Tomato chocolate spot virus, a member of a new torradovirus species that causes a necrosis-associated disease of tomato in Guatemala. Arch. Virol. 155:857-869.

Pospieszny, H., Borodynko, N., Obrępalska-Stęplowska, A., and Hasiów, B. 2007. The first report of Tomato torrado virus in Poland. Plant Dis. 91:1364.

Turina, M., Ricker, M. D., Lenzi, R., Masenga, V., and Ciuffo, M. 2007. A severe disease of tomato in the Culiacan area (Sinaloa, Mexico) is caused by a new picorna-like viral species. Plant Dis. 91:932-941.

Verbeek, M., Dullemans, A. M., van den Heuvel, J. F. J. M., Maris, P. C., and van der Vlugt, R. A. A. 2007. Identification and characterization of tomato torrado virus, a new plant picorna-like virus from tomato. Arch. Virol. 152:881-890.

Verbeek, M., Dullemans, A. M., van den Heuvel, J. F. J. M., Maris, P. C., and van der Vlugt, R. A. A. 2008. Tomato marchitez virus, a new plant picorna-like virus from tomato related to tomato torrado virus. Arch. Virol. 153:127-134.

(Prepared by O. Batuman, Y.-W. Kuo, M. Palmieri, and R. L. Gilbertson)

Tomato yellow leaf curl virus

Tomato yellow leaf curl virus (TYLCV) is a monopartite begomovirus and one of the most devastating viruses to infect tomato. A tomato crop can be severely affected by TYLCV despite daily applications of insecticides, and yield losses of 100% have been recorded in cases in which the whitefly population was high.

TYLCV was first described in the region that is now Israel and was originally found only in the eastern Mediterranean. In the late 1990s, the virus became of concern in tomato production in the western Mediterranean, as well as the Caribbean, Japan, Mexico, and the southern United States. This geographic expansion has been attributed to the movement of TYLCV-infected transplants and fruit. Commercial cluster tomatoes (i.e., fruit plus vine) harvested from TYLCV-infected plants have been shown to be a source of TYLCV for whiteflies, and the shipment of such fruit has been implicated as a means of distributing the virus. Movement of tomato transplants for both the retail and commercial sectors was significant in the rapid distribution of TYLCV throughout Florida. Moreover, in theory, the virus can be distributed via the movement of infected but asymptomatic host plants.

Symptoms

The symptoms of *Tomato yellow leaf curl virus* are severe, and consequently, the virus can have a catastrophic impact on tomato production in many locations. TYLCV symptoms appear 10 days to 3 weeks after infection (depending on the rate of plant growth) and include severe stunting, marked reduction in leaf size, upward cupping and chlorosis of leaf margins, mottling, flower abscission, and significant yield reduction (Fig. 155). Although the fruit size may be normal on infected plants, the fruit number may be dramatically reduced because of the high rate of flower abscission.

Causal Agent

TYLCV belongs to the genus *Begomovirus* and the family Geminiviridae. Both mild and severe strains of TYLCV have been identified, and their DNA sequences are approximately 94% identical. The mild isolate (TYLCV-Mld) occurs in the eastern and western Mediterranean, and although it is described as "mild," the symptoms it causes in tomato are nearly indistinguishable from those of the severe strain. The severe strain of TYLCV is the result of a recombination event between TYLCV-Mld and another *Begomovirus* ancestor. The severe strain is the most widely distributed of the two and thus the strain found outside the Mediterranean.

Disease Cycle and Epidemiology

TYLCV is transmitted in a persistent manner by adult whiteflies (*Bemisia tabaci* (Gennadius)). Approximately 15–30 min of feeding by an adult whitefly is required to acquire TYLCV. This is followed by a latent period of approximately 6 h. Transmission can then occur following a single 15- to 30-min feeding period. An adult whitefly can retain the virus for several weeks, spreading the virus as far as it ranges to feed. TYLCV has been shown to be acquired by immature whiteflies developing on infected plants. However, adults are responsible for the spread of virus to healthy susceptible host plants. Reservoirs of the vector vary somewhat across production regions. Collectively, members of *B. tabaci* have a large number of hosts on which to feed and reproduce. In addition, whiteflies are believed to be able to move long distances over land and water (i.e., at least 16 km), using wind-assisted flight.

Although known primarily as a pathogen of tomato, TYLCV has one of the largest described host ranges within the genus *Begomovirus*. It is capable of infecting more than 30 species in more than 12 plant families (although about half of these hosts have been identified in reports from greenhouse experiments). Most of the known hosts belong to the Solanaceae family and include cultivated vegetable and ornamental crops, as well as wild and weed plant species. Crop plants include bean (*Phaseolus vulgaris* L.), petunia (*Petunia* Juss.), and lisianthus or prairie gentian (*Eustoma grandiflorum* (L.) Salisb. ex. G. Don subsp. *russellianum* (Hook.) Kartesz). As such, TYLCV reservoirs may vary across production regions. Furthermore, TYLCV is able to establish symptomless infections in both wild and non-tomato crop species, so reservoirs of TYLCV may not always be obvious. For example, pepper (*Capsicum annuum* L.), a symptomless host, has been shown to be a reservoir of TYLCV for tomato in Florida but not in Spain.

Older tomato fields remain the most important source of both TYLCV and whiteflies for young tomato plants. Whiteflies move readily from old plants to young plants, and begomoviruses can spread quite rapidly in older or abandoned fields, where whitefly management has ceased. Early infections result in greater yield losses than late infections. In addition, although resistant cultivars produce acceptable yields after TYLCV infection, they still support viral replication and can act as reservoirs of TYLCV for susceptible crops.

Fig. 155. Comparison of symptoms from infection by a bipartite begomovirus on an early tomato transplant **(A)** with infection of a mature plant with *Tomato yellow leaf curl virus* (TYLCV) **(B).** (Courtesy D. P. Maxwell)

Management

Management of TYLCV is difficult, often requiring significant changes in production and management practices and reduced expectations for crop yield. TYLCV can be managed (although not always successfully) through the simultaneous application of multiple approaches.

The most economical control of TYLCV for both protected and open field production can be achieved with the use of resistant cultivars. In fact, however, cultivars are limited and not available for all production conditions, climates, and market preferences. Also, resistance in these cultivars can be overcome by early and high inoculation pressures, so growers must use additional approaches even with the use of resistance cultivars.

In protected production, growers should rely mainly on exclusion through the use of whitefly-proof screening and ultraviolet (UV)-absorbing plastic screens and covers. In open field production, growers should rely heavily on a rotation of pesticides to reduce whitefly populations, frequently using a neonicotinoid (i.e., systemic insecticide) that kills all life stages of whiteflies. However, the development of pesticide resistance and the loss of natural predators and parasites after repeated insecticide applications have contributed to control problems and environmental concerns. Additional open field practices include roguing, timing of planting to avoid high whitefly populations, allowing crop-free periods, using reflective plastic mulches, and sanitation.

Selected References

Cohen, J., Gera, A., Ecker, R., Joseph, R. B., Perlsman, M., Gokkes, M., Lachman, O., and Antignus, Y. 1995. Lisianthus leaf curl a new disease of lisianthus caused by tomato yellow leaf curl virus. Plant Dis. 79:416-420.

Cohen, S., and Antignus, Y. 1994. Tomato yellow leaf curl virus, a whitefly-borne geminivirus of tomatoes. Adv. Dis. Vector Res. 10:259-288.

Cohen, S., and Nitzany, F. E. 1966. Transmission and host range of the tomato yellow leaf curl virus. Phytopathology 56:1127-1131.

Czosnek, H., ed. 2007. Tomato Yellow Leaf Curl Virus Disease: Management, Molecular Biology, Breeding for Resistance. Springer, Dordrecht, the Netherlands.

Delatte, H., Dalmon, A., Rist, D., Soustrade, I., Wuster, G., Lett, J. M., Goldbach, R. W., Peterschmitt, M., and Reynaud, B. 2003. *Tomato yellow leaf curl virus* can be acquired and transmitted by *Bemisia tabaci* (Gennadius) from tomato fruit. Plant Dis. 87:1297-1300.

Hilje, L., Costa, H. S., and Stansly, P. A. 2001. Cultural practices for managing *Bemisia tabaci* and associated viral diseases. Crop Prot. 20:801-812.

Jones, D. R. 2003. Plant viruses transmitted by whiteflies. Eur. J. Plant Pathol. 109:195-219.

Lapidot, M., and Friedmann, M. 2002. Breeding for resistance to whitefly-transmitted geminiviruses. Ann. Appl. Biol. 140:109-127.

Navas-Castillo, J., Sánchez-Campos, S., Antonio-Díaz, J., Sáez-Alonso, E., and Moriones, E. 1999. Tomato yellow leaf curl virus-Is causes a novel disease of common bean and severe epidemics of tomato in Spain. Plant Dis. 83:29-32.

Navas-Castillo, J., Sánchez-Campos, S., Noris, E., Lourno, D., Accotto, G. P., and Moriones, E. 2000. Natural recombination between *Tomato yellow leaf curl virus*-Is and *Tomato leaf curl virus*. J. Gen. Virol. 81:2797-2801.

Navot, N., Pichersky, E., Zeidan, M., Zamir, D., and Czosnek, H. 1991. Tomato yellow leaf curl virus: A whitefly-transmitted geminivirus with a single genomic component. Virology 185:151-161.

Polston, J. E., and Anderson, P. K. 1997. The emergence of whitefly-transmitted geminiviruses in tomato in the Western Hemisphere. Plant Dis. 81:1358-1369.

Polston, J. E., Cohen, L., Sherwood, T. A., Ben-Joseph, R., and Lapidot, M. 2006. *Capsicum* species: Symptomless hosts and reservoirs of *Tomato yellow leaf curl virus*. Phytopathology 96:447-452.

Polston, J. E., McGovern, R. J., and Brown, L. G. 1999. Introduction of Tomato yellow leaf curl virus in Florida and implications for the spread of this and other geminiviruses of tomato. Plant Dis. 83:984-988.

Salati, R., Nahkla, M. K., Rojas, M. R., Guzman, P., Jaquez, J., Maxwell, D. P., and Gilbertson, R. L. 2002. *Tomato yellow leaf curl virus* in the Dominican Republic: Characterization of an infectious clone, virus monitoring in whiteflies, and identification of reservoir hosts. Phytopathology 92:487-496.

(Prepared by J. E. Polston)

Tomato yellow mosaic virus

Tomato yellow mosaic was first described in 1963 as a "yellowish mosaic" of tomato in Venezuela. By 1978, the disease had spread throughout the four main tomato-producing states of Venezuela, and the causal agent was named *Tomato yellow mosaic virus* (ToYMV). In 1981, ToYMV was reported to be an occasional pathogen of potato (*Solanum tuberosum* L.) in the state of Aragua but always in the vicinity of ToYMV-infected tomatoes.

Five years later, in independent research by British plant pathologists, ToYMV was isolated from an infected potato plant in Venezuela. These researchers were the first to characterize the virus at the molecular level. They described it as "a new virus infecting potatoes" and named it *Potato yellow mosaic virus* (PYMV). Subsequent research conducted in 2001 clearly showed that PYMV and ToYMV are the same virus, yet both names continue to appear in the literature.

Tomato yellow mosaic remains one of the most important viral diseases of tomato in the Caribbean Basin, including Colombia, Panama, Venezuela, and several islands in the Caribbean Sea. ToYMV has recently moved into the Andean region of central and western Colombia, causing serious outbreaks in tomato fields located 1,000–1,700 m above sea level. ToYMV has not been reported to infect potatoes in this region, however.

Symptoms

Tomato yellow mosaic virus induces varying degrees of systemic yellowing in infected tomato plants, ranging from intense to mild depending on the genotype and environmental conditions. Leaf curling is a consistent symptom induced by ToYMV, and plants may also show different degrees of stunting (Fig. 156).

Fruit production can be severely affected in some tomato cultivars. This is particularly likely when plants are infected at an early stage of development—for instance, in unprotected seedbeds before transplanting.

Causal Agent

ToYMV has characteristic twinned or geminate virion morphology formed by two quasi-isometric particles (approximately 22 × 38 nm in diameter) that encapsidate two single-stranded DNA components (approximately 2,500–2,600 nucleotides long). This virus is transmitted by the whitefly *Bemisia tabaci* (Gennadius) and is thus a species of the genus *Begomovirus,* the family Geminiviridae.

Potato yellow mosaic is the original and true name of (and thus a synonym for) tomato yellow mosaic. First described in the early 1960s, it was the second disease of tomato in the Americas induced by a begomovirus. ToYMV probably evolved from the original begomoviruses that infect plants in the family Malvaceae in the tropical Americas. Consequently, ToYMV is commonly found with other tropical begomoviruses belonging to related phylogenetic clusters.

Disease Cycle and Epidemiology

Tomato yellow mosaic occurs wherever the virus and its vector coexist. Thus, the disease affects tomato in tropical regions of Latin America located less than 1,200 m above sea level, where *B. tabaci* can thrive. However, ToYMV has been found in tomato fields above this altitude in Colombia, resulting from the transport of virus-infected or whitefly-infested seedlings from nurseries in the lowlands. Disease incidence under normal conditions depends on the population dynamics of the whitefly vector. *B. tabaci* thrives under warm and fairly dry conditions, which usually occur in the tropical Americas during those months of the year with a low average rainfall and cloud cover. The arrival of the more aggressive, polyphagous, and prolific silverleaf whitefly in the tropical Americas in the 1990s increased the incidence of tomato yellow mosaic in this region.

ToYMV has been reported to infect different tomato and potato species under natural conditions. The experimental host range includes jimsonweed (*Datura stramonium* L.), several species of tobacco (*Nicotiana benthamiana* Domin, *N. glutinosa* L., *N. tabacum* L.), apple of Peru (*Nicandra physalodes* (L.) Scop.), petunia (*Petunia hybrida*), and Peruvian groundcherry (*Physalis peruviana* L.)—all of which react with different degrees of yellowing. ToYMV is transmitted by mechanical means but not through seed.

Management

Insecticide abuse has been identified as the main factor conditioning development of a high whitefly population capable of transmitting begomoviruses, such as ToYMV. Moreover, despite tomato's Latin American origin, it has been improved mainly in temperate regions; thus, most tomato cultivars grown in Latin America are not resistant to ToYMV.

Current integrated pest management (IPM) projects in the affected region seek to educate farmers on the proper use of insecticides and to promote the use of new systemic insecticides at the initial stages of plant development. The physical and chemical protection of tomato seedlings is the most important step in reducing the damage caused by ToYMV. The new systemic insecticides are also recommended for the protection of tomato fields during the first 30–45 days of plant development. Providing physical protection using synthetic fabrics as barriers or row covers during all or part of the plants' growth cycle is common among tomato farmers whose crops are affected by whitefly-borne viruses, such as ToYMV.

Fig. 156. Foliar malformation, induced by *Tomato yellow mosaic virus* (ToYMV). (Courtesy F. J. Morales)

Cultural practices, such as planting tomatoes during or immediately after the rainy season, are also effective in terms of avoiding high whitefly populations, as long as other pathogens that thrive under humid conditions are controlled. Other cultural practices recommended for lowering the incidence of tomato begomoviruses include implementing a tomato or virus/vector host-free period and avoiding sequential plantings of tomato.

Selected References

Debrot, C. E. F., Herold, F., and Dao, F. 1963. Nota preliminar sobre un "mosaico amarillento" del tomate en Venezuela. (In Spanish.) Agron. Trop. (Maracay) 13:33-41.

Debrot, E. A., and Centeno, F. 1985. Infección natural de la papa en Venezuela con el mosaico Amarillo del tomate, un geminivirus transmitido por moscas blancas. (In Spanish.) Trop. Agric. 35:125-138.

Lastra, J. R., and de Uzcategui, R. C. 1975. Viruses affecting tomatoes in Venezuela. Phytopathol. Z. 84:253-258.

Morales, F. J. 2006. History and current distribution of begomoviruses in Latin America. Adv. Virus Res. 67:127-162.

Morales, F. J., Lastra, R., de Uzcategui, R. C., and Calvert, L. 2001. Potato yellow mosaic virus: A synonym of *Tomato yellow mosaic virus*. Arch. Virol. 146:2249-2253.

Roberts, E. J. F., Buck, K. W., and Coutts, R. H. A. 1986. A new geminivirus infecting potatoes in Venezuela. Plant Dis. 70:603.

(Prepared by F. J. Morales)

Minor Virus Diseases

As noted earlier, Table 6 (pages 73–75) identifies more than 90 virus diseases reported to infect tomato. Most occur annually but generally have limited geographical distribution. The diverse symptomatology associated with some of these diseases is illustrated by Figures 157 through 166.

(Prepared by T. A. Zitter)

Fig. 157. Symptoms of *Eggplant mosaic virus* (EMV) on tomato plants. (Courtesy O. Gracia)

Fig. 159. Symptoms of *Pelargonium zonate spot virus* (PZSV) on tomato foliage. (Courtesy C. Vovlas)

Fig. 158. Symptoms of *Pepper veinal mottle virus* (PVMV) on tomato plants. (Courtesy J. Lapido)

Fig. 160. Symptoms of *Pelargonium zonate spot virus* (PZSV) on tomato fruit. (Courtesy C. Vovlas)

Fig. 161. Symptoms of *Potato leafroll virus* (PLRV) on tomato foliage. (Courtesy T. A. Zitter)

Fig. 162. Symptoms of *Potato leafroll virus* (PLRV) on tomato foliage. (Courtesy J. E. Thomas)

Fig. 163. Symptoms of *Tobacco streak virus* (TSV) on tomato stems and foliage. (Courtesy J. C. Watterson)

Fig. 164. Symptoms of Tomato double-virus streak (ToMV + PVX) on tomato foliage and fruit. (Courtesy T. A. Zitter)

Fig. 165. Symptoms of *Tomato pseudo-curly top virus* (ToSCTV) on a tomato plant. (Courtesy J. P. Jones)

Fig. 166. Symptoms of Tomato necrotic dwarf virus (ToNDV), transmitted by whiteflies (unknown classification), on a tomato plant. (Courtesy W. M. Wintermantel)

Diseases Caused by Viroids

Viroids are the smallest known plant pathogens, each consisting of a single-stranded RNA molecule of 246–399 nucleotides. A viroid lacks a protein capsid, and no messenger RNA activity has been detected.

Viroids are classified into two families: Avsunviroidae and Pospiviroidae. Within the latter family, five species from the genus *Pospiviroid* have been isolated from naturally infected tomatoes (Table 7):

1. a strain of *Citrus exocortis viroid* (CEVd)
2. *Potato spindle tuber viroid* (PSTVd)
3. *Tomato apical stunt viroid* (TASVd)
4. *Tomato chlorotic dwarf viroid* (TCDVd)
5. *Tomato planta macho viroid* (TPMVd)

Columnea latent viroid (CLVd) has also been shown to infect tomato; the isolates that have been detected show genetic similarity to both the *Hostuviroid* and *Pospiviroid* genera. The *Hostuviroid* genus, with the type species *Hop stunt viroid*, infects primarily hop (*Humulus lupulus* L.) and cucumber (*Cucumis sativus* L.; syn. Cucumber pale fruit viroid) and can infect tomato without developing symptoms.

Symptoms

The symptoms caused by viroids on tomato are generally similar to those observed with PSTVd. TPMVd has caused serious losses in commercial tomato regions of Mexico; infected plants produce marble-sized fruit that are unmarketable (Fig. 167). TCDVd also causes the production of very small fruit (Fig. 168). To date, the viroid has been detected only in commercial greenhouse tomato in the Netherlands and in ornamental petunia (*Petunia* Juss.) in several countries. Stunted apical growth is characteristic of TASVd. TASVd-Is has caused losses in tomato greenhouse production in Israel since 1999. The economic effects of the other viroids on tomato production have not been documented.

Causal Organisms

Viroids are low-molecular-weight, circular-RNA molecules. The nucleotide sequence for most of the known viroids (more than 20) has been elucidated. Isolates with greater than 90% sequence similarities are considered variants of a particular viroid, whereas those with less than 90% similarity are considered distinct species.

Polyacrylamide gel electrophoresis can be used to detect subnanogram amounts of viroid in plant tissue, but it is labor intensive and thus unsuitable for routine testing of a large number of samples. Because viroids do not possess a protein capsid, enzyme-linked immunosorbent assay (ELISA) and other antibody-based assays are not applicable. Nucleic acid hybridization and reverse transcription polymerase chain reaction (RT-PCR) systems (including real-time RT-PCR) have become the methods of choice, even though preparation of samples can be challenging. Detection using biological indexing is possible but labor intensive, and molecular methods are usually superior.

Disease Cycle and Epidemiology

All viroids are mechanically transmissible. They are transmitted from plant to plant by workers and farm equipment, as well as by vegetative propagation, foliar contact, and even bumble bees (*Bombus* spp.). Attempts to demonstrate transmission of TASVd by aphids and whiteflies have been unsuccessful, but aphids have been implicated in the transmission of other viroids. Vertical transmission can occur in some cases through pollen and true seed. Pepino (*Solanum muricatum* Aiton) seed imported from New Zealand and Greece may have been the source of PSTVd for greenhouse-grown tomato in the Netherlands. The viroids vary in their ability to infect and replicate in different plant species. PSTVd can replicate in approximately 160 hosts, most of which are members of Solanaceae.

TABLE 7. Viroids (Genus *Pospiviroid*, Family Pospiviroidae) Found in Tomato

Viroid	Acronym	Year(s) Identified	Source(s)	Symptoms	Reference(s)
Citrus exocortis viroid	CEVd	1991	India	As for PSTVd.	Mishra et al., 1991
(str. Indian tomato bunchy top viroid)	ITBTVd				
Columnea latent viroid	CLVd	2004	Netherlands, United Kingdom	As for PSTVd.	Verhoeven et al., 2004
Potato spindle tuber viroid	PSTVd	1971	North America, South America, Soviet Union	Stunting, apical leaf proliferation and narrowing, chlorosis, necrosis, crinkling, and brittleness of leaf tissue. Severity of symptoms varies with strain.	Diener et al., 1971; Mumford et al., 2004; Puchta et al., 1990; Verhoeven et al., 2007
(str. Tomato strain viroid)	PSTVd-N	1988	Australia, Brazil, China, Crete, Europe (East and West), India, Netherlands, New Zealand, South Africa		
Tomato apical stunt viroid	TASVd	1981	Indonesia, Israel, Ivory Coast, Senegal, Tunisia	As for PSTVd with differences in symptomology and host range for isolates. Nucleotide sequence of Ivory Coast type strain differs from that of Israel strain. Israel and Indonesia strains similar.	Antignus et al., 2002; Candresse et al., 2007; Verhoeven et al., 2006; Walter, 1987
Tomato chlorotic dwarf viroid	TCDVd	1999	Canada, Netherlands	As for PSTVd.	Singh et al., 1999
Tomato planta macho viroid	TPMVd	1974	Mexico	As for PSTVd with very small, nonmarketable fruit that have few to no seeds.	Galindo et al., 1982; Martinez-Soriano et al., 1996
(str. *Mexican papita viroid*)	—	1996	Mexico		

Most viroid diseases are thought to result from chance transfers from endemically infected wild species to susceptible cultivars. For example, TPMVd and PSTVd are believed to have been introduced from Mexico in the late 1800s as a result of the search to identify genetic resistance to the pathogen that causes late blight of potato (*Phytophthora infestans* (Mont.) de Bary).

The most complete epidemiological studies have been made with TPMVd. This viroid has been reported only in Mexico in the state of Morelos. It appears to be restricted by climate, with disease distribution being sharply defined by the 22°C isothermal. Tomato fields in districts with an annual mean temperature above 22°C (at an altitude of 800–1,400 m) have a disease incidence of 5–45%, whereas fields in districts with an annual mean temperature below 22°C (at an altitude of 1,400–2,000 m) are not usually affected by disease.

Tomato plants infected with TPMVd normally show two well-defined syndromes. The first syndrome starts 10–15 days after infection, at which time the growth of the plants slows dramatically; epinasty of leaves and leaflets is apparent and both become brittle and display veinal necrosis. The second syndrome begins after several weeks. It is characterized by renewed but slow growth of small, upright yellow leaves that develop wavy margins. Very small fruit with few or no seeds are produced—hence, the name *planta macho* ("male plant"). In many cases, no marketable fruit are produced, which can be devastating in terms of economics. Most of the economic loss in Morelos, Mexico, occurs during the autumn and winter.

Since the initial report of TPMVd, much has changed in recognizing how viroids are spread in nature. For instance, researchers have shown that important viroids infecting tomato can be seedborne and seed transmitted (i.e., TASVd and TCDVd). Also, the symptomless ornamental ground cover common periwinkle (*Vinca minor* L.) was found to contain TCDVd, and plants were able to survive subfreezing temperatures.

Management

Breeding plants for resistance is often the desired method of disease management. Unfortunately, no useful sources of natural resistance to viroids have been identified. Symptomless reservoir hosts pose a threat to greenhouse production and highlight the importance of conducting systematic surveys and implementing quarantines. With the increased economic impact of viroids affecting tomatoes grown in large greenhouse operations, alternative control methods may be necessary.

Fig. 167. Stunted plants and unmarketable fruit—symptoms of *Tomato planta macho viroid* (TPMVd). (Courtesy J. Galindo)

Modern agricultural practices—such as the growing of widespread monocultures and the worldwide distribution of seed and planting material—have probably influenced the sudden appearance and rapid spread of some viroid diseases in tomato. Greenhouse environmental conditions also favor viroid replication, allowing for high concentrations of the pathogens in new growth. Spread of viroids is also encouraged by suckering of side shoots from the main stem and by maintaining plants for extended periods of time.

Management of viroid diseases currently depends on visual inspection of plants, followed by roguing of those that appear symptomatic. Disinfesting cutting tools, particularly in nursery production, may also reduce potential spread. Regardless, the best management strategy is to prevent introducing viroids through the use of viroid-tested seed and seedlings.

Selected References

Antignus, Y., Lachman, O., and Pearlsman, M. 2007. Spread of *Tomato apical stunt viroid* (TASVd) in greenhouse tomato crops is associated with seed transmission and bumble bee activity. Plant Dis. 91:47-50.

Antignus, Y., Lachman, O., Pearlsman, M., Gofman, R., and Bar-Joseph, M. 2002. A new disease of greenhouse tomatoes in Israel caused by a distinct strain of *Tomato apical stunt viroid* (TASVd). Phytoparasitica 30:502-510.

Boonham, N., González-Pérez, L., Mendez, M. S., Lilia Peralta, E., Blockley, A., Walsh, K., Barker, I., and Mumford, R. A. 2004. Development of a real-time RT-PCR assay for the detection of Potato spindle tuber viroid. J. Virol. Methods 116:139-146.

Candresse, T., Marais, A., Ollivier, F., Verdin, E., and Blancard, D. 2007. First report of the presence of *Tomato apical stunt viroid* on tomato in Sénégal. Plant Dis. 91:330.

Diener, T. O., ed. 1987. The Viroids. Plenum Press, New York.

Diener, T. O., and Raymer, W. B. 1971. Potato spindle tuber "virus." Descriptions of Plant Viruses, no. 66. Association of Applied Biologists, Wellesbourne, UK.

Galindo A., J., Smith, D. R., and Diener, T. O. 1982. Etiology of Planta Macho, a viroid disease of tomato. Phytopathology 72:49-54.

Hadidi, A., Flores, R., Randles, J. W., and Semancik, J. S., eds. 2003. Viroids. CSIRO, Collingwood, Australia.

Hammond, R. W., and Owens, R. A. 2006. Viroids: New and continuing risks for horticultural and agricultural crops. doi: 10.1094/APSnetFeature-2006-1106.

Kryczynski, S., Paduch-Cichal, E., and Skzeczkowski, L. J. 1988. Transmission of three viroids through seed and pollen of tomato plants. J. Phytopathol. 121:51-57.

Fig. 168. Healthy greenhouse tomato plant (cultivar Rutgers) (left) compared to plant infected with *Tomato chlorotic dwarf viroid* (TCDVd) (right). Note that the infected plant has curled, chlorotic leaves and small fruit. (Courtesy R. Hammond)

Martinez-Soriano, J. P., Galindo-Alonso, J., Maroon, C. J. M., Yucel, I., Smith, D. R., and Diener, T. O. 1996. Mexican pipita viroid: Putative ancestor of crop viroids. Proc. Natl. Acad. Sci. U.S.A. 93:9397-9401.
Mishra, M. D., Hammond, R. W., Owens, R. A., Smith, D. R., and Diener, T. O. 1991. Indian bunchy top disease of tomato plants is caused by a distinct strain of citrus exocortis viroid.. J. Gen. Virol. 72:1781-1785.
Mumford, R. A., Jarvis, B., and Skelton, A. 2004. The first report of *Potato spindle tuber viroid* (PSTVd) in commercial tomatoes in the UK. Plant Pathol. 53:242.
Owens, R. A., and Diener, T. O. 1981. Sensitive and rapid diagnosis of potato spindle tuber viroid disease by nucleic acid hybridization. Science 213:670-672.
Puchta, H., Herold, T., Verhoeven, K., Roenhorst, A., Ramm, K., Schmidt-Puchta, W., and Sänger, H. L., 1990. A new strain of Potato spindle tuber viroid (PSTVd-N) exhibits major sequence differences as compared to all other strains sequenced so far. Plant Mol. Biol. 15:509-511.
Singh, R. P. 1970. Seed transmission of potato spindle tuber virus in tomato and potato. Am. Pot. J. 47:225-227.
Singh, R. P., and Dilworth, A. D. 2009. *Tomato chlorotic dwarf viroid* in the ornamental plant *Vinca minor* and its transmission through tomato seed. Eur. J. Plant Pathol. 123:111-116.
Singh, R. P., Nie, X., and Singh, M. 1999. Tomato chlorotic dwarf viroid: An evolutionary link in the origin of pospiviroids. J. Gen. Virol. 80:2823-2828.
Verhoeven, J. Th. J., Jansen, C. C. C., and Roenhorst, J. W. 2006. First report of *Tomato apical stunt viroid* in tomato in Tunisia. Plant Dis. 90:528.
Verhoeven, J. T. J., Jansen, C. C. C., Willemen, T. M., Kox, L. F. F., Owens, R. A., and Roenhorst, J. W. 2004. Natural infections of tomato by *Citrus exocortis viroid, Columnea latent viroid, Potato spindle tuber viroid* and *Tomato chlorotic dwarf viroid.* Eur. J. Plant Pathol. 110:823-831.
Verhoeven, J. Th. J., Jansen, C. C. C., Roenhorst, J. W., Steyer, S., and Michelante, D. 2007. First report of *Potato spindle tuber viroid* in tomato in Belgium. Plant Dis. 91:1055.
Walter, B. 1987. Tomato apical stunt. Pages 325-330 in: The Viroids. J. O. Diener, ed. Plenum Press, New York.

(Prepared by T. A. Zitter)

Postharvest Diseases and Disorders

The extent of yield loss following tomato harvest is not well documented. During 1974–1977 and again in 1980, approximately 23% of the fruit marketed in the New York area were estimated to have been lost to diseases and disorders, based on market surveys. Postharvest decays accounted for 80% of the losses across the entire system and for virtually all of the losses at the consumer level. Worldwide, average losses likely exceed these values.

In most developed countries, production technologies tend to minimize fruit decays. For example, the use of plastic mulch and stakes in both field and greenhouse production minimizes fruit contact with the soil and thus reduces fruit rots overall. In addition, the cultivars that are used tend to produce firm fruit. Fruit decays are also prevented through the application of sanitation practices before and after harvest, and both pathogen growth and fruit senescence are slowed by the prompt cooling of freshly harvested fruit to ideal storage temperatures.

These practices are not usually applied, however, by growers and vendors that sell at open-air farmers' markets. Also, growers in some regions routinely harvest and handle ripe fruit, which increases the incidence of harvest injury, defects and injuries related to fruit development, field diseases and insect injuries, and latent infections by fruit decay pathogens.

Selected References

Cappellini, R. A., and Ceponis, M. J. 1984. Postharvest losses in fresh fruits and vegetables. Pages 24-30 in: Postharvest Pathology of Fruits and Vegetables: Postharvest Losses in Perishable Crops. H. E. Moline, ed. Publ. NE-87 (UC Bull. 1914), Agricultural Experiment Station, University of California, Berkeley.
Ceponis, M. J., and Butterfield, J. E. 1979. Losses in fresh tomatoes at the retail and consumer levels in the Greater New York area. J. Am. Soc. Hortic. Sci. 104:751-754.
Ceponis, M. J., Cappellini, R. A., and Lightner, G. W. 1986. Disorders in tomato shipments to the New York market, 1972-1984. Plant Dis. 70:261-265.
Tietjen, W. H. 1981. Losses incurred during marketing of fresh tomatoes in the greater New York area. In: Proceedings from the Tomato Quality Workshop, 3rd. University of Maryland, College Park.

(Prepared by J. A. Bartz)

Bacterial Soft Rot

Bacterial soft rot was first described by L. R. Jones in 1901, but many growers and consumers undoubtedly had firsthand experience with the disease before then. Bacterial soft rot in tomato is a classic progressive decay. Tissue maceration expands rapidly, such that entire fruit collapses, and the liquid associated with expanding lesions quickly spreads the pathogens to adjacent and nearby fruit.

At least four different bacteria cause soft rots in tomato fruit (see Causal Organisms, below). Each produces pectolytic enzymes that destroy the structure of the fruit tissue. In a warm, wet environment, three of the four bacteria also macerate the piths and epidermis of the stems in the most succulent parts of the plant canopy.

Symptoms

A bacterial soft rot lesion can begin at a wound, at the edge of the stem scar, beside an attached stem, at or adjacent to the stylar pore, or under the fruit surface, on occasion. This small, sunken, light- to dark-colored, water-soaked area enlarges rapidly. The epidermis over the lesion tends to rupture, and the consistency of the tissue within the lesion is liquid. As lesions expand, bacteria mixed with colonized cell sap ooze from them, dispersing the pathogen among packed or stored fruit. Lesions can also form internally, and they may not be obvious until a substantial portion of the fruit's interior has been macerated (Fig. 169).

Lesions may or may not contain gas bubbles, and they have a pH of 5.4 or higher. This pH may be contrasted with the normal pH of tomato fruit tissue, which is 4.2–4.6. Certain lesions appear fluorescent when viewed under long-wave ultraviolet (UV) light.

Causal Organisms

The primary causal organisms of bacterial soft rot are *Pectobacterium carotovorum* (Jones) Waldee (syn. *Erwinia carotovora* subsp. *carotovora* (Jones) Dye) and *Dickeya chrysanthemi* (Burkholder et al.) Samson et al. Certain strains isolated from soft rot lesions on tomato appear physiologically similar to the potato black leg pathogen (*E. carotovora* subsp. *atroseptica* (van Hall) Dye). These strains cause typical lesions when inoculated in tomato fruit but continue to develop at a higher maximum temperature (36°C) than that reported for the pathogen of potato (*Solanum tuberosum* L.).

P. carotovorum and *D. chrysanthemi* are facultatively anaerobic, strongly pectolytic, non-spore-forming rods (0.7 × 1.5 μm) that are motile by peritrichous flagella. The cells occur singly, in pairs, or in short chains, on occasion. Colonies on nutrient agar are grayish white, smooth, round, glistening, slightly raised, and translucent, and they become visible without magnification within 24 h at 25–30°C. On a medium of crystal violet pectate (CVP), colonies develop relatively deep pits and have a cross-hatched appearance when viewed under oblique lighting with a bisecting microscope.

Pectolytic strains of *Pseudomonas fluorescens* Migula biovar II (syn. *Pseudomonas marginalis* (Brown) Stevens) have been isolated from naturally occurring soft rot in commercially handled fruit. The lesions glow blue-green under fluorescent illumination, as do lesions that develop after subsequent inoculation of test fruit with isolated strains. However, lesions caused by pectolytic *Pseudomonas* spp. do not expand as rapidly as those caused by *Pectobacterium carotovorum* and *D. chrysanthemi*. Consequently, their role as a primary cause of postharvest losses in tomato is unclear. The soft rot *Pseudomonas* spp. are gram-negative rods with general dimensions of 0.7–0.8 × 2.0–2.8 μm. They do not form spores or ferment carbohydrates. They are oxidase positive, motile by polar flagella, and capable of nitrate reduction. If grown on iron-deficient media, such as King's medium B, they produce green pigments that fluoresce under UV light. These bacteria grow best at 25–30°C, do not grow at 41°C, and can produce soft rot at 5°C. They produce relatively shallow pits on CVP medium, and their colonies lack the cross-hatched appearance typical of the primary bacteria that cause soft rot.

Pectolytic strains of *Bacillus* spp. have also been reported to cause bacterial soft rots in tomato fruit but only at relatively high temperatures (i.e., greater than 30°C). These strains do not grow on CVP; are nonfermentative, gram-positive rods; and produce endospores but do not likely cause extensive disease in commercially handled fruit.

Disease Cycle and Epidemiology

Soft rot bacteria are ubiquitous and survive apart from the host in a variety of ways. *Pectobacterium carotovorum* and *D. chrysanthemi* have been detected on various fleshy plant organs and in small populations (isolated after enrichment techniques) in soil and aquatic environments. In contrast, the soft rot *Pseudomonas* spp. are soil inhabitants and can often be isolated from soil with normal plating techniques. Populations of soft rot bacteria appear to increase rapidly in tomato crops

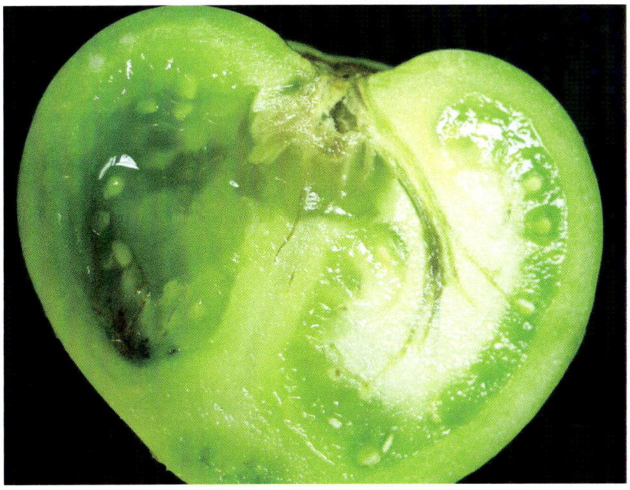

Fig. 169. Internal lesion and maceration of tissue, caused by bacterial soft rot. (Courtesy J. A. Bartz)

during wet weather and especially when the plant canopy remains wet for several days.

None of these bacteria penetrates the fruit surface directly. However, when suspended in water, the bacteria can enter any wound or naturally occurring opening, particularly if the opening is congested with cell sap or water. Bacterial multiplication, tissue colonization, and disease may occur independently of tissue injury, because internal cells are not protected against the activities of the bacterium.

Bacterial soft rot has an extremely variable latent phase, ranging from less than 18 h to more than 3 weeks. The longest latent phase occurs when a few cells have been introduced into mature-green fruit that have not been stressed and are being stored at the ideal temperature for ripening (i.e., 20°C). The shortest latent period accompanies the inoculation of green or ripe tomatoes with large populations (greater than $\log_{10}$ 8.0 cfu/ml) followed by storage at temperatures of 25–35°C.

Major postharvest outbreaks of bacterial soft rot often accompany the harvest of tomatoes from wet plants or persistently wet fields or the improper management of packinghouse water systems. During warm, wet periods, bacterial soft rot can develop in succulent stems and in fruit still attached to the plants. When harvest occurs under these conditions, many fruit will be inoculated and much inoculum will be carried on fruit surfaces. Additionally, water congestion will likely occur in fresh fruit stem scars and wounds. Tissue water congestion appears similar to the water channels reported in water-soaked tobacco leaves (*Nicotiana* L.), in which any water or aqueous suspension of particulate matter that lands on top of a water channel is rapidly sucked into the underlying tissues by capillary action.

Two forms of inoculation may occur during postharvest handling of tomato fruit: (1) potential infection courts may come in direct contact with inoculum and (2) suspended inoculum may be internalized. When direct contact occurs, bacteria in water, on equipment, or on fruit may first contaminate and then begin to grow on damaged plant cells at the surface of a wound or other opening. Alternatively, bacterial cells may diffuse in cell sap to intercellular spaces below the surface of the wound or opening and then initiate infection in cell layers beneath that surface. Also, during unloading, handling, or washing operations at the packinghouse, water containing suspended pathogens may enter openings in the fruit surface and permeate the intercellular spaces of underlying tissues. If the surface pores are dry, then trapped air appears to prevent water intrusion. Thus, some type of pressure imbalance is required to cause water to displace the air. If the pores are already water congested, then external water appears to be drawn in quite readily, perhaps by capillary forces. Inoculum internalized below the stem scar gives rise to internal lesions that often develop rapidly and lead to collapse of the entire fruit.

The pressure imbalances that result in water infiltrating the fruit include those associated with the cooling of tissues in submerged fruit (i.e., a vacuum forms in the intercellular spaces), the impact from streams of water used to dislodge piles of fruit in gondolas, the impact from dropping fruit into water during the unloading process, and the force of hydrostatic pressure when fruit are deeply submerged in water. The infiltration may force inoculum into the fruit surface or congest surface openings with water and cell sap.

Management

Internalized inoculum located in intercellular spaces is protected from the action of most contact biocides. A biocide that is mixed with inoculum prior to contact with fruit openings can prevent the internalization of viable microbes, if the infiltration occurs slowly and the biocide acts quickly. The failure of either condition creates a contamination hazard. For this reason, a rapidly acting biocide must be present in the water where fruit first enter the system. There is no reliable way to disinfect or decontaminate an infected or contaminated tomato fruit.

For more information on the management of bacterial soft rot, see later in this section, Management of Postharvest Diseases.

Selected References

Bartz, J. A. 1980. Causes of postharvest losses in a Florida tomato shipment. Plant Dis. 64:934-937.

Bartz, J. A. 1981. Variation in the latent period of bacterial soft rot in tomato fruit. Phytopathology 71:1057-1062.

Bartz, J. A., and Wei, C.-I. 2003. The influence of bacteria. Pages 519-541 in: Postharvest Physiology and Pathology of Vegetables. J. A. Bartz and J. K. Brecht, eds. Marcel Dekker, New York.

Cuppels, D., and Kelman, A. 1974. Evaluation of selective media for isolation of soft-rot bacteria from soil and plant tissue. Phytopathology 64:468-475.

Dowson, W. J. 1941. The identification of the bacteria commonly causing soft rot in plants. Ann. Appl. Biol. 28:102-106.

Johnson, J. 1947. Water-congestion in plants in relation to disease. Univ. Wis. Res. Bull. 160.

Wingard, S. A. 1924. Bacterial soft-rot of tomato. Phytopathology 14:451-459.

(Prepared by J. A. Bartz)

Sour Rot (Geotrichum Rot, Bacterial Sour Rot)

First reported in 1922, sour rot (also called watery rot and Oospora rot) occurs in all tomato-growing areas. Until 1995, any decay on tomato that smelled like pickling brine was considered sour rot, which is caused by a yeastlike organism. However, that year, K. E. Conn and co-workers reported that species of *Leuconostoc* van Tieghem—a gram-positive, coccoid bacterium—had been isolated from naturally occurring lesions on tomato and shown to cause a type of bacterial sour rot in inoculated fruit.

Sour rot is most commonly found in ripe tomatoes, where it is associated with rapid maceration. Green fruit are difficult to infect unless chilled, and the disease appears as a type of lactic acid fermentation. The sour rots have caused losses in fresh-market and processing tomatoes. Both the fungal and bacterial sour rots are progressive and likely to spread in containers of fruit.

Symptoms

Early symptoms of sour rot are greasy, water-soaked areas, which appear at wounds or the edge of the fruit stem scar or the stylar pore (Fig. 170). In green fruit, the affected tissue remains firm until the disease is well advanced, and the tissue emits an odor of lactic acid. In ripe fruit, the disease progresses rapidly, particularly under warm temperatures, and the fruit epidermis that covers the lesions usually cracks, allowing the watery contents to spill out. A white to dirty-white, crusty, yeastlike growth may develop at breaks in the fruit surface, or a cloudy bacterial suspension may drip from the lesions, depending on the pathogen.

The pH of lesions is less than or equal to 4.4. In the absence of signs of the causal organism, the lesion pH may be used to distinguish sour rot from bacterial soft rot or Rhizopus rot. Septate mycelium and barrel-shaped arthrospores or large numbers of bacterial cocci or rods can be observed in diseased tissue. Neither of the "lactic acid" bacteria is readily isolated from lesions to the standard media used for culturing most plant-pathogenic bacteria. Methods that usually prove successful include isolation to stationary broth culture (followed by selection of material from the base of the tube) or to streaks on de Man, Rogosa, and Sharpe (MRS) agar or equivalent.

Causal Organisms

Sour rot can be caused by two closely related fungi reported to cause lesions in tomato fruit: *Geotrichum candidum* Link (syns. *Oidium lactis* Fresen and *Oospora lactis* (Fresen) Sacc.) and *G. penicillatum* (Carmo Sousa) Arx. Lesions caused by *G. penicillatum* do not have the lactic acid odor, and the lesion pH has not been reported. Mycelia are hyaline and septate and branch dichotomously. Conidia form in chains by disarticulation (i.e., disjointing) at the septa of the lateral hyphae. The conidia are cylindrical, barrel shaped, ellipsoidal, or subglobose; they can be 2–8 × 3–5 µm but are more commonly 3–6 × 6–12 µm. The optimum temperature for growth of both fungi is 30°C, whereas both grow slowly at 5°C. Both fungi are much more aggressive in red fruit than in green fruit.

Three characteristics are useful for differentiating these two pathogens:

1. *G. candidum* is more aggressive in green fruit.
2. Lesions in green fruit caused by *G. penicillatum* do not have a characteristic sour odor.
3. *G. penicillatum* produces more aerial hyphae on potato dextrose agar (PDA) than does *G. candidum*.

Bacterial sour rot is caused by *L. mesenteroides* (Tsenkovskii) van Tieghem. *L. mesenteroides* cells are gram positive, are usually spherical, and occur in pairs or chains. Optimum growth occurs under low oxygen at temperatures of 20–30°C. A complex growth medium is required for culture of the organism. Isolation from infected tissue is enhanced by use of stationary cultures, as cell chains tend to sink to the bottom of the broth culture medium. Additionally, with stationary culture, low oxygen tension develops at the bottom of a container of medium. *L. mesenteroides* is commonly associated with plants and is involved with lactic acid fermentation of crucifers (Brassicaceae), green cucurbits (Cucurbitaceae), green tomatoes, and other vegetables in salt brines.

Several species of bacteria in the genus *Lactobacillus* are also found in low numbers on plant surfaces. Like *Leuconostoc* spp., *Lactobacillus* spp. grow best under reduced oxygen and have complex nutritional requirements. Cells are gram positive, are short rods 0.9–1.2 µm wide × 3–8 µm long, and occur singly, in pairs, or in short chains. It is unclear what stimulates *Leuconostoc* spp. or *Lactobacillus* spp. to infect wounds or tissues in tomato fruit or whether the strains isolated from progressive decays are different from those normally found on the fruit surface. However, given the ubiquity of these lactic acid bacteria on and around plants, both organisms—as well as *Geotrichum* spp. of fungi—may be involved in causing the disease known as sour rot.

Fig. 170. Symptoms of sour rot on green tomato fruit. (Courtesy G. W. Simone)

Disease Cycle and Epidemiology

The organisms that cause the sour rots are common inhabitants of soil, plant, and aquatic environments. The common sour rot fungus, *G. candidum,* has been isolated from soil, sewage, fruit flies (*Drosophila* spp.), buried tomato debris, and dairy products. Lactic acid bacteria (*Leuconostoc* spp. and *Lactobacillus* spp.) are normal epiphytes on plants and are found in decaying plant matter.

Tomato fruit may become contaminated with initial inoculum through contact with fruit flies and other insects, as well as machinery, splashing rain, decaying vegetation, field crews (pickers), and so on. The common fruit fly (*D. melanogaster* Meigen) can vector spores of *Geotrichum* spp. to surface cracks on healthy ripe fruit in the field. In processing plants and packinghouses, the fungus may grow on pulp residue that remains on equipment.

The sour rot organisms are considered wound pathogens and cannot directly penetrate the epidermis of the tomato fruit. However, if aqueous conidial or cell suspensions infiltrate the fruit, the sour rot microbes will cause disease. In green fruit, lesions frequently appear to originate at the edge of the stem scar of fruit still attached to the plant, suggesting that these organisms can enter the fruit under the stem attachment. Green fruit are not readily attacked, but chilled or injured green fruit are quite susceptible, with lesions being observed within 28 h of inoculation. Sour rot may advance rapidly through containers of ripe fruit, in which fruit-to-fruit spread is facilitated by the watery contents of lesions.

Some researchers have suggested that bacterial soft rot may develop following sour rot. This seems unlikely, however, as soft rot bacteria do not grow readily at the pH of sour rot lesions. The apparent presence of both bacteria in the same fruit or lesion is more likely the result of co-inoculation or the natural tendency of the sour rot organisms to colonize decaying plant matter.

Management

For information on the management of sour rot, see later in this section, Management of Postharvest Diseases.

Selected References

Butler, E. E. 1960. Pathogenicity and taxonomy of *Geotrichum candidum*. Phytopathology 50:665-672.

Conn, K. E., Ogawa, J. M., Manji, B. T., and Adaskaveg, J. E. 1995. *Leuconostoc mesenteroides* subsp. *mesenteroides,* the first report of a coccoid bacterium causing a plant disease. Phytopathology 85:593-599.

McColloch, L. P., Cook, H. T., and Wright, W. R. 1968. Market diseases of tomatoes, peppers, and eggplants. U.S. Dep. Agric. Handb. 28.

Moline, H. E. 1984. Comparative studies with two *Geotrichum* species inciting postharvest decays of tomato fruit. Plant Dis. 68:46-48.

Pritchard, F. J., and Porte, W. S. 1923. Watery-rot of tomato fruits. J. Agric. Res. (Washington, D.C.) 24:895-1003.

(Prepared by J. A. Bartz)

Rhizopus Rot

Rhizopus rot is sometimes considered a disease of ripe fruit, but in fact, it affects fruit at all stages of ripeness. Moreover, Rhizopus rot has caused extensive losses of so-called green-wrap fruit (i.e., pieces of green fruit wrapped individually in paper). This disease appears in all tomato-growing areas.

Symptoms

The lesions that are symptomatic of Rhizopus rot begin as rapidly enlarging, water-soaked areas located adjacent to wounds, open stylar pores, or the stem scar. Normally, the affected areas are not discolored but have distended surfaces. Lesions are soft and somewhat watery; even so, they usually have structure because of the coarse, threadlike strands of mycelia that spread throughout the fruit tissue.

Under high humidity, grayish-white masses of mold structures (i.e., mycelia, sporangiophores, sporangia, and occasionally zygospores) develop over the surface of a diseased fruit (Fig. 171). In packages or bins of fruit, a "nest" of fungal structures and decayed fruit may develop. The odor of extensive Rhizopus rot is similar to that of eggs and is unlike that of bacterial soft rot or sour rot.

Causal Organism

Rhizopus rot is caused by *Rhizopus stolonifer* (Ehrenb.) Vuill. (syn. *R. nigricans* Ehrenb.), a zygomycete that produces sporangiospores in a globose, columellate sporangium. The wall of the sporangium is quite fragile and ruptures easily, liberating the numerous spores. On solid media, mycelia are initially coenocytic with many branches. Septa may develop later.

Disease Cycle and Epidemiology

R. stolonifer is widespread and a common saprophyte. It may survive up to 30 years in dried deposits of decayed fruit or other plant debris. Consequently, an epidemic may be initiated by the contact of harvested fruit with containers or pieces of equipment that have not been properly disinfested from previous harvests. Spores are commonly carried long distances by air currents. Also, the pathogen has been isolated from the common fruit fly (*Drosophila melanogaster* Meigen).

Invasion of tomato fruit occurs through wounds and as a consequence of infiltration of fruit with dump tank water. Although the causal organism is considered a wound pathogen, successful penetration of the unbroken cuticle of fruit adjacent to diseased fruit in "nests" has been reported.

The optimum temperature for development of Rhizopus rot is 24–27°C, which is slightly cooler than the optimum temperatures for development of bacterial soft rot and sour rot. Little disease occurs at 10°C. The rate of disease development among infected fruit under optimal conditions appears comparable to or may even exceed that of bacterial soft rot.

Management

For information on the management of Rhizopus rot, see later in this section, Management of Postharvest Diseases.

Fig. 171. Symptoms of Rhizopus rot on ripe tomato fruit, including development of mold structures. (Courtesy G. W. Simone)

Selected References

Baker, K. F. 1946. An epiphytotic of rhizopus soft rot of green-wrap tomatoes in California. Plant Dis. Rep. 30:20-26.

Butler, E. E., and Bracker, C. E. 1963. The role of *Drosophila melanogaster* in the epiphytology of *Geotrichum, Rhizopus,* and other fruit rots of tomato. Phytopathology 53:1016-1020.

McColloch, L. P., Cook, H. T., and Wright, W. R. 1968. Market diseases of tomatoes, peppers, and egg plants. U.S. Dep. Agric. Handb. 28.

(Prepared by J. A. Bartz)

Black Mold Rot

Black mold rot occurs in all tomato production areas and can affect fruit at all stages of ripeness. This disease can be caused by at least four different fungi (see Causal Organisms, below), but the pathogens are relatively nonaggressive and normally attack stressed or senescent (ripe) tissue. Thus, an outbreak of black mold rot is most likely to occur after fruit have been stressed by heat or cold (i.e., persistent exposure to air temperatures less than 15°C), injured by sunlight, affected by calcium deficiency (i.e., blossom-end rot), or damaged by spray residue (often called "spray burn"). Fruit with a cracked cuticle or pericarp are especially vulnerable.

Under certain environmental conditions, fruit may become infected through the blossom scar or the edge of the fruit stem scar. Such lesions often develop internally. Infection through the blossom end of the fruit may be associated with a blackened, necrotic stylar pore. This is usually found among fruit that have a rough blossom scar and air pockets in the locules (see Part III, the sections Catface and Puffiness, respectively). Some of these pathogens can cause losses in processing tomatoes in the field.

Symptoms

Lesions of black mold rot begin as sunken or slightly flattened areas, which are sometimes water soaked in appearance. The lesions may or may not have definite margins but extend into the fruit as a corelike mass. The external color of the lesions soon becomes brown to black, whereas the internal color ranges from slightly darkened to black. In fruit stored under moderate to high humidity, the surfaces of lesions become covered with a dense, dark-gray to olive-green to black layer of conidia and conidiophores (Fig. 172). Lesions caused by certain strains of black mold rot may contain pimplelike structures, which are perithecia of the perfect (teleomorphic) stage.

Fig. 172. Black mold rot on tomato fruit. Note that the surfaces of lesions are covered with a dense layer of conidia and conidiophores. (Courtesy G. W. Simone)

Most lesions develop on the shoulders of tomato fruit or adjacent to the stem scar, primarily because these tissues are often injured by weather checks (i.e., small cracks in the cuticle), concentric or radial cracking, and certain pesticide tank mixes. Internal pockets of black, necrotic tissue may develop adjacent to imperfectly formed stylar pores, where the fungus appears to have colonized the style and then attacked adjacent tissues. These fungi may also grow in tissue damaged by calcium deficiency and abrupt exposure to sunlight (see Part III, the sections Blossom-End Rot and Sunscald, respectively).

Causal Organisms

Four pathogens are reported to cause black mold rot: (1) *Alternaria alternata* (Fr.:Fr.) Keissl. (syn. *A. tenuis* Nees); (2) *Stemphylium botryosum* Wallr. (teleomorph *Pleospora tarda* E. G. Simmons); (3) *Pleospora lycopersici* Marchal & É. J. Marchal (syn. *P. herbarum*); and (4) *S. consortiale* (Thüm.) J. W. Groves & Skolko. Strains designated *P. lycopersici* produce numerous perithecia in lesions, whereas strains designated *S. botryosum* do not. (The perithecia-producing strain has been reported only once, in 1919.) The *S. consortiale* strains are relatively weak pathogens and tend to be found only on stressed fruit.

Conidia produced by *A. alternata* are dilute ovoid, obclavate, obpyriform, or simple ellipsoidal (rarely) in shape, usually with an easily visible basal pore. Except for the rare ellipsoidal ones, the conidia possess a beak or apical cell that is always shorter than the remaining length of the conidium. The conidia may contain longitudinal and cross septa (myriform) and are often borne in long chains. In contrast, conidia of *S. botryosum* are never borne in chains, are globose to broadly ellipsoid, and do not have beaks. Conidia of *S. consortiale* are borne in grapelike clusters.

Disease Cycle and Epidemiology

The fungi that cause black mold rot are primarily saprophytic and cause lesions in fruit tissue that has been damaged by any of a range of conditions: frost or freezing, cracks in the cuticle or pericarp, sunscald, exposure to excessive heat or cold, attacks from other pathogens, calcium deficiency at the blossom end of developing fruit, and senescence. The disease also occurs in fruit damaged in a phytotoxic response to certain mixtures of pesticide sprays.

Green fruit are quite resistant to black mold rot, but lesions develop in chilled fruit or fruit damaged by sunscald, calcium deficiency, or impact injury to the shoulders ("black shoulder"). In ripening fruit, the pathogens infect various breaks in the pericarp that result from excessive water uptake, with or without cold weather.

Initial inocula are ubiquitous and may develop from various decaying plant debris. The conidia are dispersed primarily by wind. Early research demonstrated that the disease could develop as a result of inoculation of the stigma of the flower with conidia. In a cool to cold environment, the disease first appears on the shoulders, adjacent to the stem scar of the fruit. This region of the fruit is where cracks in the pericarp first occur.

Black mold rot progresses most rapidly at 24–28°C, particularly among overripe fruit, but lesions can develop even on fruit stored at 0°C. Fruit-to-fruit spread in storage occurs if fruit are held for long periods of time.

Management

For information on the management of black mold rot, see later in this section, Management of Postharvest Diseases.

Selected References

Bartz, J. A. 1971. Studies on the causal agent of black fungal lesions on stored tomato fruit. Proc. Fla. State Hortic. Soc. 84:117-119.

Butler, E. E. 1959. Fungi and rots in California canning tomatoes. Plant Dis. Rep. 43:187-192.

McColloch, L. P., Cook, H. T., and Wright, W. R. 1968. Market diseases of tomatoes, peppers, and egg plants. U.S. Dep. Agric. Handb. 28.
Simmons, E. G., and Roberts, R. G. 1993. Alternaria themes and variations (73). Mycotaxon 48:109-140.
Spalding, D. H. 1980. Control of Alternaria rot of tomatoes by postharvest application of imazalil. Plant Dis. 64:169-171.

(Prepared by J. A. Bartz)

Gray Mold Rot (Botrytis Fruit Rot)

Gray mold rot is one of the most important postharvest diseases of tomato fruit produced under cool, moist growing conditions. The disease has been observed in fruit virtually worldwide, but it is not often observed on fruit produced from high-calcium soils. The fungus that causes gray mold can penetrate the pericarp directly, particularly when growing from a food base (e.g., leaf or flower petal litter).

Symptoms

The lesions associated with gray mold rot begin as gray-green to gray-brown, slightly water-soaked spots, and they are usually located near the stem end of the fruit. Lesions can also originate at the point of contact between the fruit and diseased plant tissue. Gray mold rot lesions are somewhat soft but not as soft as those associated with bacterial soft rot or Rhizopus rot.

A grayish mold may develop at breaks in the fruit skin or sparsely over the surfaces of older lesions. During humid conditions, many tiny, grapelike clusters of spores borne on conidiophores are visible with the aid of a hand lens. Macroscopically, the diseased tissue appears covered with a gray-velvet coating of mold structures. Early infections in green fruit may become arrested and form circular, light-colored blemishes known as "ghost spots."

Causal Organism

The fungus that causes gray mold rot, *Botrytis cinerea* Pers.:Fr., grows well on a variety of media. It produces conidia on long conidiophores that are dichotomously branched at the tips. Conidia are borne on sterigma at the tip of each branch, and the cluster of spores resembles a bunch of tiny grapes. The conidia are colorless, single celled, and ovoid; they range in size from 7.3 × 8.0 to 9.7 × 11.1 μm. Sclerotia may form in decaying tissues and may germinate, forming mycelia, which can directly initiate lesions or produce conidia. On rare occasions, germination of sclerotia leads to development of an apothecium containing ascospores similar to those produced by *Sclerotinia* spp.

B. cinerea requires relatively cool temperatures (18–23°C) for best growth. As temperatures increase above 24°C, conidial germination decreases. Growth is inhibited at 32°C and above.

Disease Cycle and Epidemiology

B. cinerea is an excellent saprophyte and exhibits a vast host range. Disease outbreaks are particularly severe during cool, wet conditions; outbreaks are even more likely if fruit have been injured by exposure to cold temperatures.

Tomato fruit held in cold storage to prolong shelf life become extremely susceptible to *B. cinerea;* infection is particularly likely if fruit have been damaged during storage. The disease may spread in stored or packaged fruit, especially if free water exists for extended periods of time.

Management

For information on the management of gray mold rot, see later in this section, Management of Postharvest Diseases.

Selected References

McColloch, L. P., Cook, H. T., and Wright, W. R. 1968. Market diseases of tomatoes, peppers, and eggplants. U.S. Dep. Agric. Handb. 28.
Statewide Integrated Pest Management (IPM) Program. 1998. Integrated Pest Management for Tomatoes. 4th ed. Publ. 3274. Statewide IPM Program, Agriculture and Natural Resources, University of California.

(Prepared by J. A. Bartz)

Predisposition for Postharvest Diseases

Two forms of predisposition are extremely important to the severity and incidence of postharvest disease: chilling injury and cracks in the fruit surface.

Tomato fruit are injured by exposure to temperatures less than 15°C. The minimum temperature for storage of fruit without injury is not fixed but varies, decreasing as fruit ripen. Chill-injured mature-green fruit may not ripen properly and are highly susceptible to postharvest decays.

Chilling injury is usually observed on green fruit held for 2 weeks or longer at 13°C, although the recommended storage temperature for mature-green fruit ranges from 13 to 20°C (whereas that for ripe fruit ranges from 7 to 10°C). Chilling injury occurs more quickly as the temperature decreases below the recommended minimum. Additionally, the level of injury is proportional to the number of hours of continuous or discontinuous exposure to temperatures below 13°C. Field exposure is included in this formula. Thus, fruit harvested during prevailing cool weather are less tolerant of exposure to low temperatures in storage or transit than those harvested during prevailing warm weather.

Chill-injured fruit often ripen slowly, develop poor color, and have off-flavored flesh and surface pitting. As these fruit complete color development, black mold rot often forms adjacent to the fruit stem scar or blossom pore. Increased susceptibility to fruit decay (e.g., sour rot, Rhizopus rot) accompanies the onset of chilling injury. As a consequence of severe chilling injury, mature-green fruit are sometimes attacked by weakly pathogenic fungi (e.g., *Penicillium* spp.).

The second major predisposition leading to postharvest fruit rots is fruit cracking or surface cracks. Most fractures in the surface tissue of tomato fruit result when the cuticle and pericarp cannot expand sufficiently in response to the rapid uptake of water by cells. Cracks occur radially from the fruit stem scar or concentrically around the scar. Cracks may develop well into the flesh of fruit or simply involve the surface layer of cells.

Many weak pathogens colonize the tissue exposed in cracks and then become dormant, perhaps because of a resistance mechanism. This resistance is lost when fruit are chilled, are damaged by heat, or become senescent. When resistance is lost, the pathogens resume growth and initiate lesions. Tomato fruit injured by a combination of chilling, cracking, and prolonged storage may be attacked by microbes not normally associated with postharvest diseases in tomato.

Selected References

Lyons, J. M., and Breidenbach, R. W. 1987. Chilling injury. Pages 305-326 in: Postharvest Physiology of Vegetables. J. Weichmann, ed. Marcel Dekker, New York.
Segall, R. H., Hayslip, N. C., and Walter, J. M. 1963. Effect of postharvest temperatures on several tomato varieties harvested at the mature-green stage. Proc. Fla. State Hortic. Soc. 76:172-177.
Wright, R. C., Pentzer, W. T., and Whiteman, T. M. 1931. Effect of various temperatures on the storage and ripening of tomatoes. U.S. Dep. Agric. Tech. Bull. 268.

Young, P. A. 1963. Symptoms and control of tomato fruit cracking. Plant Dis. Rep. 47:523-527.

(Prepared by J. A. Bartz)

Management of Postharvest Diseases

Similar strategies are generally used for control of the different postharvest diseases. As outlined in the following sections, five strategies comprise an effective management program:
1. Use resistant cultivars.
2. Control populations of postharvest pathogens.
3. Prevent predisposition of fruit.
4. Prevent inoculation of fruit.
5. Control the postharvest environment.

None of the major postharvest diseases is reliably controlled by a single strategy. Thus, the entire range of strategies should be employed.

Use Resistant Cultivars

Resistance to postharvest diseases has been described in tomatoes, but it is not of the same magnitude or type as some of the classical host resistances. Nevertheless, certain cultivars are not as likely as others to develop decays for two general reasons: They are not as likely to become inoculated, and if they are inoculated, they are not as likely to develop a progressive decay. In certain fruit, the tissue appears to restrict growth and colonization by pathogens, thereby prolonging the latent period.

In another type of resistance, the fruit stem scar of the tomato is not as likely to be infiltrated when the fruit is exposed to water. Anecdotal evidence suggests that fruit with porous stem scars are more likely to have postharvest losses during wet weather than are fruit that resist postharvest water intrusion.

Control Populations of Postharvest Pathogens

Postharvest diseases are often directly related to the populations of pathogens in the harvest and postharvest environments. The populations may be controlled indirectly by depriving the pathogens of a food base and directly by applying general sanitation measures.

Postharvest pathogens can be deprived of a food base by controlling damage to plants caused by insects, disease, drought stress, direct sunlight, and overcrowding. Thus, tomato cultivars should be selected that are resistant to diseases such as wilts and certain leaf spots and that do not have a large or dense canopy. A large, dense canopy is difficult to protect from pests and promotes infection by postharvest pathogens—such as *Erwinia carotovora* (van Hall) Dye and *Botrytis cinerea* Pers.:Fr.—by retaining water from rainfall, irrigation, or dew and guttation. By contrast, air movement through the canopy can be promoted in a number of ways:
1. by growing plants on raised beds
2. by removing lower suckers
3. by using stakes or trellises
4. by orienting rows so they are parallel to prevailing winds
5. by properly spacing plants and rows

Overhead irrigation, if it must be used, should be scheduled to avoid prolonging natural periods of free water on plant surfaces and encouraging wet soils. Harvest, as well as various cultural practices, should not be scheduled when plants are wet.

Pathogen populations are controlled directly by applying general sanitation practices. Before the first harvest, all picking containers and equipment should be cleaned of dried deposits of plant debris, which might harbor propagules of pathogens such as *Rhizopus stolonifer* (Ehrenb.) Vuill. After the dried deposits have been removed and then periodically thereafter, containers and equipment should be cleaned and disinfested. In addition, diseased fruit should be removed from production fields, because insects and splashing rain can disperse pathogens from diseased fruit in aisles between plant rows to fruit on plants. Cull piles should be buried or removed some distance from production fields and packinghouses.

At the packinghouse, chlorine should be continuously maintained in recirculating water systems at a concentration of 150–200 ppm free chlorine and a pH of 6.5–7.5. Chlorine solutions below pH 6.5 will likely be corrosive and composed almost entirely of the highly active hypochlorous acid. By contrast, solutions at pH 8.5 or higher are both corrosive and ineffective, since most of the active ingredient is hypochlorite ion, a relatively poor sanitizer. Free chlorine gas is not a concern unless the pH drops below approximately 5.0, particularly at the recommended concentration.

The chlorine concentration can be maintained automatically through commercial systems that measure the oxidation-reduction potential (ORP) and pH of the solution. However, ORP set points should be based on direct measurement of free chlorine. Different water sources often have different ORPs, even in the absence of free chlorine. The desired ORP is that achieved by adding free chlorine to the recommended concentration and adjusting the solution to the desired pH level. Once systems have been calibrated and are in operation, managers should periodically compare the free chlorine concentration and pH level with the measured ORP. Performing this hourly maintenance will ensure that the electrode is operating properly and that the poise of the water has not changed significantly. During operation of a packinghouse water system, free chlorine will be lost because of reaction with certain chemicals dissolved in the water or on incoming fruit, microbes located on fruit and plant debris, the surfaces of incoming fruit, and organic debris (including partially decayed fruit). Additional chlorine must be added to maintain the recommended concentration. The point of addition must be designed carefully to avoid abrupt decreases in solution pH.

Sanitation should also be practiced at markets and repacking locations. As a general rule, as the incidence of postharvest disease increases, the frequency of sorting should increase, as well. Bacterial ooze and fungal mycelia from diseased fruit can quickly infect nearby healthy fruit.

Prevent Predisposition of Fruit

The progress of most postharvest diseases is greatly enhanced if fruit become injured by chilling, puncturing, or cracking, as described. The likelihood that a crop will be exposed to chilling temperatures before harvest should be a factor in scheduling the planting date. Postharvest, a crop's prior exposure to chilling temperatures in the field must be considered in determining the storage temperature.

Mechanical injuries—including cracks, punctures, bruises, and scrapes—weaken tissue and become potential infection courts. Prior to harvest, repairs should be made to all rough areas in containers and equipment that will come in contact with tomato fruit. In addition, field crews should be carefully monitored to prevent rough handling of fruit, and packinghouse crews must sort out all damaged fruit. In harvesting and marketing operations, the general philosophy should be consistent with the handling of a perishable, fragile product.

Naturally occurring cracks develop on fruit as a consequence of moisture fluctuation. These cracks appear in both cold and hot weather and increase in incidence and severity as fruit ripen from mature green to pink. Consequently, plants should be kept uniformly moist. Fruit may be harvested at the mature-green or breaker stage of ripeness to avoid some development of cracks. If the prevailing weather has been relatively dry, the harvest should be scheduled before a forecasted heavy rainfall.

Prevent Inoculation of Fruit

Postharvest pathogens are typically present on tomato fruit harvested from home gardens, greenhouses, and fields. Potential infection courts usually exist on fruit, as well. Thus, implementing measures that reduce the frequency of inoculation is quite effective in controlling postharvest fruit rots. Although the incidence of wound inoculation may be reduced by sanitation and control of wounding, as discussed earlier, practicing several key strategies will help prevent inoculation by infiltration.

Infiltration may occur if fruit soak in water for a prolonged time (i.e., more than 15 min) or cool while submerged in water. To prevent infiltration, the water in flumes, tanks, and washers should be warmed to the temperature of the incoming fruit and the duration of fruit exposure should be limited. It is preferable to restrict the period of submersion to 60 sec or less, because warming water can be expensive. However, according to anecdotal reports, cool wash or dump tank water does not dry from fruit surfaces very rapidly, and wet fruit can pick up soil from other points along the packing line after leaving the drying section.

Restricting the depth of submersion is extremely important, since the water pressure on pores in the fruit stem increases instantly with submersion. As a general policy, the maximum immersion depth should be limited to a single layer of floating fruit, except at the point where the fruit enter the system. Since most tomatoes float with the fruit stem scar up, there is not much hydrostatic pressure on pores of fruit floating in a single layer. The porosity of the stem scar increases as the fruit becomes warmer. Thus, greater restrictions must be imposed on handling fruit harvested in warm or hot temperatures compared with moderate temperatures.

The impact of the fruit with the water is a potential point of infiltration and cause of congestion. Using heavy streams of water to dislodge fruit piles in gondolas is a questionable handling practice. Even so, the gondola system is widely used and fruit handled in this manner have been successfully marketed. The fact remains, however, that using a heavy, warm stream of water drives water into fruit stem scars and other openings on the surfaces of tomato fruit, which is hazardous to the health and wholesomeness of the fruit.

Infiltration increases as the surface tension of water decreases. Given this, surfactants should never be added to the water, even if the water has been chlorinated. For example, the onset of detectable infiltration (i.e., greater than or equal to 0.1 g) occurs within 90 sec when tomatoes are submerged in water plus a surfactant, but when water alone is used, detectable infiltration does not occur until 10 min. Chlorinating the water increases the amount of water absorbed by the fruit and does not prevent the subsequent development of postharvest diseases if infiltration occurs abruptly. Thus, adequate water chlorination does not preclude the need for strategies to prevent infiltration.

Control the Postharvest Environment

Fruit stored in a warm, wet environment are more likely to become diseased and lose horticultural quality than fruit stored in a cool, dry environment. Consequently, the use of refrigeration is a key strategy for controlling postharvest diseases. Tomato fruit are susceptible to chilling injury, however, and should not be stored at temperatures below approximately 13°C or subjected to prolonged exposure to that temperature.

Moisture is the second component of the postharvest environment that may affect the incidence and severity of postharvest decay. Free water on fruit surfaces promotes the development of decay. Fruit should be adequately dried after being washed and should never be allowed to become wet with condensation. Fruit should not be wet during or after packaging.

Selected References

Bartz, J. A. 1981. Variation in the latent period of bacterial soft rot in tomato fruit. Phytopathology 71:1057-1062.

Bartz, J. A. 2006. Internalization and infiltration. Pages 75-94 in: Microbiology of Fruits and Vegetables. G. M. Sapers, J. R. Gorny, and A. E. Yousef, eds. CRC Press, Boca Raton, FL.

Bartz, J. A., and Wei, C.-I. 2003. The influence of bacteria. Pages 519-541 in: Postharvest Physiology and Pathology of Vegetables. J. A. Bartz and J. K. Brecht, eds. Marcel Dekker, New York.

Ogawa, J. M., and Manji, B. T. 1984. Control of postharvest diseases by chemical and physical means. Pages 55-66 in: Postharvest Pathology of Fruits and Vegetables: Postharvest Losses in Perishable Crops. H. E. Moline, ed. Publ. NE-87 (UC Bull. 1914), Agricultural Experiment Station, University of California, Berkeley.

Ritenour, M. A., and Bartz, J. A. 2012. Maintaining tomato fruit health after harvest. Pages 159-172 in: Tomato Health Management. R. M. Davis, K. Pernezny, and J. C. Broome, eds. American Phytopathological Society, St. Paul, MN.

Spotts, R. A. 1984. Environmental modification for control of postharvest decay. Pages 67-72 in: Postharvest Pathology of Fruits and Vegetables: Postharvest Losses in Perishable Crops. H. E. Moline, ed. Publ. NE-87 (UC Bull. 1914), Agricultural Experiment Station, University of California, Berkeley.

White, G. C. 1999. Handbook of Chlorination and Alternative Disinfectants. 4th ed. John Wiley & Sons, New York.

(Prepared by J. A. Bartz)

Diseases Caused by Nematodes

The impact of and association of plant-parasitic nematodes on tomato are among the most studied in all of nematology. In conducting nematological research, many laboratories continue to rely on tomato as a preferred plant host for testing and for culturing nematodes for population increase. Overall, more than 34 species of nematodes have been recovered from the tomato rhizosphere or shown to be hosted by tomato, but only a few species are of economic importance (Table 8).

Tomato damage thresholds are available for some root-knot (*Meloidogyne* spp.) and lesion (*Pratylenchus* spp.) nematodes, but specific information about tomato crop losses is not available. The average estimate of annual yield losses caused by nematodes for all crops worldwide is approximately 12.0%, and a 9.4% annual loss is reported for solanaceous crops in the United States.

The total economic loss caused by all species of nematodes reported to attack tomato is thought to far exceed the visible damage caused by root-knot nematodes. Little or no damage by nematodes is often observed or reported on many fresh-market tomato cultivars, which are grown extensively with plastic mulch in fumigated soil. As fumigants are eliminated from use, nematode damage to tomato is expected to increase. At the field scale, nematode-induced crop losses for tomato can be slight to severe, depending on the nematode species, initial soil population density, and edaphic and environmental conditions.

In general, root-knot nematodes pose the greatest overall nematode threat to tomato on a global basis. In subtropical and tropical cropping systems, the absence of low winter temperatures to reduce pest pressures seasonally is the primary cause of the greater diversity and severity of nematode problems, particularly those caused by root-knot nematodes. In addition to *Meloidogyne* and *Pratylenchus,* others nematode genera reported to cause crop losses in tomato are *Belonolaimus* (sting nematode) and *Rotylenchulus* (reniform nematode). Sting and

TABLE 8. Plant-Parasitic Nematodes Recovered from Tomato Rhizosphere and Documented as Parasites or Reported to Cause Damage

Aphelenchoides ritzemabosi	*Meloidogyne graminicola*
Belonolaimus gracilis	*Meloidogyne hapla*[a]
Belonolaimus longicaudatus[a]	*Meloidogyne incognita*[a]
Ditylenchus destructor	*Meloidogyne javanica*[a]
Ditylenchus dipsaci	*Meloidogyne thamesi*
Globerdera pallida	*Nacobbus aberrans*
Globodera rostochiensis	*Nacobbus batatiformis*
Helicotylenchus microbolus	*Paratrichodorus christiei*
Hemicycliophora similis	*Pratylenchus crenatus*
Heterodera cruciferae	*Pratylenchus neglectus*
Heterodera schactii	*Pratylenchus penetrans*[b]
Heterodera trifolii	*Pratylenchus projectus*
Hirschmanniella spp.	*Radopholus similis*
Longidorus maximus	*Rotylenchulus reniformis*
Meloidogyne acronea	*Tylenchorhynchus claytoni*
Meloidogyne arenaria	*Tylenchorhynchus mashoodi*
Meloidogyne chitwoodi[a]	

[a] Economically important.
[b] Economically important in sandy soils.

reniform nematodes can produce serious crop losses but appear to cause localized problems in the southeastern United States, whereas lesion nematodes appear more problematic in the northern temperate zones.

Generalized Life Cycle

Most plant-parasitic nematodes are vermiform, transparent, microscopic roundworms that attack the roots of plants. All plant-parasitic species have well-developed stylets (i.e., spearlike mouthparts), which they use to penetrate and extract cytoplasmic fluids from plant cells, and all plant-parasitic species have similar life cycles, as well. However, the shapes, means of reproduction, survival rates, and presence of males and females among nematode species can vary greatly.

The life stages of the nematode include the egg, four juvenile stages (J1–J4), and the adult male or female. Some nematodes—including the sting, lesion, and reniform nematodes—are amphimictic, requiring fertilization of the female by the male. However, other species—such as the root-knot nematode—reproduce via parthenogenesis (i.e., without fertilization). During development, the first-stage juvenile (J1) typically molts inside the egg and the second-stage juvenile (J2) hatches. In most cases, the J2 is the infective stage.

Species of root-knot and reniform nematodes exhibit sexual dimorphism; mature females become sedentary and greatly enlarged, often assuming a lemon or pear shape. The eggs are deposited into a gelatinous matrix that protrudes through a rupture of the root surface. Under optimum conditions, the nematode life cycle can be completed in 19–30 days.

Persistence in the soil is influenced by the species of root-knot nematode, prevailing environmental conditions, and population of J2 lipid reserves. The decline of the root-knot population in the soil usually occurs relatively rapidly, but the population can persist at damaging levels for 90 days or longer.

Feeding Habits

All parasitic forms feed and reproduce only on their hosts (i.e., obligate parasitism). However, some plant nematodes—including the root-knot, reniform, and cyst (*Globodera* spp.) nematodes—feed inside roots or shoots (endoparasites), whereas others feed externally (ectoparasites). Fibrous roots are typically the sites of nematode invasion. Root-knot, reniform, and cyst nematodes induce the host root tissue to develop specialized feeding cells that are essential to the nematodes' development. Some migratory endoparasites, such as those of the genus *Pratylenchus,* move and feed throughout the root tissue, causing the development of root lesions.

(Selected References appear on pages 118–119 at the end of the section Other Nematodes.)

(Prepared by J. W. Noling)

Root-Knot Nematodes

Worldwide, approximately 100 species of root-knot nematodes (*Meloidogyne* spp.) attack thousands of species of plants. The annual loss to these parasites on all food and fiber crops approaches $100 million.

Since the first report of a host–parasite relationship between the root-knot nematode and the tomato was made in 1889, 34 nematode species, belonging to 22 genera, have been recovered from the tomato rhizosphere and documented as parasites of or causing damage to tomato (see Table 8). However, root-knot nematodes are the most important genus of plant-parasitic nematodes that attack tomato. For that reason, the focus of this section on diseases caused by nematodes is on root-knot nematodes.

Symptoms

Plant symptoms induced by *Meloidogyne* spp., as well as other nematode genera and species, generally result from root dysfunction—specifically, from reductions in rooting volume and foraging and utilization efficiency of water and nutrients. Foliar symptoms of nematode infestation of tomato roots include stunting and general unthriftiness, premature wilting and slow recovery after rainfall or irrigation, leaf chlorosis (yellowing), and other symptoms characteristic of nutrient deficiency (Fig. 173). Wilting may be especially prevalent in coarse-textured soils and associated with root-knot nematodes. An increased rate of ethylene production (thought to be largely responsible for symptom expression in tomato) is also closely associated with root-knot nematode root infection and gall formation. In the field, plants exhibiting symptoms of stunting or decline usually occur in patches of nonuniform growth, rather than an overall decline.

The time required for tomato plants to express symptoms of injury from *Meloidogyne* spp. is related to the initial soil population density of nematodes, as well as the soil texture, cultivar susceptibility, and prevailing environmental conditions. For example, when nematode infestation is heavy, tomato seedlings or transplants may fail to develop, demonstrate stunted growth, or die, causing poor or patchy stand development. At a less severe level of infestation, symptom expression may be delayed until later in the season, when a number of nematode reproductive cycles have been completed. In this case, aboveground symptoms may not be readily apparent early in crop development, but with time and a reduction in root system size and function, symptoms will become more pronounced.

Root-knot nematodes cause swollen areas (galls) on the roots of infected plants (Fig. 174). The display of galls may range from a few spherical swellings to extensive areas of elongated, convoluted, and coalesced swellings that have resulted from multiple and repeated infections. The size and shape of the galls, which consist of hypertrophied cortical cells surrounding the nematodes, are often indicative of the species inducing them. Root galls induced by three common species of *Meloidogyne*—*M. arenaria, M. incognita,* and *M. javanica* (see Causal Organisms)—are rather large (often 0.5 cm or more in diameter), and they may appear continuously along the roots or, in the case of infection by *M. arenaria,* in a beadlike arrangement. In contrast, galls induced by *M. hapla* are very small (usually less than 2 mm in diameter) and are associated with an abnormally high number of lateral roots. The presence of galls is not always indicative of root-knot nematode infection, however. In most cases, an examination of tomato roots usually provides a quick, definitive diagnosis of root-knot infection, including infection severity, and if sufficient plants are observed, the field distribution and potential for crop damage can be determined, as well.

Tomato plants generally become more tolerant of nematode damage as the roots age and increase in size. Over time and with repeated infection, root symptoms may change dramatically, especially when the initial soil infestation was high and other soil-pathogenic or saprophytic organisms were present. Extensive root decay may be caused by root leakage from and invasion of the nutrient-rich root galls by a wide range of fungi and bacteria. If highly susceptible plant roots are allowed to remain in the ground, few intact galls may be present by harvest time. In general, the degree of root damage is determined by several factors: host tolerance, initial damage level, and the duration and rates of population increase. Reliable diagnostic assessments should be performed only on green, reasonably healthy plants.

Fig. 173. Symptoms of Fusarium wilt of tomato (*Fusarium oxysporum* f. sp. *lycopersici*) following root infection from southern root-knot nematode (*Meloidogyne incognita*) in tunnel (**A** and **B**) and field (**C**) production systems. (Courtesy J. W. Noling)

Causal Organisms

An undescribed species of root-knot nematode was first observed to cause severe levels of root galling in 1855. Nearly 35 years later, in 1889, S. C. Neal recorded what must be the first observation of a root-knot nematode species infecting tomato in Florida. The identity of *Meloidogyne* spp. remained ambiguous, however, and was often reported as belonging to the genus *Heterodera* or *Anguillula*. This continued until 1949, when the genus *Meloidogyne* was revised by B. G. Chitwood.

Worldwide, four common species of *Meloidogyne* are considered the most economically important pests of tomato: (1) *M. arenaria* (Neal) Chitwood; (2) *M. hapla* Chitwood; (3) *M. incognita* (Kofoid & White) Chitwood; and (4) *M. javanica* (Treub) Chitwood. Other *Meloidogyne* spp. that attack tomato are usually only of local importance. Of the four common species, *M. incognita* is the most widely distributed (occurring in every country where minimum temperatures fall below 3°C) and causes the most crop damage.

Two new *Meloidogyne* spp. have been confirmed to occur naturally on tomato in Florida: *M. mayaguensis* Rammah & Hirschmann (1988) and *M. floridensis* Handoo et al. (2004). The host range of these species is significantly different from the host ranges of other *Meloidogyne* spp., which makes the identification and distribution of these nematodes in Florida tomato fields important for disease management. Moreover, resistant tomato cultivars fail to inhibit the reproduction of these species, which means significant yield losses can occur.

All root-knot species exhibit marked sexual dimorphism, with the adult females being pyriform or saccate and the adult males being vermiform (Fig. 175). Because most species reproduce without fertilization (via parthogenesis), there are often few males. The frequency and number of males that develop varies with the species and population of *Meloidogyne,* as well as the environmental conditions. Under adverse conditions, such as crowding and suboptimal weather, developing juveniles

Fig. 174. Formation of galls on roots of tomato plants, caused by infection from southern root-knot nematode (*Meloidogyne incognita*). (Courtesy J. W. Noling)

may undergo sexual reversal and give rise to an abnormally high number of males.

Species are differentiated on various morphological characteristics and host differentials. Molecular means can be used, as well, such as monoclonal antibodies, isozyme phenotypes, polymerase chain reaction (PCR) assays, and DNA probes. Some of these tests can reveal high levels of variation in morphology, cytogenetics, physiology, and fecundity on the selected host differentials.

Although root-knot nematodes reproduce parthogenetically, they have specialized host races—including races that can attack certain tomato cultivars that have genes for resistance. Cultivars have been developed with specific resistance to *M. arenaria, M. incognita,* and *M. javanica* but not to *M. hapla.*

Disease Cycle and Epidemiology

All of the common root-knot nematode species have similar life cycles (Fig. 175). Root-knot nematodes attack almost all plant species and reproduce at a very rapid rate; many billions of eggs and juveniles can inhabit a single hectare of soil.

After hatching, the J2 moves into the soil, where it penetrates the host root to establish a permanent feeding site. The J2 completes three additional molts before becoming an adult. Upon maturity, the female deposits eggs into a gelatinous matrix; each female lays 300–2,000 or more eggs. The eggs of most nematodes, including those of *Meloidogyne* spp., can survive for a few weeks or months. A proportion of *Meloidogyne* spp. eggs within the eggmass may undergo diapause, whereas others may survive extended periods of drought conditions via anhydrobiosis to help ensure survival of the population.

Under optimum conditions, *Meloidogyne* spp. complete their life cycle in 3–4 weeks. As many as five to eight generations may be completed within a single growing season. Under tropical conditions and within a continuous cropping system, 13 to 15 generations may be completed.

M. hapla (the northern root-knot nematode) differs from the other three primary *Meloidogyne* spp. in that it thrives at moderate to low temperatures. *M. incognita* (the southern root-knot nematode) requires intermediate temperatures and is inactive at temperatures below 19°C, whereas *M. arenaria* (the peanut root-knot nematode) and *M. javanica* (the Javanese root-knot nematode) prefer temperate and tropical conditions. As noted earlier, problems caused by root-knot nematodes are generally more diverse and severe in the tropics and subtropics, where the soil moisture and soil temperature seldom deviate from optimal ranges. Regardless, most *Meloidogyne* spp. survive longer in the soil in cool winter conditions than in warm summer conditions.

The highest soil density of *Meloidogyne* spp. generally occurs near crop harvest, unless severe crop and root damage caused by a high initial population of nematodes has resulted in an early decline. In most cases, the root-knot population declines sharply right after harvest, with fewer than 10% surviving in the absence of a host. Without a host, the spring population of root-knot nematodes often declines to a nondetectable level in temperate and tropical conditions. Because most root-knot nematodes have an extensive host plant range, including many weeds, their preplant population level can remain high even in the absence of a crop. Given this, weed management can play a key role in the epidemiology of *Meloidogyne* spp. In fact, unmanaged weed growth is an important determinant of the continued need to use soil fumigation for nematode and disease management.

Root-knot nematodes are well suited to being spread by both commerce and agriculture. Spread occurs easily through soil

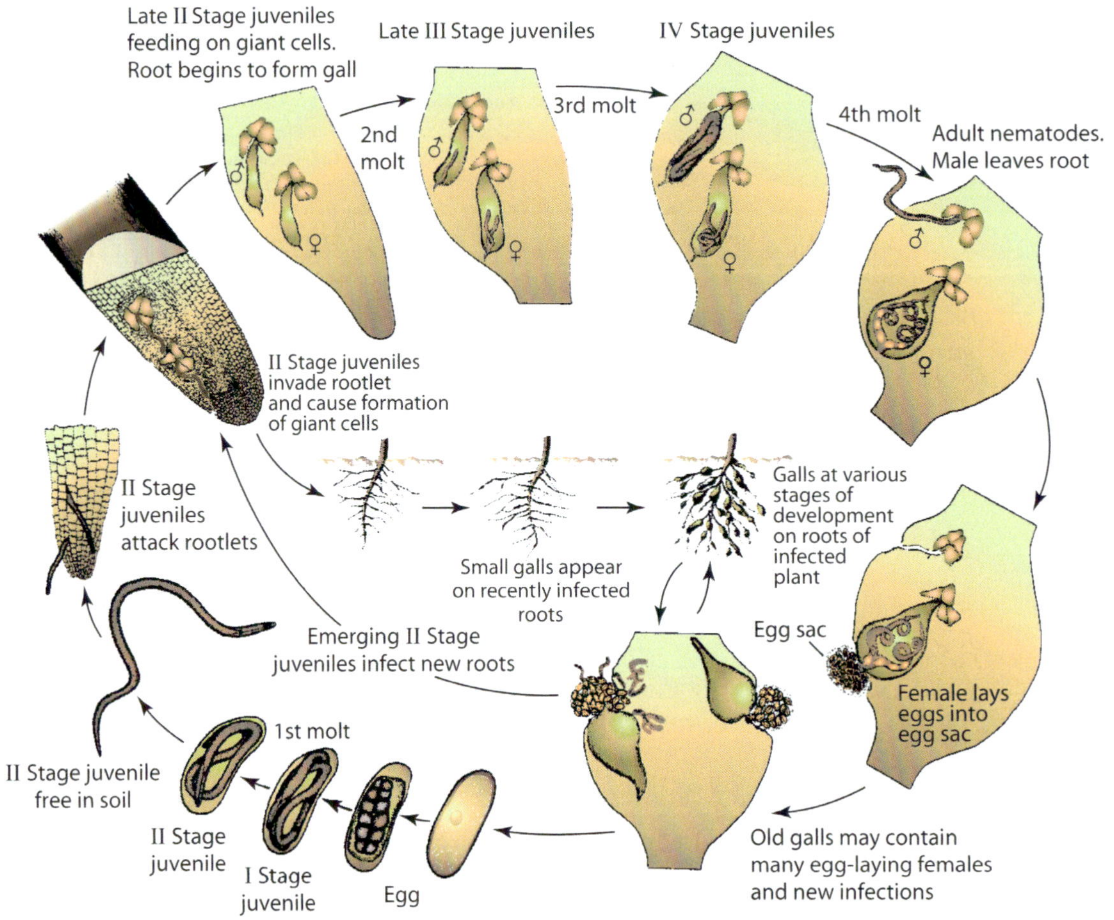

Fig. 175. Life cycle of the root-knot nematode (*Meloidogyne* spp.) and associated symptoms on infected plants. (This figure was published in *Plant Pathology*, 5th ed., George N. Agrios, Figure 15-11, page 841, Copyright Elsevier 2005.)

movement, and an epidemic caused by a single nematode species may result from moving contaminated equipment from farm to farm. Root-knot nematodes also move long distances by infected nursery plants and transplants of annual crops. This ease of spread is thought to help explain the worldwide distribution of *Meloidogyne* spp., including their rapid appearance and resulting damage within a few years after the cropping of virgin land.

Disease complexes involving root-knot nematodes and other fungal or bacterial pathogens are well documented and cause significant crop losses in tomato. The classic example is *M. incognita* predisposing wilt-susceptible tomato cultivars to infection by *Fusarium oxysporum* Schltdl.:Fr. f. sp. *lycopersici* (Sacc.) W. C. Snyder & H. N. Hansen (see Fig. 173, page 115). *M. incognita* is also associated with an increased incidence and severity of bacterial wilt of tomato, caused by *Ralstonia solanacearum* (Smith) Yabuuchi et al.

In contrast, the presence of *M. incognita* does not appear to alter resistance to Fusarium wilt in field-grown tomatoes, as can occur in other crop varieties. The wide host range, short generation times, and high reproductive rates of root-knot nematodes, coupled with their interaction among various pathogens and pests, pose significant challenges in developing successful crop and nematode management strategies in tomato.

Management

The management of plant-parasitic nematodes of tomato currently relies on integrated pest management (IPM), or the combined use of various chemical, cultural, and biological strategies. The high crop value and susceptibility of tomato to nematode damage make it critical to minimize the nematode population before planting, particularly since postplant control strategies are of limited value. Thus, making reliable preplant sample estimates of the nematode population density and consequent potential for damage forms the basis for selecting the appropriate management strategy.

In general, a soil sample of 10 to 30 composited soil cores is used to estimate the soil population densities of vermiform stages in fields up to 2 ha in size. Given the possible inefficiency and imprecision involved in sampling, however, soil samples are not particularly useful for quantifying or detecting a low or aggregated population of nematodes. Different techniques, such as bioassays, are required to estimate egg densities in soil, and end-of-season measurements of root gall severity are useful for characterizing patterns of field distribution and potential damage in current and subsequent crops.

Damage thresholds for *Meloidogyne* spp. are derived from crop damage functions, where the tomato crop yield is usually demonstrated to be inversely related to the initial preplant soil population density of nematodes. Tomato damage thresholds, expressed as juveniles per 100–200 cm^3 of soil, are generally reported in the ranges of 2–100, 20–100, and 2–100 for *M. arenaria, M. hapla,* and *M. incognita,* respectively. For *M. incognita* eggs, the preplant population threshold for damage can be extremely low (approximately 0.06 eggs per gram of soil).

Considerable variability in the level of damage observed in the field can occur because of differences in soil texture and other prevailing biotic and environmental conditions. In some instances, the damage thresholds estimated for tomato are at or below the limit of field detection of nematodes. Given the variability in damage function relationships to specific edaphic and environmental conditions, the potential for multiple pest interactions, and the imprecision that can occur in sampling, the mere presence of root-knot nematodes in preplant samples usually suggests a potentially serious problem for tomato production. In production areas with warm climates and sandy soils (where soil temperatures favor high levels of nematode reproduction and activity), nematode management decisions are often made on the basis of the mere presence or absence of the parasites.

Cultivars with race- and species-specific resistance to *M. incognita, M. arenaria,* and *M. javanica* are currently available and used extensively for nematode management in commercial tomato production on a global basis. Resistance but not immunity to these three *Meloidogyne* spp. is conferred by a single dominant gene, designated *Mi,* which invokes a cellular hypersensitive response to invasive juveniles. *Mi* gene resistance is not conferred to *M. hapla, M. mayaguensis,* and *M. floridensis* or to some races of *M. chitwoodi* Golden et al. that attack and reproduce on the *Mi*-resistant tomato cultivars.

Under suitable soil temperatures, a tomato plant's resistance to *Meloidogyne* spp. is determined within the first 24–48 h after infection by juveniles. Under high soil temperatures (27–33°C), however, functionality of the *Mi* gene can be diminished or broken, rendering the resistant plant susceptible to nematode infection and damage. The natural occurrence and resistance-breaking ability of various *Meloidogyne* populations has been well documented. With resistance-breaking populations, loss of resistance has been proven unrelated to thermal sensitivity of the *Mi* gene. Moreover, selection is not immediate but typically occurs progressively over many generations of nematodes, and completely virulent natural populations occur. The loss of resistance has also been determined to be durable over many years, in that virulence of the resistance-breaking population is retained even if the population is reproduced for several generations on other susceptible plants. Alternating the use of resistant and susceptible tomato cultivars is currently recommended to preserve durability and to minimize population shifts toward proliferation of resistance-breaking races.

The *Mi* gene has been largely durable and productive across more than 40 years of commercial use. In addition to the direct benefits to the tomato crop, multicrop benefits have been demonstrated. Reduced soil density, root gall severity, and yield losses are frequently reported in crops planted after crops of tomato with resistance to root-knot nematodes.

Crop rotation can be very useful for nematode management when the nematode species can be identified and suitable poor or nonhost crops can be substituted into the rotation scheme. Identifying a nonhost can be difficult, however, because *M. arenaria, M. incognita,* and *M. javanica* have such immense host ranges—including nearly every plant family and the majority of the estimated 250,000 species of flowering plants.

When nonhost crops are available, rotations of 1–2 years are usually adequate for control of root-knot nematodes but seldom economic. Cover crops of sunn hemp (*Crotalaria juncea* L.), hairy indigo (*Indigofera hirsute* L.), American jointvetch or shyleaf (*Aeschynomene americana* L.), cowpea (*Vigna unguiculata* (L.) Walp.), and sorghum bicolor (*Sorghum bicolor* (L.) Moench) have all been determined to be poor hosts of *Meloidogyne* spp. and can return large amounts of nitrogen and organic matter back to the soil. Overall, reduced juvenile invasion of roots, developmental rates, and fecundity and increased generation times appear to be common characteristics of poor hosts. Compared to good hosts, poor and resistant hosts are also usually much more tolerant of damage by root-knot nematodes.

Many of the small grains—such as rye (*Secale* L.), barley (*Hordeum* L.), wheat (*Triticum* L.), and oats (*Avena* L.)—are often used effectively as winter rotation crops for management of root-knot nematodes. However, these crops can support limited reproduction of root-knot nematodes if planted when soil temperatures exceed 18°C, a threshold temperature for nematode activity. Such a high temperature threshold appears to preclude the effective use of these grain covers in temperate and subtropical regions. Pasture land rotations with pangola digitgrass (*Digitaria eriantha* Steud.), bahiagrass (*Paspalum notatum* Flueggé), and bermudagrass (*Cynodon dactylon* (L.) Pers.) have significantly reduced but not eliminated the population of root-knot nematodes.

In the United States and in some European countries, rotations of certain *Brassica* spp., containing low dosages of glucosinolates or decomposition products containing isothiocyanates, are known to be effective cover crops for nematode management. In studies, the biofumigant effect derived from a green-manured *Brassica* rotation crop was synergized when used in combination with other disease management tactics, such as soil solarization. It is generally assumed that certain saprophytic microflora and fauna associated with the decomposition of soil amendments play a role in nematode suppression. As part of an IPM system, use of bare fallow for even a short period (3–5 months) can also substantially reduce the density of the nematode population, particularly when coupled with periodic tillage and when the soil temperature and moisture level are conducive for egg hatch.

Soil treatment with broad-spectrum fumigants is used effectively for soilborne pest and disease control in many regions worldwide. Where broad-spectrum fumigants are not used, nematicides such as metam sodium and 1,3-dichloropropene (1,3-D) are often used for nematode control. To be effective, these products must diffuse through the soil as gases; thus, the most efficient use of fumigants occurs when the soil is well drained, in seedbed condition, and at a temperature above 12°C. Nonfumigant nematicides have generally performed inconsistently as preplant and postplant treatments, with the loss of efficacy often greatly affected by various physical, edaphic, and biological factors. The use of nematicides in combination with plant resistance is a strategy that can be used in heavily infested fields, in situations when key nematode pests co-exist, or to preserve the durability of resistance.

Innovative production technologies and ecology-based nematode management systems have also been developed. In many areas of the world, the integration of soilless media into greenhouse and sheltered production systems has gained acceptance as a means of maintaining high yields and avoiding issues with many soil and root pathogens. In Florida, a tomato production system has been developed using strip tillage into existing bahiagrass pasture. By minimizing inputs and disruption to the pasture, growers are able to reap the rotational benefits of bahiagrass pasture land without extensive cultivation and land preparation for the rotational tomato crop. Clearly, the development of these integrated crop, nematode, and pest management systems requires further testing and economic analysis.

Many predators and parasites have been demonstrated to have potential as agents for biological control of plant-parasitic nematodes. For example, the bacterial parasite *Pasteuria penetrans* (ex Thorne) Sayre & Starr has been shown to be an effective biocontrol agent for a number of *Meloidogyne* spp. under experimental conditions within greenhouse and field microplot settings, but its performance and practicality have yet to be determined under commercial field production systems. Commercial nematode biocontrol products are increasingly becoming available.

Despite the availability of these products, it seems reasonably clear that inundative applications to soil will not ensure activity or efficacy. Additional agroecological research is needed to increase the future utility of biological control as a viable tactic for nematode management. It also seems likely that biocontrol agents will be implemented as an integral component of nematode management, rather than a stand-alone tactic.

(Selected References appear on pages 118–119 at the end of the section Other Nematodes.)

(Prepared by J. W. Noling)

Other Nematodes

In addition to *Meloidogyne* spp., many other nematode species can attack tomato, including species within the genera *Helicotylenchus, Heterodera, Longidorus, Paratrichodorus, Trichodorus,* and *Xiphinema.* Numerous species of *Pratylenchus*—such as *P. crenatus* Loof; *P. neglectus* (Rensch) Filipjev & Schuurmans Stekhoven; *P. penetrans* (Cobb) Chitwood & Oteifa; *P. scribneri* Steiner; *P. thornei* Sher & Allen; and *P. vulnus* Allen & Jensen—have been described as pathogenic to tomato, particularly in the northern climatic regions of the United States and Canada. Of the *Pratylenchus* spp., the lesion nematode (*P. penetrans*) is considered the most important, with a nematode damage threshold reported as low as 200–400 per 100 cm^3 of soil.

Another nematode that is pathogenic to tomato is the ectoparasitic sting nematode (*Belonolaimus longicaudatus* Rau). Although the sting nematode is limited largely to the sandy soils of the coastal plain in the southeastern United States, it can be a damaging pathogen of tomato (Fig. 176). Root symptoms induced by the sting nematode cause infected plants to form a tight mat of short roots, often displaying a swollen appearance. New root initials are generally killed by heavy infestations of the sting nematode, a symptom reminiscent of fertilizer salt burn.

The reniform nematode (*Rotylenchulus reniformis* Linford & Oliveira) is sometimes problematic in the rock-based and organic soils of southern Florida. This parasite can attain a high population and is, at times, an economically significant pest of tomato.

(Prepared by J. W. Noling)

Selected References

Abawi, G. S., and Chen J. 1998. Concomitant pathogen and pest interactions. Pages 135-159 in: Plant-Nematode Interactions. K. R. Barker, G. A. Pederson, and G. L. Windham, eds. American Society of Agronomy, Madison, WI.

Bakre, A. V. 1999. Ammonium accumulation and ethylene evolution by tomato infected with root-knot nematode and grown under different regimes of plant nutrition. Commun. Soil Sci. Plant Anal. 30:175-182.

Barker, K. R., and Koenning, S. R. 1998. Developing sustainable systems for nematode management. Annu. Rev. Phytopathol. 36:165-205.

Barker, K. R., Schmitt, D. P., and Imbriani, J. L. 1985. Nematode population dynamics with emphasis on determining damage potential to crops. Pages 135-149 in: An Advanced Treatise on Meloidogyne. Vol. 2, Methodology. K. R. Barker, C. C. Carter, and J. N. Sasser, eds. North Carolina State University Graphics, Raleigh.

Castagnone-Sereno, P., Bongiovanni, M., and Dalmasso, A. 1993. Stable virulence against the tomato resistance *Mi* gene in the par-

Fig. 176. Field distribution of the sting nematode (*Belonolaimus longicaudatus*) on grape tomato. Note the well-defined infestation boundaries and severe plant stunting. (Courtesy J. W. Noling)

thenogenetic root-knot nematode *Meloidogyne incognita*. Phytopathology 83:803-805.

Chellemi, D. O., Olson, S. M., Mitchell, D. J., Secker, I., and McSorley, R. 1997. Adaptation of soil solarization to the integrated management of soilborne pests of tomato under humid conditions. Phytopathology 87:250-258.

Chellemi, D. O., Rhoads, F. M., Olson, S. M., Rich, J. R., Murray, D., Murray, G., and Sylvia, D. M. 1999. An alternative, low-input production system for fresh market tomatoes. Am. J. Altern. Agric. 14:59-68.

Colyer, P. D., Kirkpatrick, T. L., Vernon, P. R., Barham, J. D., and Bateman, R. J. 1998. Reducing *Meloidogyne incognita* injury to cucumber in a tomato-cucumber double-cropping system. J. Nematol. 30:226-231.

Curran, J., and Robinson, M. P. 1993. Molecular aids to nematode diagnosis. Pages 545-565 in: Plant Parasitic Nematodes in Temperate Agriculture. K. Evans, D. L. Trudgill, and J. M. Webster, eds. C.A.B. International, Wallingford, UK.

Duncan, L. W., and Noling, J. W. 1998. Agricultural sustainability and nematode IPM. Pages 251-287 in: Plant-Nematode Interactions. K. R. Barker, G. A. Pederson, and G. L. Windham, eds. American Society of Agronomy, Madison, WI.

Eisenback, J. D., and Triantaphyllou, H. H. 1991. Root-knot nematodes: *Meloidogyne* species and races. Pages 191-274 in: Manual of Agricultural Nematology. W. R. Nickle, ed. Marcel Dekker, New York.

Evans, K., Trudgill, D. L., and J. M. Webster, eds. 1993. Plant Parasitic Nematodes in Temperate Agriculture. C.A.B. International, Wallingford, UK.

Fernandez, M., and Ortega, J. 1998. An overview of nematological problems in Cuba. Nematropica 28:151-164.

Glazer, I., Orion, D., and Apelbaum, A. 1985. Interrelationships between ethylene production, gall formation, and root-knot nematode development in tomato plants infected with *Meloidogyne javanica*. J. Nematol. 15:539-544.

Goodey, J. B., Franklin, M. T., and Hooper, D. J. 1965. The Nematode Parasites of Plants Catalogued Under Their Hosts. 3rd ed. Commonwealth Agricultural Bureaux, Farnham Royal, UK.

Hussey, R. S., and McGuire, J. M. 1987. Interaction with other organisms. Pages 294-320 in: Principles and Practices of Nematode Control in Crops. R. H. Brown and B. R. Kerry, eds. Academic Press, New York.

Johnson, A. W., and Campbell, G. M. 1980. Managing nematode populations densities on tomato transplants using crop rotation and a nematicide. J. Nematol. 12:6-18.

Koenning, S. R., Overstreet, C., Noling, J. W., Donald, P. A., Becker, J. O., and Fortnum, B. A. 1999. Survey of crop losses in response to phytoparasitic nematodes in the United States in 1994. J. Nematol. 31:587-618.

Mai, W. F., Mullin, P. G., Lyon, H. H., and Loeffler, K. 1996. Plant Parasitic Nematodes: A Pictorial Key to Genera. Cornell University Press, Ithaca, NY.

McSorley, R. 1987. Extraction of nematodes and sampling methods. Pages 13-48 in: Principles and Practices of Nematode Control in Crops. R. H. Brown and B. R. Kerry, eds. Academic Press, New York.

McSorley, R. 1998. Population dynamics. Pages 109-135 in: Plant-Nematode Interactions. K. R. Barker, G. A. Pederson, and G. L. Windham, eds. American Society of Agronomy, Madison, WI.

McSorley, R. 1999. Host suitability of potential cover crops for root-knot nematodes. J. Nematol. 31:619-623.

Neal, J. C. 1889. The root-knot disease of the peach, orange and other plants in Florida due to the work of *Anguillula*. Bull. U.S. Bur. Entomol. 20:1-31.

Nickle, W. R. 1991. Manual of Agricultural Nematology. Marcel Dekker, New York.

Noe, J. P. 1992. Variability among populations of *Meloidogyne arenaria*. J. Nematol. 24:404-414.

Noling, J. W. 2002. The practical realities of alternatives to methyl bromide: Concluding remarks. Phytopathology 92:1373-1375.

Noling, J. W., Duncan, L. W., and Gilreath, J. P. 2003. Grower conducted surveys for field diagnosis and sampling for root-knot nematode based on root gall indices. Pages 431-434 in: Annual International Research Conference on Methyl Bromide Alternatives and Emissions Reductions, Methyl Bromide Alternatives Outreach, Fresno, CA.

Noling, J. W., and Gilreath, J. P. 2002. Weed and nematode management: Simultaneous considerations. Pages 61-64 in: Annual International Research Conference on Methyl Bromide Alternatives and Emissions Reductions, Methyl Bromide Alternatives Outreach, Fresno, CA.

Potter, J. W., and Olthof, T. H. A. 1993. Nematode pests of vegetable crops. Pages 171-207 in: Plant Parasitic Nematodes in Temperate Agriculture. K. Evans, D. L. Trudgill, and J. M. Webster, eds. C.A.B. International, Wallingford, UK.

Rich, J. R., Dunn, R. A., and Noling, J. W. 2004. Nematicides: Past and present uses. Pages 468-481 in: Nematology: Advances and Perspectives. Vol. 2, Nematode Management and Utilization. Z. X. Chen, S. Y. Chen, and D. W. Dickson, eds. C.A.B. International, Beijing.

Roberts, P. A. 2002. Concepts and consequences of resistance. Pages 23-41 in: Plant Resistance to Parasitic Nematodes. J. L. Starr, R. Cook, and J. Bridge, eds. C.A.B. International, Wallingford, UK.

Sasser, J. N., and Carter, C. C., eds. 1985. An Advanced Treatise on *Meloidogyne,* vol. 1. North Carolina State University Graphics, Raleigh.

Siddiqui, I. A., Sher, S. A., and French, A. M. 1973. Distribution of Plant Parasitic Nematodes in California. State of California, Department of Food and Agriculture, Division of Plant Industry, Sacramento.

Sorribas, F. J., Ornat, C., Verdejo-Lucas, S., Galeano, M., and Valero, J. 2005. Effectiveness and profitability of the *Mi*-resistant tomatoes to control root-knot nematodes. Eur. J. Plant Pathol. 111:29-38.

Starr, J. L. 1993. Recovery and longevity of egg masses of *Meloidogyne incognita* during simulated winter survival. J. Nematol. 25:244-248.

Statewide Integrated Pest Management (IPM) Program. 1998. Integrated Pest Management for Tomatoes. 4th ed. Publ. 3274. Statewide IPM Program, Agriculture and Natural Resources, University of California.

Trudgill, D. L., and Blok, V. C. 2001. Apomictic, polyphagous root-knot nematodes: Exceptionally successful and damaging biotrophic root pathogens. Annu. Rev. Phytopathol. 39:53-77.

Williamson, V. M. 1998. Root-knot nematode resistance genes in tomato and their potential for future use. Annu. Rev. Phytopathol. 36:277-293.

Part II. Arthropod Pests

Arthropods constitute a large group of animals with exoskeletons and jointed appendages. However, only mites and insects are described in this part of the compendium.

Mites have piercing–sucking mouthparts, and the damage they cause is sometimes confused with that caused by environmental and nutritional disorders of tomato. Insects have various types of mouthparts, and the damage they cause to tomato is determined by the type of mouthpart that characterizes the damaging stage of the life cycle. Many insects have chewing mouthparts or adaptations of them and produce holes in leaves or fruit, as well as leaf mines, leaf rolling, and so on. Many other insects have piercing–sucking mouthparts or adaptations of them, and the damage they cause to tomato is sometimes confused with that caused by biotic and abiotic disorders. In addition, insects with sucking mouthparts may transmit many important diseases of tomato, primarily those incited by viruses and phytoplasmas. Because these viral and phytoplasmal diseases are described elsewhere in this book (see Part I), their insect vectors are not described here unless they inflict direct or indirect damage themselves or the damage they inflict is of unknown etiology.

A thorough review of the arthropod pests that attack tomato is beyond the scope of this discussion. For more information on these pests and the damage they cause to tomato, readers should consult other works (see Selected References). For monitoring and considerations for integrated pest management (IPM) regarding arthropod pests of tomato, readers should consult *Tomato Health Management* (Davis et al., 2012).

Selected References

Capinera, J. L. 2001. Handbook of Vegetable Pests. Academic Press, San Diego, CA.

Davis, R. M., Pernezny, K., and Broome, J. C., eds. 2012. Tomato Health Management. American Phytopathological Society, St. Paul, MN.

Pernezny, K. 2008. Florida Tomato Scouting Guide: Solutions for Your Life. IFAS Extension, University of Florida, Gainesville. Available online at http://erec.ifas.ufl.edu/tomato-scouting-guide

Sorensen, K. A., and Baker, J. R., eds. 1994. Insect and Related Pests of Vegetables. Publ. AG-136. North Carolina Agricultural Extension Service, Department of Agricultural Communications, North Carolina State University, Raleigh.

Statewide Integrated Pest Management (IPM) Program. 1998. Integrated Pest Management for Tomatoes. 4th ed. Publ. 3274. Statewide IPM Program, Agriculture and Natural Resources, University of California.

(Prepared by D. J. Schuster, F. Zalom, and D. Gilrein)

Mites

Spider Mites

Biology

The most common species of spider mite attacking tomato in the field or greenhouse is the two-spotted spider mite (*Tetranychus urticae* Koch), but a closely related species, which is sometimes called the tomato red spider mite (*T. evansi* Baker & Pritchard), is often encountered in Florida. Both species usually have two dark spots on the dorsal surface of the body, which is paler in color; in some cases, the spots are absent or the body color is a uniform reddish orange.

Spider mites are 0.3–0.5 mm long and live on the undersides of the lower leaves on tomato plants. Adult females are oval, whereas adult males have a pointed abdomen (Fig. 177); females are also larger than males. Females can deposit up to 200 spherical eggs, attaching them to the undersides of leaves (Fig. 178).

The larvae that hatch from the eggs have three pairs of legs, while the two subsequent nymphal stages and the adults have four pairs of legs. Development from egg to adult takes 5–19 days, depending on the temperature. A population increase of spider mites is favored by hot, dry conditions.

Fig. 177. Adult spider mite. (Courtesy D. J. Schuster)

Damage

Feeding damage from spider mites begins as small, chlorotic spots (stippling) on the upper surfaces of affected leaves (Fig.

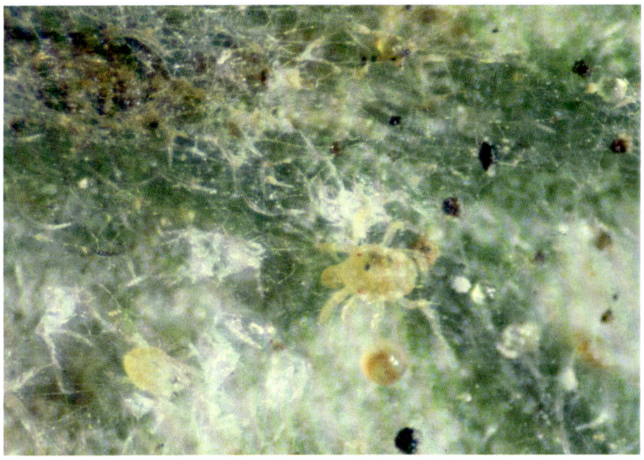

Fig. 178. Spider mite eggs and nymphs on the underside of a leaf. Note the silken threads covering the leaf. (Courtesy D. J. Schuster)

Fig. 179. Stippling on upper surfaces of leaves, caused by feeding of spider mites. (Courtesy D. J. Schuster)

Fig. 180. Aggregations of spider mites at the top of the tomato plant, where they collect before being dispersed by the wind ("ballooning"). (Courtesy D. J. Schuster)

179) and as silken threads covering the lower surfaces of leaves (see Fig. 178). However, as the density of the mite population increases, the mites move to the top leaves of the tomato plant (Fig. 180), where they produce copious amounts of silk webbing. Migrating individuals use silk strands to "balloon" (float) in the air. In a severe infestation, the leaves become generally chlorotic, necrotic, and desiccated.

(Prepared by D. J. Schuster, F. Zalom, and D. Gilrein)

Tomato Russet Mite

Biology

The tomato russet mite (*Aculops lycopersici* (Massee)) is an eriophyid mite; it is so small that a dissecting microscope or a hand lens of at least 14× is needed to detect its presence. In addition to being smaller than spider mites, the tomato russet mite is more elongated and tapers toward the rear end (Fig. 181, top). The tomato russet mite is pale and has four legs as both a nymph and an adult. Nymphs and adults also have long hairs at the end of the abdomen (Fig. 181, bottom).

Development from egg to adult occurs in 8–15 days, depending on the temperature, and is favored by hot, dry conditions. The tomato russet mite is not known to overwinter outdoors in cold climates.

Fig. 181. Top, Adult tomato russet mite. Bottom, Both nymphs and adults have long hairs at the end of the abdomen. (Courtesy D. J. Schuster)

Fig. 182. Progressive damage (left to right) to tomato leaves caused by feeding of tomato russet mites. (Courtesy D. J. Schuster)

Damage

The tomato russet mite causes far more damage in field production than the two-spotted spider mite. The tomato russet mite attacks the stems, leaf axils, leaf petioles, and leaves. Damaged stems and leaves appear silvery or greasy and become bronzed (russeted). Leaves develop chlorosis, curl at the edges, and then desiccate and drop from the plant (Fig. 182). The infestation begins at the bottom of the plant, and if the mite population is not controlled, the infestation progresses up the plant until all of the foliage turns brown and desiccates. Tomato plants may be killed in only a few days in hot, dry weather.

Foliar damage by the tomato russet mite may be confused with that caused by moisture stress or by nutrient deficiencies or imbalances. Off-color tomato foliage should be examined with a powerful hand lens to check for russet mites.

(Prepared by D. J. Schuster, F. Zalom, and D. Gilrein)

Insects

Aphids

Biology

Aphids are small, pear-shaped insects that have a pair of tubercles (i.e., cornicles) projecting up and back from the rear of the dorsal body surface. The two species of aphids that most commonly attack tomato are the potato aphid and the green peach aphid. The potato aphid (*Macrosiphum euphorbiae* (Thomas)) is large (approximately 3 mm long) and may be green or pink. The green peach aphid (*Myzus persicae* (Sulzer)) is smaller (approximately 1.5 mm long); it is usually light to dark green but sometimes orange, pink, or red.

Aphids invade tomato fields as winged adults but produce unwinged offspring parthenogenetically. Aphids are found on the undersides of plants' upper leaves, where they suck out juices with their piercing–sucking mouthparts. Green peach aphids are especially damaging, because they can disseminate many important virus diseases of tomato, including *Alfalfa mosaic virus* (AMV), *Potato virus Y* (PVY), and *Tobacco etch virus* (TEV) (see Part I, the section Diseases Caused by Viruses).

Damage

Even when aphids do not serve as vectors of virus diseases, they may damage plants directly and indirectly when they exist in a large population. Through direct feeding, aphids cause chlorotic spotting, general chlorosis and distortion of leaves, stunting and wilting of plants, and abscission of blooms. Bloom abscission can be caused by other nutritional, physiological, and environmental factors, but when the density of the aphid population is high enough to cause this type of damage, their culpability is obvious from their sheer number.

Indirectly, aphids consume more plant juices than they need for sustenance and excrete the excess as a sugary substance called "honeydew." When heavy secretions of honeydew are deposited on foliage and fruit, sooty mold can grow, resulting in reduced photosynthesis and fruit quality.

(Prepared by D. J. Schuster, F. Zalom, and D. Gilrein)

Whiteflies

Biology

Two species of whitefly are most often seen on tomato. The first species, the greenhouse whitefly (*Trialeurodes vaporariorum* (Westwood)), is most common in the greenhouse but can reach damaging numbers in the field. It is known to vector *Tomato infectious chlorosis virus* (TICV) to both greenhouse and fresh-market tomatoes (see Part I, the section Diseases Caused by Viruses). The second species, *Bemisia tabaci* (Gennadius), is actually a complex of morphologically indistinguishable biotypes or species. (Because of the fluidity of the taxonomy, the complex will be referred to as *B. tabaci* in this discussion.) *B. tabaci* is more common in the field than in the greenhouse and vectors several important virus diseases worldwide. Begomoviruses, including *Tomato yellow leaf curl virus* (TYLCV), are transmitted by *B. tabaci.*

Whiteflies are closely related to aphids, with all life stages occurring on the undersides of tomato leaves. Development from egg to adult requires 15–30 days, depending on the temperature. Unlike adult aphids, however, adult whiteflies are winged and have a white, powdery wax on their wings and bodies. They appear as small, white flecks and are approximately 1.2 mm long (including wings). Adult whiteflies are easily disturbed and make frequent, short flights, usually ending on the same or an adjacent plant. Immature whiteflies are scalelike and range in length from approximately 0.3 to 0.7 mm. The first nymphal stage (called a "crawler") is mobile, whereas the subsequent nymphal and pupal stages are sessile.

The greenhouse whitefly and the sweetpotato whitefly can be easily distinguished from one another using pupal characteristics. The greenhouse whitefly has a fringe of waxy filaments along the upper rim of the pupa, whereas the silverleaf whitely has a flatter pupa with little ornamentation.

Like aphids, whiteflies have piercing–sucking mouthparts at both the nymphal and adult stages. Whiteflies attack the undersides of leaves by sucking out the plants' juices.

Damage

The damage to tomato plants is inflicted primarily by nymphal feeding and is similar to that inflicted by aphid feeding. Common damage includes chlorotic spotting and chlorosis of leaves (Fig. 183), spotting of fruit, stunting and wilting of plants, and production of "honeydew" (and the sooty mold that grows on it).

The silverleaf whitefly produces a disorder characterized by reduced or incomplete ripening of fruit, often in longitudinal sections (Fig. 184). This disorder has no foliar symptoms, unlike similar disorders in some cucurbits (Cucurbitaceae), in which infestations of silverleaf whitefly are associated with leaves having a silvery appearance. On mildly affected fruit, this ripening disorder may produce a pronounced yellow star on the blossom end, with the points of the star extending up the sides of the fruit toward the calyx. The amount of internal white tissue that may occur in some commercial cultivars increases with the density of the whitefly population. Less severe external symptoms may disappear as the fruit ripens, but the internal white tissue remains. Similarly, a fruit that appears to have ripened normally may have extensive internal white tissue.

This irregular ripening disorder is not evident on fruit until ripening occurs. It is most severe in hot weather and is related to nymphal density, rather than adult density. Although the etiology of the disorder is not known, it is physiological in nature and not caused by a pathogen.

(Prepared by D. J. Schuster, F. Zalom, and D. Gilrein)

Fig. 183. Foliar damage caused by nymphal stages of the sweetpotato whitefly. (Courtesy D. J. Schuster)

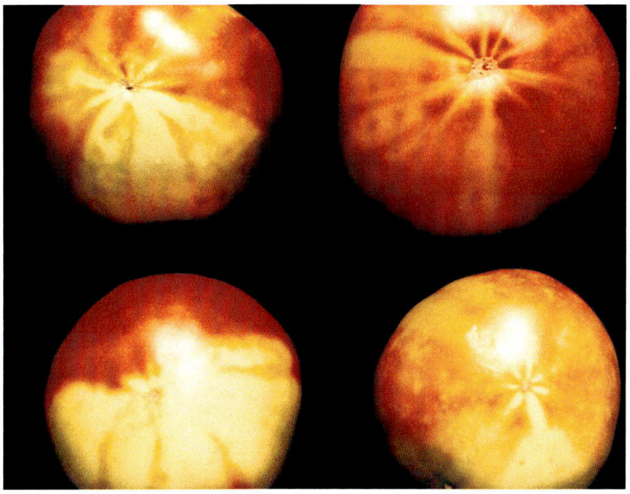

Fig. 184. Irregular ripening of tomato fruit caused by the silverleaf whitefly. (Courtesy D. J. Schuster)

Thrips

Biology

Thrips are small, elongated insects (approximately 0.2 mm long). Some species, such as the greenhouse thrips (*Heliothrips haemorrhoidalis* (Bouché)), attack the foliage of tomato plants, but other species attack primarily the flowers and small fruit. The species that commonly attack tomato flowers include the flower thrips (*Frankliniella tritici* (Fitch)), the western flower thrips (*F. occidentalis* (Pergande)), the tobacco thrips (*F. fusca* (Hinds)), and the onion thrips (*Thrips tabaci* Lindeman).

Adult thrips are usually yellow or brown and have slender, featherlike wings fringed with long hairs. The first two larval stages resemble small, wingless adults, whereas the last two (nonfeeding prepupal and pupal stages) have wing pads. Eggs are inserted into succulent plant tissue, including leaves, stems, pistils, and small fruit.

Adults and immature thrips have rasping–sucking mouthparts, which are used to rupture plant cells and suck out their contents. Thrips move frequently while feeding, thus causing damage over an area disproportionate to their size.

Damage

Tissue damaged by the feeding of thrips initially appears silvery but may turn necrotic or black. Damage to leaves is not usually confused with that produced by noninsect causes. However, damage to flower parts (particularly pistils) may cause bloom abscission, which may be attributed erroneously to a number of nutritional, environmental, and physiological factors. Thrips feeding may interact with these other factors in causing or increasing bloom drop. Slight feeding damage on one side of the pistil may result in deformed or catfaced fruit (see Part III, the section Catface). The feeding or oviposition of thrips on small fruit may result in the development of small pits as the fruit develop, mostly on the blossom end.

Oviposition by the western flower thrips in young fruit may produce subepidermal white spots surrounding the oviposition sites (Fig. 185). The white spots may extend longitudinally on either side of the oviposition site or produce small, pale halos surrounding the site.

(Prepared by D. J. Schuster, F. Zalom, and D. Gilrein)

Fig. 185. Oviposition injury caused by the western flower thrips. (Courtesy D. J. Schuster)

True Bugs

Biology

Three families of true bugs (order Hemiptera) commonly attack tomatoes: Pentatomidae (stink bugs), Coreidae (leaf-footed bugs), and Miridae (tarnished plant bugs). Adults of these families have a thickened basal area on the front pair of wings and a membranous apical area.

Among the Pentatomidae, the most common species that attack tomato are the brown stink bug (*Euschistus servus* (Say)) (Fig. 186A), the green stink bug (*Acrosternum hilare* (Say)) (Fig. 186B), the southern green stink bug (*Nezara viridula* (L.)), the consperse stink bug (*E. conspersus* Uhler), the red-shouldered stink bug (*Thyanta accerra* McAtee), the Say stink bug (*Chlorochroa sayi* Stål), and the brown marmorated stink bug (*Halyomorpha halys* (Stål)). Other *Euschistus* and *Chlorochroa* spp. may also affect tomato. Stink bug eggs are barrel shaped and typically multicolored; they are laid side by side in groups on the undersides of leaves. Nymphs, which are not winged, are usually multicolored and appear nearly round from the top; older nymphs may be less brightly colored and appear more like adults. Adult stink bugs are shield shaped (approximately 1.3 cm long) and winged, and they are usually gray, green, or brown. Both nymphs and adults have piercing–sucking mouthparts, and although they may feed on foliage, they produce the most damage when they feed on fruit.

Within the Coreidae, three species are recognized as most commonly causing damage to tomato: the eastern leaf-footed bug (*Leptoglossus phyllopus* (L.)), the western leaf-footed bug (*L. zonatus* (Dallas)), and the coreid bug (*Phthia picta* (Drury)). The eggs of leaf-footed bugs are oval, somewhat flattened, and often metallic in appearance; they are laid in lines or clusters on the undersides of leaves. Compared to stink bug nymphs, leaf-footed bug nymphs appear more elongated from the top and more swaybacked from the side. Leaf-footed bug nymphs do not have wings and may be brightly colored, although older bugs may be less colorful, appearing more like adults. Compared to adult stink bugs, adult leaf-footed bugs are larger (approximately 2 cm long) and more slender. The tibiae of the hind legs of leaf-footed bugs may be expanded and leaflike, but not all species have this characteristic. Nymphs and adults have piercing–sucking mouthparts and may feed on tomato foliage. However, like stink bugs, leaf-footed bugs cause the most damage when they feed on tomato fruit.

The third family of true bugs that attack tomato is the Miridae. Within this family, two species from the genus *Lygus* feed on tomato fruit: the tarnished plant bug (*L. lineolaris* (Palisot)) and the western tarnished plant bug (*L. hesperus* Knight). Both species feed on the fruit or seeds of many flowering plants, including cultivated crops such as alfalfa (*Medicago sativa* L.) and safflower (*Carthamus tinctorius* L), and they move to tomato when their primary hosts dry or are disturbed. *Lygus* spp. adults are yellowish, brownish, or greenish in color and have a conspicuous triangle in the center of the back that is tinged with brown, red, or yellow. Adult tarnished plant bugs are smaller (approximately 0.75 cm in length) than either stink bug adults or leaf-footed bug adults.

Fig. 186. Stink bugs use piercing–sucking mouthparts to probe and feed on green tomato fruit. **(A)** Brown stink bug. **(B)** Green stink bug. (Courtesy D. Gilrein)

Damage

Stink bugs and leaf-footed plant bugs usually attack green tomato fruit. Damage appears as small, dark pinpricks on the surface, which are surrounded by white, corky tissue below the epidermis. The white tissue may appear crystalline or like frost on glass. This tissue remains firm and as the fruit ripens, the surface at the damaged area turns yellow (Fig. 187). Penetration of tomato fruit by the mouthparts of leaf-footed bugs is apparently deeper than that of stink bugs, and the fruit often becomes distorted and deformed as it enlarges.

Surface damage caused by tarnished plant bugs is similar to that caused by stink bugs immediately after feeding. However, the area below the feeding site dries instead of remaining white and corky, and the surface of the fruit cracks slightly.

Stink bugs may also vector the yeast *Nematospora* (syn. *Eremothecium*), which causes fruit rot.

(Prepared by D. J. Schuster, F. Zalom, and D. Gilrein)

Fig. 187. Feeding damage caused by stink bugs on green fruit turns yellow when the fruit ripen. (Courtesy D. J. Schuster)

"Worm" Pests

Some of the most conspicuous feeding injury on tomato foliage and fruit is caused by larval stages of insects in the order Lepidoptera: the moths and butterflies. Adult lepidopterans are winged, whereas the immature "worm" stages feed on tomato through several instars after hatching from eggs. The final resting stage is called the pupa; it typically resides in the soil and allows overwintering of the pest.

This category of so-called worm pests includes several important pests of tomato: tomato fruitworm, armyworms, tomato pinworm, cutworms, hornworms, and loopers.

Tomato Fruitworm

Biology

The tomato fruitworm (*Helicoverpa* (*Heliothis*) *zea* Boddie), which is also known as the corn earworm and cotton bollworm, has a broad host range on both crops and weeds. In areas with long growing seasons, as many as three generations of fruitworm may be produced each season.

Fig. 188. Older larva of tomato fruitworm boring into fruit. (Courtesy Jack Kelly Clark, University of California Statewide IPM Program)

Fig. 189. Frass-filled cavities inside tomato fruit, caused by tomato fruitworm larvae. (Courtesy Jack Kelly Clark, University of California Statewide IPM Program)

Tomato fruitworm eggs are sphere shaped, and a female can lay 500–3,000 eggs during her lifetime. Young larvae are white with black heads and have rows of dual-bristled, black spots on their bodies; older larvae change color to yellowish green or dark brownish/black but maintain their dark spots (Fig. 188). Newly hatched larvae begin to feed on the leaves of tomato plants, but later instars bore into the fruit, reaching a length of approximately 4.4 cm before dropping to the soil to pupate.

Adults are moths with a body 2.5 cm long and a wingspan of 3.2 cm. Moths often move to flowering tomato plants after the end of the silking stage of corn (*Zea mays* L.), their favorite host.

Damage

Some injury is caused to tomato foliage by the chewing mouthparts of the youngest larvae, which generally go on to attack green tomato fruit. The burrowing larvae form deep, water-soaked, frass-filled cavities inside the tomato fruit (Fig. 189). Often, the invasion of secondary microorganisms compounds the injury.

(Prepared by D. J. Schuster, F. Zalom, and D. Gilrein)

Armyworms

Biology

The beet armyworm (*Spodoptera exigua* Hübner) is a significant lepidopteran pest of tomato in Mexico and in specific regions of the United States, including California and the southern and mid-Atlantic states. The western yellow-striped armyworm (*S. praefica* Grote) is also problematic in California.

Coloring of the larvae is different for the two *Spodoptera* spp. The beet armyworm larva is light green with fine, pale stripes along the back and a single wider stripe along each side (Fig. 190). The western yellow-striped armyworm larva is dark gray to black and has two bright-yellow stripes on each side. The adult armyworm has a wingspan of 2.5 cm.

Armyworms often move to tomato fields after other host crops have been harvested. Weeds may also be a source of infestation. Three to five generations of armyworms are produced per year.

Fig. 190. Larva of beet armyworm. (Courtesy Jack Kelly Clark, University of California Statewide IPM Program)

Fig. 191. Damage to tomato surfaces caused by armyworms. Note that some of the damage has scarred over. (Courtesy Jack Kelly Clark, University of California Statewide IPM Program)

Fig. 192. Mature larva of the tomato pinworm. (Courtesy Jack Kelly Clark, University of California Statewide IPM Program)

Fig. 193. Blotch-type mines on tomato foliage, caused by feeding of tomato pinworm larvae. (Courtesy Jack Kelly Clark, University of California Statewide IPM Program)

Fig. 194. Small piles of frass beside the calyx, indicating feeding by tomato pinworm larvae. (Courtesy Jack Kelly Clark, University of California Statewide IPM Program)

Damage

Armyworm larvae feed on tomato fruit and foliage. On fruit, the damage they cause is not as deep as that caused by the tomato fruitworm. Armyworm damage later scars over, so the fruit is sometimes still acceptable for processing (Fig. 191). Armyworms do not complete their life cycle within the fruit.

(Prepared by D. J. Schuster, F. Zalom, and D. Gilrein)

Tomato Pinworm

Biology

The tomato pinworm (*Keiferia lycopersicella* Busck) is a common and serious threat to tomato crops in Mexico, the U.S. state of California, and the U.S. Southeast. Other production areas of the U.S. South and West experience infestations more sporadically. The impact of this pest is derived from its ability to complete as many as seven or eight overlapping generations in areas where it overwinters. Solanaceous plant debris, including old tomato vines and weed hosts, provides shelter for overwintering in warm climates.

Eggs of the tomato pinworm are laid singly. The larva is pale in color, and its body is segmented but appears smooth. As the larva matures, it becomes gray or yellowish and has a red or purple band across each segment (Fig. 192). After feeding on tomato leaves, the larva enters the fruit, primarily under the calyx, but if the pinworm population is large, the larva may enter from anywhere on the fruit surface. The adult tomato pinworm is a light-colored moth with small, black flecks on its wings. The pinworm moth is smaller than an armyworm moth or a fruitworm moth.

Damage

On tomato foliage, pinworm larvae create blotch-type mines as they feed (Fig. 193). Within tomato fruit, the larvae bore dry tunnels from the calyx to the core. Small piles of frass may be visible beside the calyx (Fig. 194); when frass is not present, fruit infestation may remain undetected at harvest.

(Prepared by D. J. Schuster, F. Zalom, and D. Gilrein)

Cutworms

Biology

The broad host range and geographic distribution of cutworms makes them an occasional problem wherever tomatoes are grown. The moth species in this group that are most likely to affect tomato are the variegated cutworm (*Peridroma saucia* Hübner), the black cutworm (*Agrotis ipsilon* Hufnagel), and the granulate cutworm (*Feltia subterranea* (F.)).

Eggs are usually laid singly or in small clusters on foliage at or near the soil surface. Cutworm larvae are roughly 2.5–5.0 cm at maturity. Larvae of the variegated cutworm and the black

Fig. 195. Larva of the cutworm. (Courtesy Jack Kelly Clark, University of California Statewide IPM Program)

Fig. 196. Larva of the tobacco hornworm. (Courtesy D. Gilrein)

cutworm are dark and appear smooth, whereas larvae of the granulate cutworm are light and have subdorsal chevron-like markings (Fig. 195). Cutworm larvae are most active at night and likely to be near the soil surface during the day, sometimes burrowing down a few inches in dry soil conditions. When disturbed, a cutworm larva curls into a C-shape.

Damage

Emerging tomato seedlings or young transplants may be cut off at the soil line by the activity of cutworm larvae. Direct feeding injury to the surface of the fruit may also occur; most commonly, larvae bore irregular holes in fruit that are near to or in contact with the ground. The variegated cutworm is also known as the "climbing cutworm" because of its ability to climb staked plants. Cutworm infestations are usually clumped within fields.

(Prepared by D. J. Schuster, F. Zalom, and D. Gilrein)

Hornworms

Biology

Hornworms are conspicuous pests, and the voracious feeding of hornworm larvae causes noticeable injury to tomato. The two hornworm species found in tomato-growing areas across North America are the tobacco hornworm (*Manduca sexta* L.) and the tomato hornworm (*M. quinquemaculata* (Haworth)).

Hornworm eggs are spherical and approximately 1.5 mm in diameter; they are laid singly, usually on lower leaf surfaces. The larvae of both species are large, reaching 7.6–10.2 cm at maturity, and they have a single upright horn at the end of the abdomen. Larvae of the tobacco hornworm are green and have seven white stripes (Fig. 196); larvae of the tomato hornworm are also green but have eight chevrons on each side of the body. Hornworm adults are known as hawk moths or sphinx moths; they have a wingspan of 11.4–12.7 cm and are strong fliers.

Damage

Hornworm larvae chew obvious holes in tomato foliage, but they can also consume pieces of green fruit. The damage caused by hornworms is most extensive in home gardens, rather than commercial production fields, where infestations tend to be localized and kept under control by natural infestations of

Fig. 197. Larva of the looper. (Courtesy Jack Kelly Clark, University of California Statewide IPM Program)

parasitic wasps (e.g., *Polistes* spp., *Trichogramma* spp., *Cotesia* spp., *Hyposoter* spp.).

(Prepared by D. J. Schuster, F. Zalom, and D. Gilrein)

Loopers

Biology

Injury to tomato by loopers is usually incidental, because tomato is not their preferred host. Even so, the cabbage looper (*Trichoplusia ni* Hübner) and the alfalfa looper (*Autographa californica* Speyer) sometimes feed on tomato foliage. Cabbage loopers are found across North America, and alfalfa loopers are most common in the western United States.

The moths are nocturnal, laying up to 300 eggs on the undersurfaces of leaves. The larvae, which are usually light green, have three pairs of legs at the front and three pairs of prolegs at the rear of the abdomen (Fig.197). This structure leads to the typical "looping" of the body as a larva crawls.

Damage

Looper larvae feed on tomato leaves, creating holes, but rarely damage tomato fruit.

(Prepared by D. J. Schuster, F. Zalom, and D. Gilrein)

Part III. Noninfectious Diseases, Disorders, and Damage

Physiological Disorders

Physiological disorders that affect the appearance and thus the marketability of tomato fruit usually develop while the crop is being grown. At harvest, fruit that are observed to be damaged are culled and discarded. Other physiological disorders develop after harvest and become apparent as the fruit ripen or are being shipped to market. Temperature extremes are associated with many of these postharvest disorders.

Selected References

Abbott, J. D., Peet, M. M., Willits, D. H., and Gough, R. E. 1985. Water management of greenhouse tomatoes. HortScience 20:688-690.

Banuelos, G. S., Offermann, G. P., and Seim, E. C. 1985. High relative humidity promotes blossom-end rot on growing tomato fruit. HortScience 20:894-895.

Barten, J. H. M., Scott, J. W., Kedar, N., and Elkind, Y. 1992. Low temperatures induce rough blossom-end scarring of tomato fruit during early flower development. J. Am. Soc. Hortic. Sci. 117:298-303.

Brooks, W. M., Alban, E. K., Bauerle, W. L., Jr., George, W. L., Jr., Kretchmann, D. W., Linquist, R. K., Short, T. H., and Walker, R. J. 1973. Growing greenhouse tomatoes in Ohio. Ohio State Univ. Ext. Bull. SB-19.

Emmons, C. L. W., and Scott, J. W. 1998. Ultrastructural and anatomical factors associated with resistance to cuticle cracking in tomato (*Lycopersicon esculentum* Mill.). Int. J. Plant Sci. (Chicago, IL, U.S.) 159:14-22.

Francis, D. M., Barringer, S. A., and Whitmoyer, R. E. 2000. Ultrastructural characterization of shoulder disorder in a uniform ripening tomato genotype. HortScience 35:1114-1117.

Hayslip, N. C. 1979. Effect of potassium fertilizer rates on severity of white tissue in fruits of graywall resistant and susceptible tomato varieties. Ft. Pierce Agric. Res. Cent. Res. Rep. RL1979-6.

Jones, J. B., Weber, G. F., and Kelbert, D. G. A. 1969. Tomato diseases in Florida. Fla. Agric. Exp. Stn. Bull. 731.

Karracher, P. E., and Bortner, C. E. 1934. Studies of frenching of tobacco. Ky. Agric. Exp. Stn. Res. Bull. 349:63-109.

Kuo, C. G., Tsay, J. C., Chen, B. W., and Lin, P. Y. 1982. Screening for flooding tolerance in the genus *Lycopersicon*. HortScience 17:76-78.

Peet, M. M. 1992. Fruit cracking in tomato. HortTechnology 2:216-223.

Rabinowitch, H. D., and Sklan, D. 1980. Superoide dismustase; a possible protective agent against sunscald in tomatoes (*Lycopersicon esculentum* Mill.). Planta 148:162-167.

Ricker, M. D. 2002. Association of zebra-stem symptoms on processing tomato with the *Pto* gene. (Abstr.) Phytopathology 92(Suppl.):S69.

Saltveit, M. E., Jr., and Cabrera, R. M. 1987. Tomato fruit temperature before chilling influences ripening after chilling. HortScience 22:452-454.

Scott, J. W., and Barten, J. H. M. 1992. Inheritance of spider track incidence, a tomato fruit disorder. Proc. Fla. State Hortic. Soc. 105:349-352.

Sikes, J., and Coffey, D. L. 1976. Catfacing of tomato fruits as influenced by pruning. HortScience 11:26-27.

Smith, O. 1932. Relation of temperature to anthesis and blossom drop of the tomato, together with a histological study of the pistils. J. Agric. Res. (Washington, D.C.) 44:183-190.

Taylor, M. D., Locascio, S. S., and Alligood, M. R. 2004. Blossom-end rot incidence of tomato as affected by irrigation quantity, calcium source, and reduced potassium. HortScience 39:1110-1115.

Woltz, S. S. 1978. Nonparasitic plant pathogens. Ann. Rev. Phytopathol. 16:403-430.

Yakir, D., Sadovski, A., Rabinowitch, H. D., and Rudich, J. 1984. Effect of high temperature on quality of processing tomatoes of various genotypes ripened on the vine. Sci. Hortic. 24:33-43.

Blossom Drop

Symptoms

Blossom drop involves yellowing of the pedicels and calyx followed by abortion of the flower. It can take place before or after anthesis. In jointless tomato varieties, the flower withers and turns brown but does not actually abscise.

Causes

The abortion of flowers can result from a variety of causes generally associated with stress conditions. Blossom drop can be caused by any factor that inhibits pollination/fertilization, such as prolonged low temperatures (less than 20/10°C daytime/nighttime), prolonged high temperatures (greater than 30/21°C daytime/nighttime), high relative humidity (greater than 90%), and excessive wind.

Improper nutrition, such as deficient fertilization or excessive nitrogen, can also contribute to blossom drop. Foliar diseases and insects may also cause flowers to abort, and plants with a heavy fruit load may abort later flowers.

Management

Proper fertilization and water management can minimize blossom drop. However, if stress conditions occur during these critical stages, little can be done.

In the greenhouse, temperature and humidity control are important for managing blossom drop. In the field, overhead watering during hot weather may reduce plant temperatures and thus limit blossom drop. However, the value of overhead watering decreases under prolonged periods of high temperatures. Overhead watering can also promote the development of foliar diseases.

Heat-tolerant tomato cultivars have been developed. These cultivars may have less blossom drop associated with high temperatures and sometimes with low temperatures.

(Prepared by J. W. Scott)

Blossom-End Rot

Symptoms

The initial symptom of blossom-end rot is the development of light-tan, water-soaked lesions on tomato fruit. These lesions then enlarge and turn black and leathery, and they often become overrun with secondary black mold (Fig. 198). The lesions generally form at the blossom end of the fruit, but they can form at the side of the fruit. In some cases, internal black lesions form, which are not visible from the exterior of the fruit (Fig. 199). Fruit affected by blossom-end rot ripen more rapidly than normal.

Causes

A localized calcium deficiency in the distal end of the fruit causes blossom-end rot (see Nutritional Disorders later in Part III). The most sensitive fruit are those on rapidly growing plants 1–3 weeks past anthesis.

Because calcium is not a highly mobile element, a deficiency can occur with a fluctuation in water supply—even a short-term fluctuation. Thus, moisture extremes promote the likelihood that blossom-end rot will develop. The incidence of this disease also increases under high relative humidity, which reduces plants' evapotranspiration and thus water uptake. Other conditions that reduce calcium uptake can also make plants more susceptible to blossom-end rot, including high levels of salts (especially cations) and the use of ammonium nitrogen.

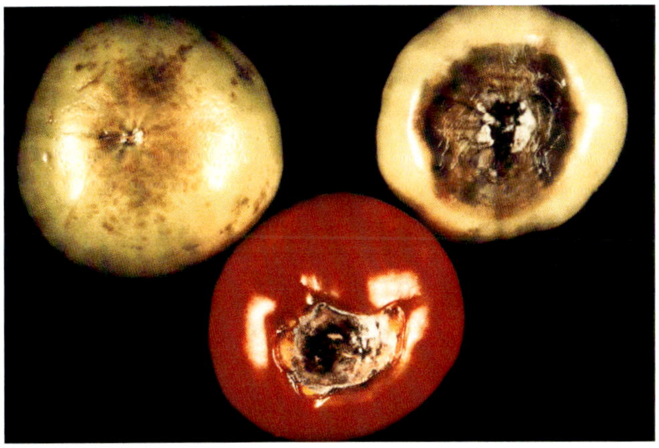

Fig. 198. External lesions caused by blossom-end rot. (Courtesy J. W. Scott)

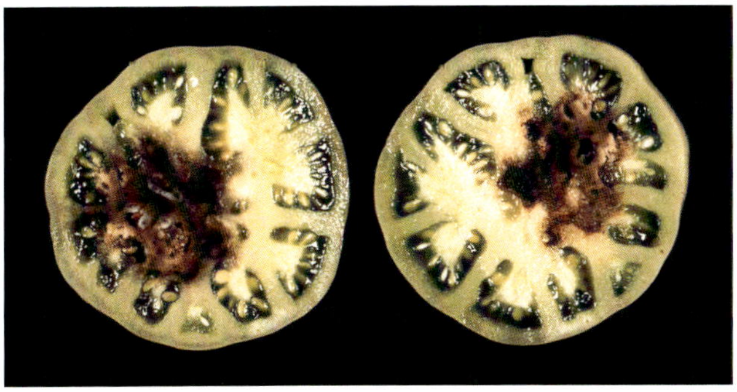

Fig. 199. Internal lesions caused by blossom-end rot. (Courtesy J. P. Jones)

Management

Soil testing is recommended to determine if there is a shortage of calcium. Liming with dolomitic or high-calcium limestone 2–4 months before planting can alleviate blossom-end rot. If a calcium deficiency or high level of salts occurs during the growing season, foliar spraying with anhydrous calcium chloride may help alleviate the problem. Plants should be properly fertilized, avoiding products with high potassium, sodium, magnesium, and ammonium, which compete with calcium for uptake.

Other management strategies include maintaining an even water supply via irrigation to avoid defects and providing drain fields to prevent excessively wet conditions. Use of cultivars tolerant to blossom-end rot will minimize this defect, as well.

(Prepared by J. W. Scott)

Catface

Symptoms

Catfaced tomatoes are generally misshapen and have enlarged scars and holes in the blossom end (Fig. 200). Fruit are sometimes kidney shaped and have elongated blossom scars, but they can also be distorted into other shapes. Even normally shaped fruit are considered catfaced if the blossom scar is enlarged or perforated.

Causes

Although catface is a serious disease of large-fruited, fresh-market tomatoes, little has been published about it. Cold weather is one cause that has been documented. Flowers are susceptible to blossom-end roughness when exposed to cold temperatures approximately 3 weeks before anthesis. Extreme flucations in day and night temperatures also likely increase the incidence of catface, as does pruning tomato plants, under some conditions. High levels of nitrogen can enhance the development of catface, as well.

Management

In the field, excessive pruning and excessive nitrogen fertilization should be avoided. In the greenhouse, low tempera-

Fig. 200. Misshapen and scarred tomato fruit, caused by catface. (Courtesy J. W. Scott)

tures should be avoided by heating the structure. Temperatures should also be controlled when growing transplants for field planting, because low temperatures might affect early fruit. The best management strategy is to use cultivars that are not subject to catface.

(Prepared by J. W. Scott)

Chilling Injury

Symptoms

Symptoms of chilling injury are the failure of tomato fruit to ripen, irregular ripening, premature softening, surface pitting, and browning of seeds, along with increased decay. Symptoms usually appear after chilling and upon exposure to higher ripening temperatures. Green tomatoes are more sensitive to this type of injury than ripe tomatoes. Flavor is adversely affected because of the reduction of aromatic volatiles.

Cause

Chilling injury can be caused when temperatures remain above the freezing point but below 12.5°C for a period of time. The amount of time needed to cause injury to tomato fruit depends on the temperature: Shorter times are required with lower temperatures, and longer times are required with higher temperatures. Tomatoes subjected to higher temperatures before chilling are more sensitive to injury than those exposed to lower temperatures before chilling.

Management

Little can be done about cool weather in the field. However, after harvest, maintaining storage temperatures above 12.5°C will prevent postharvest chilling injury.

(Prepared by J. W. Scott)

Cuticle Cracking (Rain Check, Weather Check, Crazing, Russetting)

Symptoms

Cuticle cracking is characterized by the development of numerous tiny, concentric cracks on the shoulders of tomato fruit, which feel rough or bumpy to the touch. These small cracks sometimes coalesce into larger, lenticel-like cracks (Fig. 201). The affected area may become leathery and remain green when the fruit ripens, or the area may become blackish. In some cases, the affected area turns red, but the rough, tiny cracks remain visible.

Fig. 201. Cuticle cracking, characterized by development of numerous tiny, concentric cracks on the shoulders of tomato fruit. (Courtesy T. A. Zitter)

Ripe tomatoes are not affected by cuticle cracking. It occurs most frequently on exposed, mature-green, and possibly breaker fruit after rain, although unexposed fruit and small, immature-green fruit are sometimes severely affected. In addition, cuticle cracking occurs more in large, fresh-market tomatoes than in plum and cherry cultivars.

Symptoms are often seen under dead leaves that stick to fruit or near adjacent flowers that have not set. The affected area tends to become wrinkled in storage, presumably because of water loss. Cuticle cracking can result in the development of black shoulder under favorable conditions (see Part IV, Diseases of Unconfirmed Etiology).

Causes

The exact cause of cuticle cracking is not known. One theory is that the development of the shoulder epidermis is disrupted after rain alters the fruit temperature or water uptake. The physical impact of the rain may also promote development of the disease. Often, cuticle cracking is more severe when a dry period is followed by a period of heavy rain than when light rain falls frequently throughout the season.

Cultivars vary widely in their susceptibility to cuticle cracking. Tolerance is based on good leaf cover to protect the tomato fruit and on the characteristics of the fruit epidermis. Resistant fruit have a thicker epicarp (i.e., epidermis and cuticle) than susceptible fruit. Glossy fruit tend to crack less than fruit with a dull surface, but little is known about the chemical or morphological differences between resistant and susceptible genotypes.

Management

Cultural procedures that reduce fruit exposure can be used to manage cuticle cracking. Excessive pruning, damaging vines during harvest, and foliar diseases should all be avoided. The use of tolerant cultivars is recommended, as well.

(Prepared by J. W. Scott)

Growth Cracks

Symptoms

Tomato fruit are vulnerable to two types of large cracking during growth: (1) concentric cracking, which is a splitting of the epidermis in a circular pattern around the fruit stem scar, and (2) radial cracking, which is a splitting that radiates outward from the stem scar (Fig. 202).

Fig. 202. Growth cracks may be concentric (top fruit) or radial (bottom fruit). (Courtesy J. P. Jones)

Growth cracks occur on tomatoes as they near maturity, depending on the cultivar. When conditions are conducive, susceptible cultivars crack while tomatoes are at the mature-green stage. Less susceptible cultivars do not crack until the breaker stage, more tolerant cultivars do not crack until the red-ripe stage, and resistant cultivars rarely crack at all or only under extreme conditions. (Epidermal splitting of ripe fruit is referred to as radial bursting.) The earlier in the fruit's development the crack occurs, the deeper and longer it often becomes.

Causes

Susceptibility to cracking relates to the strength and stretching ability of the fruit epidermis. Alterations in growth rate promote the disease, and rapidly growing fruit tend to be most susceptible. Thus, succulent plants (provided with high-nitrogen, low-potassium nutrition) tend to be more susceptible than less vigorous plants.

Cracking is also induced by rain and wide fluctuations in temperature. Exposed fruit crack more readily than those protected by foliage, probably because greater temperature fluctuations result from exposure to the direct rays of the sun. During periods of rain, exposed fruit cool more rapidly. Both rain and temperature fluctuations likely alter the water uptake of tomato fruit, which may promote cracking.

Management

Cultural measures that reduce cracking include proper water management, a good nutritional program to prevent plants from becoming overly succulent, proper pruning to limit fruit exposure, and prevention of defoliation from foliar diseases to limit fruit exposure. Some cultivars are tolerant to cracking.

(Prepared by J. W. Scott)

Internal White Tissue

Symptoms

The expression of white tissue varies widely, depending on the cultivar and environmental conditions. Sometimes, only a few white fibers are spread through the pericarp, but usually, the most prevalent white tissue formation is in the outer wall of the fruit (Fig. 203). White tissue can vary from a few white vascular bundles to a coalescence of these strands into a solid mass of wall tissue.

Fig. 203. Tomatoes with internal white tissue and puffiness. (Courtesy T. A. Zitter)

The expression can be variable, with tomatoes on a single cluster of fruit often differing dramatically in the amount of white tissue. Some cultivars are prone to white tissue formation in the placental area, proximal to the locules (see Fig. 203). In some cases, white tissue extends from the fruit stem scar (core) into the fruit.

Causes

The formation of white tissue is caused by potassium deficiency and high temperatures. Other causes have not been documented by research, but stress factors seem to be involved. Feeding of the sweetpotato whitefly (*Bemisia tabaci* (Gennadius)) can also cause severe white tissue formation.

Some tomato cultivars are more susceptible to white tisue formation than others, and late fruit on plants with a heavy fruit load often have increased white tissue, based on observations. In addition, plants grown under cool conditions and then later under hot conditions have more white tissue than plants grown under consistently high temperatures.

Management

Proper nutrition (especially with potassium) reduces white tissue formation. The incidence of affected fruit can also be reduced by preventing stressful conditions, using tolerant cultivars, and controlling the sweetpotato whitefly population.

(Prepared by J. W. Scott)

Puffiness (Boxiness, Pockets)

Symptoms

From the exterior, a tomato fruit affected with puffiness appears slab sided or angular. When the tomato is cut transversely, hollow cavities are revealed in one or more locules between the locular contents and the outer wall (see Fig. 203). Thus, the depressed areas seen from the exterior are above the locules, and the protruded areas are above the transverse walls. Tomato fruit with this disease are less dense than normal fruit.

Causes

Puffiness can be caused by anything that contributes to inadequate pollination, fertilization, or seed development. Likely factors include improper nutrition (e.g., high nitrogen and/or low phosphorus or potassium), insufficient light, and low or high temperatures; wet soils and drought may be factors, as well. The use of auxin growth regulators for fruit set often results in puffiness. Some cultivars are more susceptible to this disease than others.

Management

Puffiness can be minimized with the use of tolerant cultivars, proper nutrition, and water control. If possible, fruit should be harvested before a predicted rain.

(Prepared by J. W. Scott)

Spider Track

Symptoms

The symptoms of spider track resemble those of zippering somewhat, but there are several differences (see the section Zippering, later in Part III). Spider tracks are generally more numerous than zippers and are confined to the shoulder area of the

Fig. 204. Brown, necrotic scars and rays of epidermal tissue, caused by spider track. (Courtesy J. W. Scott)

Fig. 205. Areas of white, necrotic tissue with yellow halos, caused by sunscald. (Courtesy T. A. Zitter)

fruit. Brown, necrotic scars generally radiate from the fruit stem scar but do not always connect with it. In addition, rays of off-colored epidermal tissue extend from the stem scar (Fig. 204). In green-shouldered fruit, the rays are generally lighter in color; in fruit with a uniformly green shoulder, the rays can be darker green. When the tomato turns red, the rays remain yellowish.

Cause

No cause has been documented for spider track, but cultivars are known to vary in susceptibility. Spider track is often more prevalent on some large-fruited and jointless cultivars, but plum cultivars can be affected, as well.

Management

Resistant cultivars should be used.

(Prepared by J. W. Scott)

Sunscald

Symptoms

Mature-green and breaker tomatoes are most susceptible to sunscald. An affected fruit develops areas of white, necrotic tissue surrounded by yellow halos; these lesions often become sunken and wrinkled. When the fruit turns red, the necrotic areas remain white or become covered by black, secondary fungi (Fig. 205).

The damage caused by sunscald generally appears on the side or top half of the tomato. Damage occurs when a fruit that has been covered by leaves is suddenly exposed to sunlight.

Cause

Sunscald occurs when sunlight heats the tomato pericarp to greater than 40°C. Fruit that are protected by foliage but suddenly exposed to the sun are more prone to sunscald than fruit that are partially exposed as they develop.

Sudden exposure of tomato fruit can be caused by careless pruning and harvesting. Defoliation caused by foliar diseases and insects also predisposes fruit to sunscald.

Management

Care in pruning and harvesting will minimize the occurrence of sunscald, as will implementation of a program for foliar disease control. The use of protective sprays of clay or diluted white paint is sometimes effective in preventing sunscald on exposed fruit. In ground culture, cultivars should be used that have good foliage cover, which will not open up and expose the fruit.

(Prepared by J. W. Scott)

Tomato Little Leaf

Tomato little leaf was first diagnosed in the fall of 1986, when several U.S. tomato producers in the northern Florida–Quincy production region observed plants in their fields with unusual growth characteristics. Since then, the disease has been observed sporadically in fall crops in this production region, but in some cases, it has affected large areas and caused significant yield reductions (as high as 80%). Reports of tomato little leaf have also been made from the U.S. states of Georgia, Maryland, North Carolina, Ohio, and Texas and from the nations of Italy and Poland and possibly Indonesia.

Symptoms

Early symptoms of tomato little leaf consist of interveinal chlorosis in young leaves, with the veins remaining dark green (Fig. 206). Symptoms progress to increased leaflet distortion, in which the margins become smooth, and to more pronounced interveinal chlorosis (Fig. 207). Subsequently, the top growth of the plant becomes severely distorted. The leaflets along the midrib fail to expand properly (Fig. 208), producing a "little-leaf" appearance. Additional symptoms included cessation of terminal growth, leaflets with twisted and brittle midribs, and axillary buds with reduced and distorted growth. The latter symptoms can be confused with those of *Cucumber mosaic virus* (CMV). (See the section on CMV in Part I.)

Tomato fruit that set when plants are mildly affected with tomato little leaf are distorted and thus unmarketable. Specifically, the fruit are flattened and may have radial cracks extending from the calyx to the blossom scar (Fig. 209). There is virtually no seed set in affected fruit. In more severely affected plants, blooms are distorted and fail to set fruit. At times, affected plants grow next to normal plants.

Cause

After the 1986 diagnosis of tomato little leaf, several growers from the northern Florida–Quincy production area noted that the same set of symptoms had been observed before but on very few plants. Following the designation of this set of symptoms as tomato little leaf, numerous attempts were made to determine its cause.

Suspecting that a nutritional deficiency was responsible, researchers treated plants with sprays of iron, calcium, zinc, and boron but without success; extensive tissue sampling also failed to reveal an abnormal nutritional situation. Rigorous testing of soil and plant tissue for herbicide residues proved negative, as well, and the problem was shown not to be virus related when cuttings made from distorted plants tops developed normal growth habits after rooting. One common observation was made, however: Affected plants were associated with wet areas of fields.

Researchers also considered frenching—a nonparasitic disease of tobacco (*Nicotiana* L.)—in exploring the cause of tomato little leaf. (Frenching is considered a nonparasitic disease, because the organism or organisms that cause it do not live within the tissue of the plant.) Tomato had been mentioned as being sensitive to frenching in research literature from more than 70 years earlier, but unfortunately, symptoms were not described. Chrysanthemum (*Chrysanthemum* × *morifolium* Ramat. (pro. sp.), alt. *Dendranthema* × *grandiflorum* Kitam.) is affected by a similar disorder, which is referred to as yellow strap leaf.

In tobacco, frenching is more severe with warm soil temperatures. Symptoms can appear in 8–15 days with soil temperatures of 35°C but rarely appear with soil temperatures of 21°C or less. Symptoms are more severe on neutral or alkaline soils and are rarely observed with a soil pH of 6.3 or less. In addition, frenching of tobacco and yellow strap leaf of chrysanthemum are more prevalent with wet, poorly aerated soils.

The current hypothesis on the etiology of tomato little leaf is that one or more amino acid analogs are synthesized by certain soil microorganisms and then released into the rhizosphere. These compounds, which are structurally similar to the amino acid leucine, are taken up by the plant, causing morphological changes and stunting in susceptible plants at very low concentrations. These compounds are believed to act as antimetabolites of leucine.

There is no definitive test for tomato little leaf; thus, diagnosis occurs via a process of elimination. Diagnosis is made after viruses, phytoplasmas, and factors such as herbicide damage have been eliminated as causing the symptoms characteristics of little leaf. (Fortunately, not many sets of symptoms mimic this disease.) Cuttings can also be made to confirm diagnosis.

In sum, it is apparent that for tomato little leaf to develop, a fairly unique set of soil conditions must occur in combination with an appropriate soil microorganism. Researchers have not been able to determine the exact circumstances under which tomato little leaf develops, because they have not been able to produce the symptoms on demand.

Fig. 206. Interveinal chlorosis of young leaves, an early symptom of tomato little leaf. (Courtesy S. M. Olson)

Fig. 207. Increased leaflet distortion, a more advanced symptom of tomato little leaf. (Courtesy S. M. Olson)

Fig. 208. Severe leaf distortion at the top growth of the plant, caused by tomato little leaf. (Courtesy S. M. Olson)

Fig. 209. Fruit distortion caused by tomato little leaf. (Courtesy S. M. Olson)

Management

The development of tomato little leaf can be reduced in fields where soil moisture is monitored and waterlogged conditions do not occur. If symptoms are caught early in the growth cycle, drying out a field will usually result in subsequent growth being normal and plants will resume setting normal fruit. Controlling the soil pH may help prevent the disease, as well; high soil pH has been implicated in tobacco but not chrysanthemum. In most cases, tomato little leaf does not seem to carry over from one season to the next.

(Prepared by S. M. Olson)

Water Wilt

Symptoms

A plant affected by water wilt has waterlogged roots. These roots cannot respire, which produces an oxygen deficiency and thereby alters the plant's hormone levels. This results in epinasty and wilting, beginning at the growing tip of the plant.

Initially, the wilt occurs with little color change in the foliage, and the symptoms are similar to those of bacterial wilt (caused by *Ralstonia solanacearum* (Smith) Yabuuchi et al.). However, if the waterlogged conditions are not corrected, the wilted areas become chlorotic and necrotic and the stem tends to collapse. Adventitious roots may form at the base of the stem. Although these symptoms appear similar to those of bacterial wilt, there is no discoloration of the vascular system.

Cause

As the name implies, water wilt results when soils become saturated with water, generally after excessive rain. The effects are more pronounced and damage occurs more rapidly under high temperatures. If plants are already infected by soilborne pathogens, they will be stressed more readily by waterlogging, and disease symptoms will be manifested.

Management

Fields should be well drained and cultivated as soon as possible after a heavy rain to break water tension in the soil.

(Prepared by J. W. Scott)

Yellow Shoulder (Yellow Top, Persistent Green Shoulder)

Symptoms

Yellow shoulder affects the shoulder of tomato fruit exposed to the sun. The chlorophyll in this area is slow to break down as ripening occurs, resulting in a patch that remains green or eventually turns yellow but not red (Fig. 210). The disease may affect the entire shoulder or only a small, irregular patch. The outer pericarp in the affected area is hard and white.

Cause

The physiological cause of yellow shoulder is not known, but apparently, fruit exposed to high temperatures during maturation and ripening can develop this disorder. It is usually observed on the sun-exposed fruit of cultivars with green shoulders, but these cultivars vary considerably in susceptibility. Cultivars that have uniform ripening (shoulders) are generally not as prone to this disorder but can be susceptible to it.

Fig. 210. Persistent green shoulder (top) and yellow shoulder or yellow top (bottom). (Courtesy T. A. Zitter)

If the fruit of susceptible cultivars are picked green and ripened in the dark, the disorder is not expressed. This indicates that damage from yellow shoulder is manifested by light.

Management

The most effective management strategy is to use resistant cultivars with or without green shoulders. In addition, practices should be avoided that expose tomato fruit to the sun, such as improper nutrition, excessive pruning, and defoliation from foliar diseases.

Yellow shoulder is not a concern if tomatoes are picked mature green and ripened by exposure to ethylene in the dark.

(Prepared by J. W. Scott)

Zebra Stem

Symptoms

Symptoms of zebra stem vary between tomato plants in the field versus those in the greenhouse. In the field, leaves become brown, necrotic, and desiccated, and the lower and middle portions of stems become covered with light-colored lesions in a "bull's eye" or zebra pattern. In the greenhouse, foliar lesions begin as small spots near the veins at the bases of expanding leaflets. The lesions coalesce and the leaflets desiccate, twist, and die. Stem lesions on greenhouse plants are black and sometimes coalesce into wide bands. Seedlings are often stunted but rarely die.

Cause

Zebra stem is associated with tomato cultivars that are homozygous for the *Pto* gene, which confers resistance to bacterial speck (caused by *Pseudomonas syringae* pv. *tomato* (Okabe)

Young, Dye, & Wilkie). Zebra stem is likely caused by a gene linked to *Pto*.

Management

When growing cultivars with resistance to bacterial speck, heterozgyotes resistant for the *Pto* gene should be used.

(Prepared by J. W. Scott)

Zippering

Symptoms

Tomato fruit affected by zippering have thin, brown, necrotic scars, which start at the fruit stem scar and extend part or all of the way to the blossom end. Each longitudinal scar has small, transverse scars along it, resembling a zipper (thus the name). Often, a tomato fruit has only one scar, but it can have several. Sometimes, a hole open to the locule forms in addition to the zipper scar (Fig. 211).

Cause

Zippers are caused by anthers that are adnate (i.e., attached) to the ovary wall of a newly forming fruit. The scarring is most pronounced in cool weather but occurs under other weather conditions, as well. Cultivars vary in their susceptibility to zippering.

Fig. 211. Zippering on mature fruit, accompanied by a hole open to the locule. (Courtesy T. A. Zitter)

Management

The most effective management strategy is to grow tolerant cultivars. To prevent zippering among greenhouse-grown plants, proper temperatures should be maintainted.

(Prepared by J. W. Scott)

Nutritional Disorders

Optimal amounts of essential mineral elements are required to produce the greatest yields of high-quality tomato fruit. Essential nutrients are most often supplied from the native fertility of the soil or from fertilizers, but they can also be supplied from some crop chemicals, such as certain fungicides and foliar nutritional sprays. Inadequate nutrition of tomato plants results in reduced plant growth and fruit yield, and nutrient deficiency can lead to certain fruit disorders or impaired fruit quality.

A key to diagnosing nutrient disorders is presented in Table 9. Some of these disorders are direct effects of nutrient deficiencies, whereas others may be indirect nutritional effects from environmental factors (e.g., drought, low temperatures), root diseases, or root system damage caused by nematodes (Table 10).

Leaf nutrient analysis should be used to confirm a nutrient deficiency diagnosis made visually in the field (Table 11). This analysis examines the *most recently matured leaf,* which is the leaf nearest the tip of the plant or branch that has reached full size most recently. This leaf is typically the one that has turned from a juvenile light-green color to a darker green. For tomato, this leaf is usually the fifth or sixth expanded leaf from the tip of the plant or branch.

The Selected References at the end of this section provide sources of information about collecting and analyzing plant samples and interpreting the results. Information about the absorption, transport, and function of the essential elements for tomato plant nutrition is also provided in the Selected References (see Marschner, 1995).

Nitrogen

Nitrogen-deficient tomato plants are smaller and less vigorous than healthy plants (Fig. 212). In foliage, nitrogen deficiency is manifested as a general yellowing of the entire leaf and sometimes a slight purpling of the leaf veins. Affected plants also have thinner stems and smaller fruit; the overall diminished size of the fruit renders them less marketable. Fruit of nitrogen-deficient plants may be misshapen and lighter red at maturity.

Nitrogen deficiency first appears on older leaves, because under nitrogen stress, plant nitrogen is translocated from the older leaves to sustain growth in the upper parts of the plant. The most recently matured leaf of a nitrogen-deficient tomato plant at early flowering will have a total nitrogen concentration of less than 2.8% on a dry weight basis (see Table 11). The fresh petiole sap nitrogen concentration of a nitrogen-deficient leaf at the time of early flowering will be less than 600 ppm nitrate-nitrogen (Table 12).

Fig. 212. Reduced growth and general yellowing, caused by nitrogen deficiency in the tomato crop. (Courtesy T. Hartz)

TABLE 9. Key to Nutritional Diseases[a]

Criterion	Symptom(s)	Additional Criterion or Deficient Nutrient
A.	Symptoms appear on leaves, stems, or petioles.	B
	Flowering or fruiting is affected.	M, B
B.	Youngest leaves are affected first.	C
	Entire plant is affected or oldest leaves affected first.	I
C.	Chlorosis appears on youngest leaves.	D
	Chlorosis is not a dominant symptom. Growing points eventually die, and storage organs are affected.	H
D.	Leaves are uniformly light green, followed by yellowing and poor, spindly growth.	Sulfur
	Uniform chlorosis does not occur.	E
E.	Leaves wilt, become chlorotic, then necrotic.	Copper
	Wilting and necrosis are not dominant symptoms.	F
F.	Distinct yellow or white areas appear between veins and eventually become chlorotic. Symptoms are rare on mature leaves. Necrosis is usually absent.	Iron
	Yellow or white areas are not as distinct, and veins remain green.	G
G.	Chlorosis is less marked near veins. Some mottling occurs in interveinal areas. Chlorotic areas eventually become brown, transparent, or necrotic. Symptoms may appear later on older leaves.	Manganese
	Leaves are abnormally thick, small, and necrotic and curl downward.	Zinc
H.	Tissues are brittle. Young, expanding leaves may be necrotic or distorted, followed by death of growing points. Internodes may be short, especially at shoot terminals. Stems may be rough and cracked.	Boron
	Brittle tissues are not a dominant symptom. Growing points are usually damaged or dead (dieback). Margins of leaves developing from the growing point are first to turn brown or necrotic; old leaves remain green.	Calcium
I.	Chlorosis.	J
	Chlorosis is not a dominant symptom.	L
J.	Interveinal or marginal chlorosis.	K
	General chlorosis. Chlorosis of older leaves progresses from light green to yellow. Entire plant becomes yellow under prolonged stress. Growth is immediately restricted, and plants soon become spindly and drop older leaves.	Nitrogen
K.	Marginal chlorosis or chlorotic blotches, which later merge. Leaves show yellow chlorotic interveinal tissue, progressing to necrosis. Younger leaves are affected with continued stress. Chlorotic areas may become necrotic and brittle and curl upward. Symptoms usually occur late in growing season.	Magnesium
	Interveinal chlorosis, with early symptoms resembling nitrogen deficiency. Older leaves are chlorotic or blotched, with veins remaining pale green. Leaf margins become necrotic and may roll or curl. Symptoms appear on younger leaves as deficiency progresses.	Molybdenum
L.	Leaf margins are tanned or scorched or have necrotic spots (may be small black dots, which later coalesce). Margins become brown and cup downward. Growth is restricted, and dieback may occur. Mild symptoms appear first on recently matured leaves; become pronounced on older leaves and finally on young leaves. Symptoms may be more common late in the growing season from translocation of potassium to developing fruits.	Potassium
	Leaves appear dull, dark green, blue green, or red purple, especially on underside and at midrib and veins. Petioles may also exhibit purpling. Restriction in growth may be noticed.	Phosphorus
	Terminal leaflets wilt with slight water stress. Wilted areas become bronzed and finally necrotic. Observed very infrequently.	Chlorine
M.	Fruit appear rough, cracked, or spotted. Flowering is greatly reduced. Fruit show open locules, internal browning, blotchy ripening, or stem-end russeting.	Boron
	Cracking and roughness are not dominant symptoms. Fruit exhibits water-soaked lesions at blossom end, which later become sunken, dark, or leathery (blossom-end rot).	Calcium
	Ripening diseases (such as blotchy ripening or gray wall) are prominent. Fruit may be puffy and have low acidity.	Potassium

[a] Based on Maynard, 1991.

Too much nitrogen can lead to excessive vegetative growth, delayed fruit set, misshapen fruit, and an increase in fruit-ripening disorders. Tomato transplants that are overfertilized with nitrogen in the transplant production greenhouse produce reduced yields in the production field compared to plants grown with optimum nitrogen fertilization in the transplant greenhouse. Under field conditions, excessively fertilized plants are more attractive to thrips (*Frankliniella, Heliothrips,* and *Thrips* spp.).

In most commercial plantings of tomato, nitrogen deficiency typically results from inadequate nitrogen fertilization. In coarse, sandy soils, nitrogen can be lost from the root zone by leaching after heavy rainfall or excessive irrigation (see Table 10).

When diagnosed early enough, nitrogen deficiency can be corrected by timely applications of nitrogen fertilizer, by soil-directed side dressings, and through a drip irrigation system (i.e., fertigation).

Phosphorus

Phosphorus deficiency is characterized by reduced growth in tomato seedlings (Fig. 213) and smaller but darker-green (sometimes purplish) leaves. Older plants are stunted and have upright leaves that are light green on the upper sides and purple on the lower sides. Phosphorus is mobile within the plant, so the deficiency shows up first on older leaves.

Phosphorus deficiency is rarely observed in commercial field-produced tomato crops, for several reasons. Most agricultural soils contain adequate reserves of phosphorus, and most growers add at least a small amount of phosphorus to fertilizer. The most likely field situations in which phosphorus deficiency might occur include a crop planted on newly cleared land that has not been in agricultural production and that has had very little phosphorus applied and a crop planted in limestone soil, where phosphorus can become "fixed" in unavailable forms. Phosphorus deficiency might occur in soilless greenhouse cul-

TABLE 10. Essential Nutrients for Tomato and Conditions Favoring Deficiencies[a]

Nutrient	Conditions for Deficiency
Nitrogen (N)	Inadequate fertilization in field Low concentration of N in soilless nutrient solution Sandy soils with excessive irrigation or leaching rains
Phosphorus (P)	Inadequate fertilization in field Low concentration of P in soilless nutrient solution Low soil temperature Soil compaction Acidic or basic (high pH) soil reaction
Potassium (K)	Inadequate fertilization in field Low concentration of K in soilless nutrient solution Sandy soils with excessive irrigation or leaching rains
Calcium (Ca)	Low concentration of Ca in soilless nutrient solution High concentration of K, Mg, or NH_4N^b in soil solution Low or fluctuating soil moisture conditions High atmospheric humidity Acidic soils
Magnesium (Mg)	Inadequate fertilization in field Low concentration of Mg in soilless nutrient solution Sandy soils with excessive irrigation or leaching rains Acidic soils High concentration of K, Ca, or NH_4N^b in soil solution
Sulfur (S)	Inadequate fertilization in field Low concentration of S in nutrient solution Sandy soils with low organic matter Leaching rains or excessive irrigation
Boron (B)	Inadequate fertilization in field Low concentration of B in nutrient solution Acidic or basic soil reactions Sandy soils with low organic matter Low soil moisture
Copper (Cu)	Inadequate fertilization in field Low concentration of Cu in nutrient solution Some organic (muck) soils Acidic, highly leached sandy soils Soil with basic reaction
Iron (Fe)	Inadequate fertilization in field Low concentration of Fe in nutrient solution Soils with basic reaction Compacted soils Root damage Excessive concentrations of other heavy metals
Manganese (Mn)	Inadequate fertilization in field Low concentration of Mn in nutrient solution Soils with basic reaction
Molybdenum (Mo)	Inadequate fertilization in field Low concentration of Mo in nutrient solution Acidic soils
Zinc (Zn)	Inadequate fertilization in field Low concentration of Zn in nutrient solution Certain organic (muck) soils Soils with low organic matter Acidic, highly leached soils Soils with basic reaction Excessive P fertilization

[a] Based on Maynard, 1991.
[b] Ammonium-N.

Fig. 213. Phosphorus deficiency of a tomato plant grown in a poor soil mix (left). The same plant 1 month later (right), after the soil ball was loosened and the plant was repotted in new mix and fertilized with a balanced fertilizer. (Courtesy T. A. Zitter)

ture if inadequate phosphorus is supplied in the nutrient solution (see Table 10). A plant exhibiting phosphorus deficiency will have a phosphorus concentration of less than 0.2% in the most recently matured whole leaf at early flowering (see Table 11).

Soil testing should be conducted to determine the phosphorus needs of field-grown crops. Provisions for adequate phosphorus nutrition should be made before planting or very early in the season, since in-season applications of phosphorus are unlikely to correct the deficiency in time to produce normal yields.

Potassium

A deficiency of potassium results in a type of leaf discoloration sometimes referred to as "bronzing." Older leaves of potassium-deficient plants turn tan and then brown at the margins (Fig. 214). In cases of prolonged deficiency, the leaves become necrotic at the margins, with some interveinal yellowing or tanning—hence, the descriptor "bronzing" (Fig. 215).

A tomato plant with potassium deficiency at early flowering will have a potassium concentration of less than 2.5% in the most recently matured whole leaf on a dry-weight basis (see Table 11). The fresh petiole sap potassium concentration in a deficient plant at early flowering will be less than 3,500 ppm potassium (see Table 12).

Fig. 214. Early stages of potassium deficiency on a tomato plant in field culture. (Courtesy G. J. Hochmuth)

TABLE 11. Interpretations of Leaf Tissue Analyses for Most Recently Matured Whole Tomato Leaves[a]

		Concentration (% by dry weight)						Concentration (ppm by dry weight)					
Growth Stage	Interpretation	N	P	K	Ca	Mg	S	B	Cu	Fe	Mn	Mo	Zn
First flower	Deficient level (<)	2.8	0.2	2.5	1.0	0.3	0.3	20	5	40	30	0.2	25
	Adequate range	2.8–4.0	0.2–0.4	2.5–4.0	1.0–2.0	0.3–0.5	0.3–0.8	20–40	5–15	40–100	30–100	0.2–0.6	25–40
	High level (>)	4.0	0.4	4.0	2.0	0.5	0.8	40	15	100	100	0.6	40
	Toxic level	—	—	—	—	—	—	250	—	—	1,500	—	300
First ripe fruit	Deficient level (<)	2.0	0.2	2.0	1.0	0.3	0.3	20	5	40	30	0.2	20
	Adequate range	2.0–3.5	0.2–0.4	2.0–4.0	1.0–2.0	0.3–0.5	0.3–0.6	20–40	5–10	40–100	30–100	0.2–0.6	20–40
	High level (>)	3.5	0.4	4.0	2.0	0.5	0.6	40	10	100	100	0.6	40
	Toxic level	—	—	—	—	—	—	250	—	—	1,500	—	300

[a] Based on Hochmuth et al., 2004.

TABLE 12. Interpretations of Fresh Tomato Petiole Sap Analyses[a]

Stage of Growth	Petiole Sap Nitrate-N Concentration (ppm)	Petiole Sap Potassium Concentration (ppm)
First flower buds	1,000–1,200	3,500–4,000
First open flowers	600–800	3,500–4,000
Fruit: 1-inch[b] diameter	400–600	3,000–3,500
Fruit: 2-inch[c] diameter	400–600	3,000–4,000
First: Harvest	300–400	2,500–3,000
Fruit: Second harvest	200–400	2,000–2,500

[a] Based on Hochmuth, 2004.
[b] 2.5 cm.
[c] 5.0 cm.

Fig. 215. Yellowing and tanning of foliage associated with potassium deficiency. (Courtesy G. J. Hochmuth)

Fig. 216. Potassium deficiency is sometimes associated with the fruit disorders graywall and blotchy ripening. (Courtesy T. Hartz)

Plants under potassium stress are smaller than normal and have reduced fruit production and smaller fruit. Potassium is mobile within the plant; thus, deficiency occurs initially on the lower leaves and advances to the middle and upper leaves. Mild symptoms of potassium deficiency might appear on a mature plant with a heavy fruit load, in which potassium is being rapidly translocated from the leaves to the developing fruit.

Potassium deficiency is sometimes associated with the fruit disorders gray wall and blotchy ripening, in which areas on the sides and upper parts of the fruit remain grayish or yellow and fail to develop red color (Fig. 216). These fruit disorders are particularly noticeable under cool growing conditions (such as winter in southern Florida) and in fields where soil potassium is inadequate. Fruit from potassium-deficient plants are often soft and have low acidity.

Potassium deficiency is most often the result of inadequate potassium fertilization. Potassium is mobile in coarse, sandy soils, and deficiency can result from potassium leaching following heavy rainfall or excessive irrigation (see Table 10). Like nitrogen deficiency, potassium deficiency can be corrected by timely applications of soil-test-predicted potassium fertilizer, by soil-directed side dressing, and through a drip irrigation system (i.e., fertigation).

Calcium

Symptoms of calcium deficiency appear at the tops of tomato plants in the newest growing points and consist of stunting and necrosis of the youngest leaves and growing points (Fig. 217). Sometimes, affected leaves take on a curled, straplike appearance.

Foliar symptoms of calcium deficiency are rarely observed in tomato production fields, because calcium is usually in ample supply from liming materials, fertilizers, and native soil calcium. Also, in many production regions of the world, significant calcium is supplied by irrigation water drawn from wells in limestone aquifers.

In commercial hydroponic culture, calcium deficiency can occur when calcium is inadvertently omitted from the nutrient solution (see Table 10). The most recently matured leaf with less than 0.1% calcium is considered low in calcium (see Table 11). Severe deficiency symptoms will most likely be reflected in plants with less than 0.5% calcium.

Calcium deficiency in tomato crops can lead to the fruit disorder known as blossom-end rot (see the section Blossom-End Rot, earlier in Part III), in which the distal end of the fruit develops a sunken, dark lesion (see Figs. 198 and 199, page 129). The lesion might reside just under the surface of the fruit, a form known as internal blossom-end rot (Fig. 218).

Magnesium

A tomato crop needs adequate magnesium for normal plant growth and development of dark-green leaves. In a magnesium-deficient plant, the older leaves, followed by the middle-aged leaves, show a marked interveinal chlorosis; a mild purpling of the affected leaves sometimes accompanies the chlorosis. As the deficiency progresses, leaves can appear almost white and have green veins (Fig. 219). Magnesium

Fig. 217. Signs of calcium deficiency on tomato leaflet, including interveinal chlorosis and necrosis. (Courtesy T. A. Zitter)

Fig. 218. Internal blossom-end rot, resulting from calcium deficiency. (Courtesy T. A. Zitter)

Fig. 219. Interveinal chlorosis of foliage of a tomato plant with magnesium deficiency. (Courtesy J. Arboleya)

deficiency will likely be manifested in plants with whole, dried leaves having a magnesium concentration of less than 0.25–0.30% (see Table 11).

Magnesium deficiency is observed occasionally in commercial tomato crops in the field but more often in greenhouse soilless culture. Deficiency is most often observed in coarse, sandy soils (from which magnesium can leach from the root zone), in crops not routinely treated with magnesium fertilizer, and in soils of limestone origin (in which calcium can dominate the exchange sites in the soil) (see Table 10). Mild deficiency might be observed on maturing plants with a heavy fruit load, from which magnesium is rapidly translocated to the fruit. In the case of a severe magnesium deficiency, tomato plants fail to grow and yield is reduced.

Magnesium deficiency can be prevented by applying dolomite lime, if a liming material is required. Magnesium can be applied to the soil with a magnesium-containing fertilizer (e.g., magnesium sulfate, potassium-magnesium sulfate) or with a soluble magnesium fertilizer via drip irrigation (i.e., fertigation).

Sulfur

Sulfur deficiency in tomato plants begins with light-green coloration of the younger leaves (Fig. 220). As the deficiency progresses, older leaves also become light green, until the whole plant takes on a light-gray/light-green appearance (Fig. 221). Early sulfur deficiency will be observed in leaves with a

Fig. 220. Light-green coloration of tomato foliage, an early sign of sulfur deficiency. (Courtesy G. J. Hochmuth)

Fig. 221. Sulfur deficiency in tomato plants in the field, indicated by light-gray/light-green coloration. (Courtesy G. J. Hochmuth)

sulfur concentration of less than 0.3% in dried, whole leaves (see Table 11). Symptoms of sulfur deficiency will be evident as the sulfur concentration falls to 0.15–0.20%.

The effects of well-developed sulfur and nitrogen deficiencies can appear similar to a casual observer, but the two deficiencies can be distinguished by examining their ontogenies. Nitrogen deficiency appears first on older leaves and progresses over the whole plant, whereas sulfur deficiency appears first on younger leaves and moves to older leaves.

Deficiencies of sulfur are rare in field tomato crops, primarily because many fertilizers are formulated from sulfate salts (e.g., potassium sulfate, potassium-magnesium sulfate, ammonium sulfate, magnesium sulfate, etc.). Sulfur deficiency is most likely to occur in extremely coarse, sandy soils, because sulfur leaches from these soils and they typically contain little sulfur-containing organic matter (see Table 10). The sulfur needs of a tomato crop can be satisfied with the application of sulfur-containing fertilizer materials.

Sulfur deficiency can be corrected by soil-directed side dressing with sulfur or by applying a soluble sulfur fertilizer using drip irrigation (i.e., fertigation).

Boron

Signs of boron deficiency begin with yellowing of the lower parts of the youngest leaf tips and progresses across the leaf. Petiolules (i.e., leaflet petioles) can become brittle, and growing points can become necrotic. Brown discoloration of the veins in leaves and fruit can also occur under boron deficiency. In addition, fruit can develop corky areas near the calyx end, become puffy, and have uneven ripening. Boron deficiency symptoms will likely appear in younger leaves containing less than 20–25 ppm boron (see Table 11).

Boron deficiency is rare in field tomato production. It is more commonly observed in greenhouse hydroponic culture in which the boron concentration in the nutrient solution is low (see Table 10).

Boron toxicity is indicated by yellowing, scorching, and necrosis of leaf margins of older leaves in the tops of plants. Toxicity has been observed in cases in which foliar sprays contained an excessive concentration of boron.

Chlorine

Documented cases of chlorine deficiency in tomato plants are extremely rare, especially in crop production situations. In controlled growing systems, chlorine deficiency can be produced by omitting chlorine from the nutrient solution and environmental surroundings. Chlorine deficiency appears as leaf wilting and restricted leaf growth, followed by chlorosis.

Copper

Symptoms of copper deficiency in tomato include stunted plant growth, wilting and distortion of youngest leaves, and necrosis of growing points. Copper is a micronutrient required by tomato plants in an extremely small concentration (4–6 ppm copper in dried, whole leaves) (see Table 11).

A deficiency of copper is rarely observed in commercial tomato production. Most soils contain adequate copper for tomato production, and copper-containing bactericides are used routinely for disease control in most U.S. tomato production areas (see Table 10). When a copper deficiency exists, it can be corrected by applying foliar sprays of copper or by adding copper-containing fertilizers to the soil.

Copper can be toxic to tomato plants if the leaf concentration rises above 20–30 ppm in dried, whole leaves. Toxicity can become a concern in vegetable-producing areas in which copper has built up in the soil from the long-time use of copper-containing fungicides and bactericides.

Fig. 222. Interveinal chlorosis of tomato leaves, caused by iron deficiency. Note that the chlorosis is worse on the lower leaflets. (Courtesy J. Arboleya)

Iron

Iron deficiency is indicated in the youngest, expanding leaves at the tips of the tomato plant or branches. Interveinal yellowing begins with the lower portion of the leaf and progress to the tip, and in the case of a severe deficiency, the interveinal area can become white (Fig. 222). Iron deficiency is difficult to diagnose with leaf analysis, but deficient young leaves will likely have iron concentrations of less than 40–50 ppm (see Table 11).

Iron deficiency is most often observed in tomato crops growing in soils with a very basic reaction (greater than pH 7.0), in which iron can be unavailable to the plant (see Table 10). These soils may exist in areas of limestone geology or where the vegetable grower has applied excessive amounts of liming materials, leading to the so-called lime-induced chlorosis of plants. Iron deficiency might be observed in tomato crops that have a heavy fruit load and that grow in the cooler part of the season (e.g., winter tomato production in southern Florida); however, this transient iron deficiency does not appear to reduce fruit yield or quality. Iron deficiency might also be observed periodically in soilless tomato culture under winter growing conditions of low light and cool temperatures (especially cool soil temperatures in the root zone).

Application of iron-containing fertilizers to high-pH soils is usually not very effective in correcting iron deficiency. Repeated foliar sprays of iron are more effective. Tomato plants can tolerate a wide range of iron concentrations, so iron toxicity is rare.

Manganese

Manganese deficiency occurs under conditions similar to those described for iron deficiency: soils with a high pH and containing high levels of free carbonates (see Table 10). In most situations, symptoms of manganese deficiency are similar to but not as striking as those of iron deficiency.

Manganese deficiency of tomato appears as interveinal chlorosis and speckling of the youngest, expanding leaves (Fig. 223). Young leaves with less than 20–30 ppm manganese will likely be deficient (see Table 11).

Manganese deficiency is rare in most tomato production fields, in part because some commonly used crop fungicides contain manganese. Deficiency can be prevented by applying soil-test-predicted amounts of manganese with preplant fertilizer mixtures.

Symptoms of manganese toxicity include brown lesions on stems and petioles, especially near the nodes. In addition, leaf blades develop bleached-appearing, necrotic patches.

Fig. 223. Interveinal chlorosis and speckling of tomato leaves, caused by manganese deficiency. (Courtesy G. E. Wilcox)

Molybdenum

Symptoms of molybdenum deficiency begin with the bright-yellow or orange mottling of lower, older leaves; the leaf margins then become scorched and rolled. These leaf symptoms progress to younger leaves, and flowers drop from the plants. Healthy tomato plant leaves contain 0.2–0.6 ppm molybdenum (see Table 11).

Molybdenum deficiency is extremely rare under typical field conditions but can occur in fields with acidic soil conditions (see Table 10). Deficiency is also more likely to occur in hydroponic culture when molybdenum is inadvertently omitted from the nutrient solution.

Zinc

Zinc deficiency symptoms begin with interveinal chlorosis of the lower leaves, which then spreads up the plant. In addition, brown spots can form on the veins of leaflets and their petiolules. Severely affected leaflets can curl downward and petioles can twist and coil. Deficiency will likely develop in plants with a leaf-zinc concentration of less than 20 ppm (see Table 11).

Zinc uptake by tomato plants can be suppressed by high concentrations of other heavy metals (e.g., manganese) and by high concentrations of phosphorus in the medium (see Table 10). Zinc deficiency has been reported in greenhouse culture but is rare in the field.

Zinc toxicity has been observed in greenhouse tomatoes. Plants demonstrate spindly growth, and younger leaves remain small and have interveinal yellowing. Toxicity will likely occur in plants with a leaf-zinc concentration greater than 300 ppm (see Table 11).

Selected References

Bould, C., Hewitt, E. J., and Needham, P. 1983. Diagnosis of Mineral Disorders in Plants. Vol. 1, Principles. Her Majesty's Stationary Office, London.

Hochmuth, G. J. 2003. Plant Petiole Sap-Testing for Vegetable Crops. Fla. Coop. Ext. Serv. Publ. CIR1144. Available online at http://edis.ifas.ufl.edu/pdffiles/CV/CV00400.pdf

Hochmuth, G. J., Maynard, D. N., Vavrina, C. S., and Hanlon, E. A. 2004. Plant Tissue Analysis and Interpretation for Vegetable Crops in Florida. Fla. Coop. Ext. Serv. Publ. HS964/EP081. Available online at http://edis.ifas.ufl.edu/pdffiles/ep/ep08100.pdf

MacNab, A. A., Sherf, A. F., and Springer, J. K. 1983. Identifying Diseases of Vegetables. The Pennsylvania State University, University Park.

Marschner, H. 1995. Mineral Nutrition of Higher Plants. 2nd ed. Academic Press, London.

Maynard, D. N. 1979. Nutritional disorders of vegetable crops: A review. J. Plant Nutr. 1:1-23.

Maynard, D. N., and Hochmuth, G. J. 1997. Knott's Handbook for Vegetable Growers. 4th ed. John Wiley and Sons, New York.

Roorda van Eysinga, J. P. N. L., and Smilde, K. W. 1981. Nutritional Disorders in Glasshouse Tomatoes, Cucumbers, and Lettuce. Center for Agricultural Publishing and Documentation, Wageningen, Netherlands.

Watterson, J. C. 1985. Tomato Diseases: A Practical Guide for Seedsmen, Growers, and Agricultural Advisors. Petoseed, Saticoy, CA.

Weir, R. G., and Cresswell, G. C. 1993. Plant Nutrient Disorders 3: Vegetable Crops. NWS Agriculture and Inkata Press, Sydney, Australia.

Wilcox, G. E. 1993. Tomato. Pages 137-140 in: Nutrient Deficiencies and Toxicities in Crop Plants. W. F. Bennett, ed. American Phytopathological Society, St. Paul, MN.

Winsor, G., and Adams, P. 1987. Diagnosis of Mineral Disorders in Plants. Vol. 3, Glasshouse Crops. Her Majesty's Stationary Office, London.

(Prepared by G. J. Hochmuth and D. N. Maynard)

Herbicide Damage

Pesticide-related disorders of tomato generally involve herbicides. Injury related to pesticides often appears similar to injury from other sources, such as excessive fertilizer salts and air pollutants. Also, symptoms of pesticide-related disorders are not always definitive, as similar damage may be caused by different pesticides.

Pesticide injury can result from misapplication, the presence of residual material in soil, and the movement of a pesticide through the soil into the crop root zone. Misapplication can take several forms, including spray or vapor drift, contamination of spray equipment, application of the wrong pesticide, and application at the wrong rate.

The types of herbicides most commonly used on tomato are discussed in the following sections. Discussion includes the conditions under which each herbicide is likely to cause damage and the nature of the damage that may occur.

Amides

Amides—such as bensulide, diphenamid, isoxaben, napropamide, naptalam, pronamide, and propanil—interfere with cell division and inhibit growth in tomato leaves, roots, and shoot terminals. Amides are soil-applied herbicides and are not readily translocated within tomato plants. However, contact with an excessive quantity of amides in the soil may produce plants with stubby roots (since root growth is inhibited), similar to the effects of dinitroaniline herbicides (see Dinitroanilines, later in this section). Foliar symptoms include stunted growth, malformed leaves, and downward and inward curling of leaves, similar to the epinasty associated with phenoxy and benzoic acid herbicides (see Phenoxy and Substituted Benzoic Acids, later in this section).

Naptalam may cause tomato plant roots to grow up and out of the soil. Napropamide and diphenamid are registered for use on tomato and do not cause damage at normal application rates, but excessive application rates and use in soils with high accumulated amide concentrations may produce damage.

Amino Acid Derivatives

Amino acid derivatives interfere with the synthesis of amino acid. This family of herbicides includes glufosinate, glyphosate, and sulfosate.

Glyphosate damage is usually associated with spray drift from the application of directed sprays on tomatoes or from applications made in adjoining areas, such as pastures. However, isolated instances of damage have been observed from residual glyphosate in soil and on polyethylene mulch.

The degree of injury from glyphosate varies, depending on several factors, including the stage of plant growth, the glyphosate application rate or concentration, and the cultivar. Injury can range from a temporary growth cessation without foliar symptoms to the death of the affected plant. A lethal dose of glyphosate produces chlorosis of terminals followed by necrosis; symptoms progress slowly from the apical to the basal portion of the plant. Wilting may be associated with this chlorosis. A sublethal dose of glyphosate can produce symptoms ranging from temporary growth cessation without foliar symptoms to varying degrees of chlorosis (Fig. 224). Chlorosis of terminal buds and developing leaves may eventually disappear and normal growth may resume. Other symptoms may include crinkling and cupping of foliage, development of some strap-shaped or "shoestring" leaves, pseudoepinasty, and leaf mottling similar to the effects caused by a virus.

One of the most common effects of a sublethal dose of glyphosate on tomato is delayed bloom or fruit set. This delay often results in a large number of flowers developing late in the season and the subsequent production of a large number of very small, unmarketable fruit. In addition to being small, fruit may be irregular, pearlike, elongated, or flattened ("button") in shape; have a rough surface; and ripen in an uneven, blotchy fashion.

Tomato seedlings have been injured by glyphosate vapor when the herbicide has been applied to weeds growing on the greenhouse floor during periods of high temperatures. This scenario is not common, however.

Soil residues of glyphosate occasionally damage tomato plants. This damage is generally associated with freshly set transplants, and the symptoms are similar to those already described but expressed more rapidly. With this source of damage, the plant first develops chlorotic lesions around the stomate. Necrosis follows fairly rapidly, followed by the death of the plant. Root absorpton of glyphosate can sometimes cause veinal chlorosis, as well.

Residual activity is observed for as long as 2 days following application of glyphosate to the soil surface. Residues of glyphosate applied to polyethylene mulch photodegrade very slowly, if at all, and may remain in a phytotoxic concentration for several weeks. Damage may result if the residue contacts tomato foliage by some means, such as the splash of a raindrop, or if plants are brought into contact with the mulch film while wet from dew or rainfall.

Arsenicals

Monosodium methanearsonate (MSMA) and disodium methanearsonate (DSMA) are arsenical herbicides to which tomato plants are normally exposed as a result of spray drift from another crop (e.g., cotton (*Gossypium* L.)) or from pasture or sod. Arsenicals interfere with phosphorus metabolism and produce foliar symptoms on tomato plants that include chlorosis, development of a purplish cast, and necrosis. In many cases, necrosis occurs rapidly after exposure.

Even a small dose of an arsenical herbicide causes severe injury to tomato. Plants seldom recover from exposure.

Aryl Triazinones

Aryl triazinones have a similar mode of action as diphenyl ethers, disrupting cell membranes in grasses and broadleaf weeds. Sulfentrazone and carfentrazone are the most well known members of this group. Aryl triazinones are used mainly in postemergence applications.

When tomato leaves sprayed with aryl triazinones are exposed to sunlight, they rapidly become necrotic. When light herbicide drift occurs, leaves become speckled and localized necrosis and death of affected tissue follows (Fig. 225).

Benzothiadiazoles

Bentazon belongs to the family of herbicides called benzothiadiazoles. On tomato plants, bentazon makes leaves become chlorotic, and they later turn brown and die. This herbicide translocates poorly in tomato plants. Therefore, when a plant has been partially exposed to a low rate of bentazon, symptoms characteristic of the triazines develop, in which the affected part of the plant is destroyed but the other growing points remain green (Fig. 226) (see Triazines, later in this section).

Bipyridyliums

The bipyridylium family of herbicides includes paraquat and diquat. These herbicides destroy the integrity of cellular membranes, including those around chloroplasts. Following the

Fig. 224. Injury to a tomato plant exposed to a sublethal rate of glyphosate. (Courtesy B. M. Santos)

Fig. 225. Speckling and necrosis on foliage, caused by exposure to sulfentrazone. (Courtesy B. M. Santos)

disruption of membranes, the cells quickly die. Bipyridyliums require sunlight for activity; thus, applications made on cloudy days produce symptoms more slowly than applications made on sunny days.

Under optimum conditions, symptoms on tomato occur in a matter of hours, beginning with wilting and followed by the develoment of water-soaked areas where spray droplets have contacted tissue. These areas darken rapidly, becoming necrotic and desiccated. Under most conditions, contacted tissue dies within 24–48 h. Limited soil activity of bipyridyliums has occasionally been observed but is not common.

Paraquat damage is fairly common in tomato production with polyethylene mulch, usually occurring as the result of spray drift or misapplication. Paraquat is subject to photodegradation. However, the rate of photodegradation of residues on polyethylene mulch varies with the season of the year, and phytotoxic levels may remain for as long as 5 days after application (Fig. 227).

Chloroacetamides

Chloroacetamides, such as alachlor and metolachlor, inhibit cell division, much like amides (see Amides, earlier in this section). Unlike amides, however, chloroacetamides are readily translocated within the plant, moving in the transpirational stream of water from the roots to the leaves. Exposure of tomato plants to a low concentration of these herbicides may be noninjurious, but exposure to a higher concentration causes stunted overall growth and distortion of foliage, with leaf crinkling, cupping, and twisting.

Dinitroanilines

Dinitroaniline herbicides—such as ethalfluralin, oryzalin, pendimethalin, and trifluralin—inhibit cell division, similar to amides and acetanilides (see Amides, earlier in this section). Dinitroanilines inhibit growth of the entire tomato plant by limiting root growth, especially the development of lateral and secondary roots. Moreover, the roots that do develop are somewhat short, thick, and stubby.

Many of the dinitroanilines are prone to volatilization and photodecomposition when applied to the soil surface and must be incorporated with another product, but some are not (e.g., oryzalin). When the less volatile dinitroanilines are applied to the soil surface, they may cause the development of callus tissue on the stem of the tomato plant at the soil line. This callusing can result in lodging and stem breakage.

Diphenyl Ethers

Acifluorfen, lactofen, and oxyfluorfen belong to the herbicide family known as diphenyl ethers. These herbicides rupture cellular membranes and inhibit photosynthesis. All have foliar activity, and preemergence activity occurs at high rates.

Tomato plants exposed to diphenyl ethers may display a range of symptoms, from leaf crinkling, marginal leaf curling, and downward cupping of leaves at lower rates of exposure to stunting, contact burning, and necrosis at higher rates of exposure (Fig. 228). In addition to the leaf malformations produced by vapor drift, occasional interveinal chlorosis may develop with slightly higher exposure rates.

Even when visible damage occurs to tomato plants, fruit yield and quality are not always affected. Plants frequently grow out of the effects of exposure to diphenyl ethers.

Fumigants

Fumigants, such as methyl bromide and 1,3-dichloropropene (1,3-D), are another source of damage to tomato plants. Damage sometimes occurs when tomato transplants are planted in soil that has been treated with a fumigant before the fumigant has completely dissipated. In such a case, general chlorosis develops rather quickly on the foliage and is usually followed by wilting, necrosis, and death of the plant. Damage may be localized, affecting only certain areas of the field, or it may be spread throughout the field, depending on how soon after fumigant application the planting occurred.

Fig. 226. Chlorosis on tomato foliage, caused by bentazon exposure. (Courtesy B. M. Santos)

Fig. 227. Paraquat damage to tomato stems and leaves in plants grown with polyethylene mulch. (Courtesy J. P. Gilreath)

Fig. 228. Symptoms of lactofen exposure on tomato foliage and fruit. (Courtesy B. M. Santos)

Imidazolinones and Sulfonylureas

Imidazolinones and sulfonylureas are meristematic inhibitors and inhibit acetolactate synthase, which is involved in the synthesis of the essential amino acids leucine, isoleucine, and valine. The imidazolinones include imazamox, imazapic, imazapyr, imazaquin, and imazethapyr, and the sulfonylureas include halosulfuron, nicosulfuron, and rimsulfuron. In certain U.S. states, halosulfuron and rimsulfuron are labeled for selective use on tomato.

Symptoms of damage from these herbicides on tomato are the cessation of growth or shortening of internodes and the development of a compact growth habit, followed by foliar chlorosis and necrosis. Chlorosis usually develops first in the terminal buds.

Imidazolinones and sulfonylureas have both foliar and soil activity and may remain active in the soil for long periods of time. Most cases of injury from these herbicides result from soil residues. Damage from volatilization may occur in the greenhouse, and it is seldom uniform.

Phenoxy and Substituted Benzoic Acids

Perhaps the most classic symptoms of herbicide injury on tomato plants are those associated with phenoxy herbicides (e.g., 2,4-dichlorophenoxyacetic acid (2,4-D)) and substituted benzoic acids (e.g., 3,6-dichloro-2-methoxybenzoic acid (dicamba)). These herbicides have auxinlike properties (which cause excessive cellular growth) and are true growth regulators.

Damage from these herbicides usually results from spray drift, but volatilization and movement of vapor can occur. Mild symptoms range from light to moderate crinkling of young leaves to epinasty, frenching of leaves, and temporary cessation of growth for up to approximately 2 weeks. Exposure to high concentrations of these herbicides causes long-term growth cessation and considerable growth distortion, including epinasty, thickening and increased brittleness of stems and petioles, and formation of adventitious roots (Fig. 229).

The extent of damage is related not only to the concentration of phenoxy or benzoic acid herbicide but also to the plant's stage of growth at the time of exposure, the duration of exposure, and the growing conditions. At a given herbicide concentration, young, actively growing plants are damaged more than older plants.

Fig. 229. Growth distortion caused by exposure to 2,4-dichlorophenoxyacetic acid (2,4-D). (Courtesy T. A. Zitter)

The effects of herbicide injury on tomato fruit are varied. The shape may be affected, so that fruit are elongated (resembling some processing cultivars), large or blocky, flattened ("button" fruit), or otherwise irregularly shaped. Another effect of herbicide injury is a roughening of the fruit surface, with small pimples forming. Fruit may also ripen prematurely or unevenly, creating a blotchy appearance, or an excessive number of small fruit may be produced. Finally, herbicide injury may result in increased incidences of several disorders of tomato, including blossom-end rot (see Blossom-End Rot, earlier in Part III), the formation of hollow locules or the absence of gel, the limited production of seeds, and the emergence of radicles from seeds within the fruit prior to harvest.

Pyridazinones and Pyridinones

The group of herbicides known as pyridazinones and pyridinones includes the compounds dithiopyr, fluridone, and norflurazon. These herbicides kill plants by inhibiting the synthesis of carotenoid pigments, which protect chlorophyll molecules from destruction by ultraviolet (UV) light. The dramatic effect they produce on tomato foliage is complete chlorosis, with the leaves developing a bleached appearance. In the early stages of injury from these herbicides, foliage may have areas of pink tissue as well as white tissue (Fig. 230). Sublethal doses of these herbicides may result in extended chlorosis, but lethal doses eventually cause necrosis. The pyridazinones and pyridinones have prolonged soil and foliar activity. Tomato plants may be exposed to these herbicides from spray drift or soil residues.

A closely related herbicide is clomazone (a member of the isoxazolidinones). Like dithiopyr, fluridone, and norflurazon, clomazone kills plants by disrupting carotenoid synthesis, leaving chlorophyll molecules vulnerable to sunlight. On tomato plants, the loss of chlorophyll produces white foliage within 1 week. Plants are sometimes damaged by clomazone in soil residues, following use of the herbicide on a previous crop (e.g., soybean (*Glycine max* (L.) Merr.)), but most instances of clomazone injury result from spray and vapor drift. Although the visible effects of clomazone drift on tomato plants are quite dramatic, yield losses do not often occur. The chlorosis usually disappears over time if good cultural practices are maintained.

Generally, norflurazon and fluridone cause much greater damage to tomato plants than clomazone.

Fig. 230. Foliar damage caused by exposure to norflurazon. Note the areas of pink and white tissue. (Courtesy J. P. Gilreath)

Substituted Ureas

Substituted ureas, such as diuron and linuron, are photosynthesis inhibitors. Most of these herbicides are applied to the soil, but some have foliar activity.

On tomato, these herbicides initially produce leaf chlorosis, followed by necrosis of affected plants. In most cases, chlorosis develops first in the leaf veins and then spreads to the interveinal tissue. Older leaves tend to be affected worse than younger leaves, and dieback of leaf tips is often observed. Thus, the substituted ureas produce symptoms that are similar to those produced by triazine and uracil herbicides (see Triazines and Uracils, later in this section).

Phenylureas can have rather long residual lives in soil and are fairly mobile in soils with low cation-exchange capacities. Under these conditions, tomato plants may not be damaged until their roots grow into the layers of soil where the herbicide has accumulated or until the herbicide moves up in the soil profile, as occurs with some types of irrigation.

Although diuron is usually considered nonvolatile, it may volatilize and injure tomato seedlings in greenhouses under high temperatures. In this case, injury results in variable chlorosis, including the formation of a chlorotic band around the leaf margins.

Thiocarbamates

The thiocarbamate family of herbicides includes EPTC, pebulate, thiobencarb, and vernolate. These are soil-active herbicides that inhibit cell division and elongation by what is believed to be an inhibition of protein synthesis.

Several thiocarbamates are registered in the United States for use on tomato plants. However, when applied at excessive rates, these products can cause a range of damage. Mild to severe stunting may occur, and leaves may be twisted, cupping upward and inward. In addition, foliage may develop a crinkled appearance and have necrotic margins; over time, the necrosis spreads across the leaves and they abscise. Foliar application or spray drift of thiocarbamates produces mostly leaf crinkling, rapid necrosis of affected foliage, or both effects (Fig. 231).

Triazines

Triazine herbicides—such as atrazine, metribuzin, and simazine—are photosynthetic inhibitors and produce leaf chlorosis followed by necrosis (Fig. 232). Chlorosis usually develops first in the leaf veins (often on lower leaves) and then spreads to the interveinal areas, but in some cases, it may be restricted to interveinal tissue. The response of tomato plants varies with the cultivar and triazine herbicide involved. Some tolerance exists for metribuzin but not for atrazine and other triazines.

Triazine injury may result from shoot or root uptake. Injury from root uptake may result from soil residues from a previous season or the movement of soil-applied material into the root zone. Residual levels of triazines depend on the application rate, the history of triazine use in a field, the qualities of the soil (e.g., pH, organic matter and clay content, drainage characteristics), and environmental factors (most notably, rainfall and temperature).

Although metribuzin is registered for use on tomato in the United States, it occasionally causes damage when applied before or after transplanting during a period of cloudy, cool weather, when plants are stressed and photosynthesis rates are low. In fact, because the triazines inhibit photosynthesis, anything that might stress plants or slow photosynthesis enhances the effects of these herbicides, if only slightly, and thus increases the chance of injury.

Uracils

Like the phenylureas and triazines (see previous sections), the uracils are photosynthesis inhibitors. Uracil herbicides, such as bromacil and terbacil, are generally applied to the soil and have a long residual life in most soils (5 years or longer).

Symptoms produced by the uracils on tomato are similar to those produced by the phenylureas and triazines. Foliage usually develops veinal chlorosis, which may later spread to interveinal tissue and be followed by necrosis. Exposure to small doses of uracils may produce only veinal chlorosis.

Selected References

Ashton, F. M., and Crafts, A. S. 1973. Mode of Action of Herbicides. John Wiley & Sons, New York.

Baumann, P. A., Dotray, P. A., and Prostko, E. P. 1999. Herbicides: How They Work and the Symptoms They Cause. Tex. Agric. Ext. Serv. Bull. B-6081.

Eagle, D. J., and Caverly, D. J. 1981. Diagnosis of Herbicide Damage to Crops. Chemical Publishing, New York.

Ferrell, J. A., Stall, W. M., and MacDonald, G. E. 2007. Diagnosing Herbicide Injury—2007. Univ. Fla. Coop. Ext. Serv. SS-AGR-15.

Klingman, G. C., and Ashton, F. M. 1982. Weed Science: Principles and Practices. John Wiley & Sons, New York.

(Prepared by B. M. Santos, J. P. Gilreath, and P. R. Gilreath)

Fig. 231. Crinkling and necrosis of foliage resulting from thiocarbamate exposure. (Courtesy J. P. Gilreath)

Fig. 232. Necrosis of tomato foliage caused by exposure to atrazine. (Courtesy B. M. Santos)

Genetic Diseases

Tomatoes are subject to numerous genetic diseases. Even so, only a few cause serious problems in crop production, because plant breeders eliminate the causal factors in the development of tomato cultivars.

Two factors are responsible for the existence of genetic diseases in cultivars: incomplete penetrance and variable expressivity. *Penetrance* refers to the frequency with which a gene produces a recognizable effect in the individuals that carry it, and *expressivity* refers to the degree of manifestation of a genetic character. Thus, the gene system for a genetic disease may exist but be expressed to a serious degree only rarely under most environmental conditions. Breeders may never see the defect or see it so rarely that they consider it too trivial to detract from the positive qualities of the cultivar. Also, since symptoms are not always expressed, such a defect can be cumbersome to eliminate by selection. The defect may arise by mutation in the seed production field or the grower's field, but careful roguing by seed producers should reduce this problem. In sum, for tomato cultivars, the occurrence of genetic defects by mutation should be rare in growers' fields.

Genetic mutations are classified as genetic diseases when they adversely affect the quality of tomato fruit directly by producing undesirable symptoms on fruit or indirectly by reducing fruit yield and quality through their effects on other plant characteristics. Given the large number of reported mutations and the low frequency of their occurrence, this compendium will not describe all of the genetic diseases of tomato. The most common genetic diseases are described in the following sections. (Extensive descriptions of genetic mutations can be found in the *Report of the Tomato Genetics Cooperative,* published annually.[1])

As a general rule, diagnostic plant pathologists, extension specialists, and farmers should not be overly concerned about the occurrence of genetic diseases in tomato crops, since the frequency of occurrence is normally quite low. Regardless, the appearance of only a few genetically diseased plants in a large field can be quite alarming to the farmer if the cause and nature of the problem are not understood. The best management results from understanding the genetic nature of the problem—in other words, realizing that the disease is not contagious and will not spread to healthy plants in the field. If the incidence of genetic disease is quite high, then the farmer should work with a producer who can provide seed free of genetic diseases.

Autogenous Necrosis

Autogenous necrosis (AN) is caused by the incompatible reaction of a dominant Cladosporium leaf mold resistant gene (*Cf–*) with a recessive necrosis gene (*ne*), which produces necrotic spots on tomato foliage under some environmental conditions. The development of AN is favored by high temperatures and light intensity.

Symptoms of AN start at the base of the tomato plant and progress upward as the plant grows. Lesions first form at the distal ends of leaflets, and new lesions form proximally. In severe cases of AN, fruit size is greatly reduced.

Foliage symptoms are more prominent on the lower leaf surfaces than on the upper leaf surfaces. On the upper surfaces, bright-yellow spots appear, which coalesce and turn brown and necrotic over time. On the lower surfaces, the yellow spots are often densely crisscrossed by fine, light-brown streaks (Fig. 233). The leaves also tend to cup downward and inward, eventually shriveling and dying.

AN generally does not occur in finished cultivars, because it is selected against by the breeders. On occasion, however, a cultivar that was developed in a cool climate and never displayed symptoms will develop severe AN when grown in a warm climate. Thus, AN may be more likely to develop in experimental cultivars that are being tested commercially.

The best management strategy for AN is to grow cultivars that are not subject to the disease.

Blind Plants

A genetic abnormality that occurs in tomato plants grown in seedbeds is referred to as a "blind plant." Blind plants germinate, have cotyledons, and sometimes develop one or two true leaves, but they do not have a normal meristem. These plants therefore fail to grow (except for an expansion of the cotyledon) and produce fruit.

Blind plants are generally caused by genetic mutations and should occur infrequently enough not to cause real concern. These plants are more likely to be produced from old seed, so if their growth becomes prevalent, a newer lot of seed should be used. The prevalent growth of blind plants may also result from using seed that has been excessively heated (80°C) or from exposing seedlings to volatile toxic materials (e.g., fumigants, disinfectants, creosote).

Bull Plants

A genetic disease that occurs in approximately one out of 1,000 tomato plants is an unfruitful plant, often referred to as a "bull plant." Late in the season, when normal plants are weighted down with fruit, bull plants can be easily identified from across the field by their vigorous, bushy appearance. In greenhouse production, bull plants stand out as dark-green plants with enlarged leaves.

In 1945, C. M. Rick reported that bull plants were caused primarily by chromosome abnormalities. Many were the result of triploidy (i.e., the presence of one extra set of chromosomes), while others were caused by other chromosome aberrations or single gene mutations (e.g., male sterility).

Fruit Pox

Fruit pox was first reported in Texas in 1937, and over the years, it has been reported on various cultivars and in vari-

Fig. 233. Symptoms of autogenous necrosis on lower leaf surfaces. (Courtesy J. P. Jones)

[1] Available online at http://tgc.ifas.ufl.edu/onlinevo.htm

ous tomato production regions of the United States and other countries. Environmental conditions have been discounted as having a primary effect on the development of fruit pox, yet observations in North Carolina over several seasons have indicated that the disease is more severe under conditions of high temperature and rapid plant and fruit growth and on fruit lacking good foliage cover.

Symptoms of fruit pox begin on immature fruit as small, white lesions that are slightly elongated to oval in shape. As the fruit approaches maturity, the lesions enlarge, rupture, and become necrotic. In some cases, the lesions do not rupture and assume a tan color as the fruit ripens (Fig. 234). Fruit pox lesions tend to be concentrated on the shoulder area of the fruit and can vary in number from very few to very many.

Cultivars differ greatly in susceptibility to fruit pox, but none can be considered immune. Within susceptible cultivars, fruit on individual plants may be severely affected, whereas fruit on other plants may show few symptoms.

Fruit pox susceptibility is inherited and mutation is fairly common, resulting in severe expression of the disease in susceptible cultivars. Growers should be cautious when increasing seed of open-pollinated cultivars and inbred parent lines of hybrids by roguing plants that express symptoms of fruit pox.

Gold Fleck

Gold fleck often occurs on tomato cultivars that are susceptible to fruit pox and for many years was considered part of that disease. However, more detailed study has distinguished gold fleck from fruit pox.

Symptoms of gold fleck begin on the surface of immature fruit as small, dark-green spots; the spots are typically round but are sometimes elongated or irregular in shape. As the fruit ripens, the spots become lighter in color, changing to a light tan and then a golden yellow by the time the fruit is red and ripe (see Fig. 234). The spots that are symptomatic of gold fleck are often more concentrated on the blossom end of the fruit than on the shoulder area. In terms of number, very few to very many spots may occur on a single fruit.

The growing conditions that favor the expression of fruit pox also favor the expression of gold fleck. The two diseases often occur on the same fruit, and severe expression of one is often accompanied by severe expression of the other.

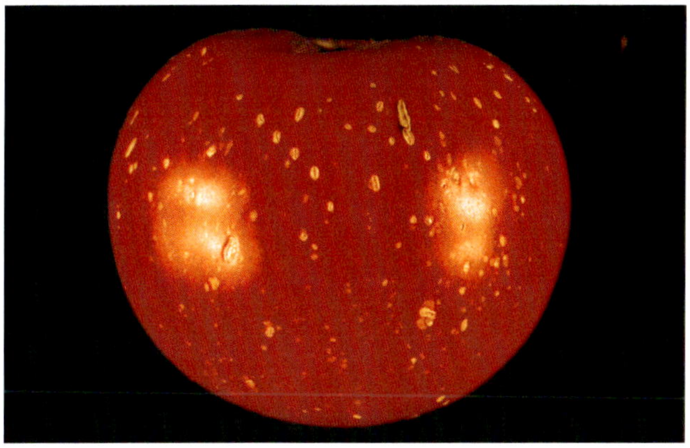

Fig. 234. Small, tan lesions on ripe tomato fruit are symptomatic of both fruit pox and gold fleck. (Courtesy R. G. Gardner)

Gold fleck susceptibility is inherited and mutation is common; severe expression of the diseases is likely in susceptible cultivars. Growers should be certain to avoid saving seed from plants with the disease when making selections for increasing seed.

Selected References

Crill, J. P., Burgis, D. S., Jones, J. P., and Strobel, J. W. 1973. The fruit pox and gold fleck syndromes of tomato. Phytopathology 63:1285-1287.

Langford, A. N. 1948. Autogenous necrosis in tomatoes immune from *Cladosporium fulvum* Cooke. Can. J. Res. 26:35-63.

Rick, C. M. 1945. A survey of cytogenetic causes of unfruitfulness in the tomato. Genetics 30:347-362.

Wagner, R. P., and Mitchell, H. K. 1955. Genetics and Metabolism. John Wiley and Sons, New York.

(Prepared by J. W. Scott, J. P. Crill, and R. G. Gardner)

Part IV. Diseases of Unconfirmed Etiology

Black Shoulder

Symptoms

Black shoulder affects primarily fresh-market tomato fruit. Early symptoms are gray spots or striations on the shoulder of green fruit. These spots later develop into larger, irregularly shaped, dark-gray to black areas surrounding the fruit stem scar. As the fruit ripens, the diseased tissue typically shrivels and becomes tough, although in severe cases, the lesions may continue to enlarge and become sunken. Under favorable conditions, the lesions coalesce, producing the characteristic symptom of black shoulder.

Etiology and Epidemiology

Black shoulder is caused by *Alternaria alternata* (Fr.:Fr.) Keissl. Avenues for invasion by this fungus are believed to be provided by small cuticle cracks in green fruit, which are caused by sudden environmental changes and mechanical injuries.

A. alternata remains quiescent and symptoms will not develop if the fruit temperature stays above 15°C. However, at temperatures below 10°C, the fungus will continue to colonize exposed epidermal cell layers, producing numerous small, black lesions. The extent of damage caused by this disease is proportional to the length of time that mature-green fruit are exposed to cool temperatures.

Periods of rapid plant growth followed by cool, rainy weather also seem to be associated with the appearance of black shoulder (see the section Cuticle Cracking in Part III). In addition, severe black shoulder (as well as Alternaria stem canker) can develop in harvested tomato fruit approaching the mature-green stage that are kept chilled (at 0°C) for 6 days or longer.

Only green fruit are susceptible to initial infection by *A. alternata,* although ripe fruit may exhibit symptoms of black shoulder. Cultivars with fruit that have dark-green stem ends are more prone to cracking and black shoulder than cultivars that have uniformly light-green fruit. Tomato plants with lush growth resulting from heavy fertilization also appear more prone to this disease.

Management

Given the importance of sudden environmental changes in the development of black shoulder, management of the disease is difficult. One effective strategy is to plant tomato cultivars that are more tolerant to shoulder cracking. In addition, careful attention should be paid to supplying adequate but not excessive fertilizer and to avoiding injuries to developing fruit (e.g., those caused by high-pressure spraying of pesticides).

Selected References

McColloch, L. P., and Worthington, J. T. 1952. Low temperature as a factor in the susceptibility of mature-green tomatoes to Alternaria rot. Phytopathology 42:425-427.

Morton, D. J., Worley, R. E., and Jaworski, C. A. 1964. Black shoulder of mature-green tomatoes in Georgia. Plant Dis. Rep. 48:956.

(Prepared by J. C. Watterson)

Graywall

Graywall can be a devastating disease of tomato, for two reasons: The affected fruit are unmarketable, and the disease occurs after all of the costs of production have been expended. The percentage of fruit affected by graywall depends on the tomato cultivar planted and the environmental conditions in which the crop is grown. However, with a susceptible cultivar, it is not unusual to find that 40% of the fruit have been affected.

Note that graywall is the name adopted for a disease of tomato fruit in the U.S. state of Florida. There is much uncertainty about the cause of the disease, however, and thus uncertainty about the distribution of the disease around the world. The disorders referred to as "blotchy ripening" in greenhouses in the northern United States and southern Canada and as "cloud" in New Zealand are probably the same disease as graywall. Diseases involving blotchy ripening of field tomatoes in New York and California and internal necrosis of tomatoes in New Jersey and Pennsylvania also have some similarities to graywall. Even so, the latter two diseases should not likely be assumed to be graywall and so are not included in this section.

Symptoms

Graywall is expressed as the formation of black to dark-brown necrotic tissue in the walls of tomato fruit. The necrosis is sometimes present in the crosswalls but seldom in the center column of the fruit; in most cases, only the outer walls are affected. The necrosis causes a partial collapse of the wall tissue, and the outer surface of an affected fruit may appear wrinkled. The necrosis gives the epidermis a grayish appearance—hence, the name of the disease (Fig. 235).

Often, the necrosis in the wall is associated with the vascular tissue (Fig. 236), and the necrotic area in the walls can usually be traced to the stem end of the fruit. However, the necrosis does not begin at the vascular bundles but rather at the corky ring associated with the stem end.

Graywall typically develops on green fruit prior to harvest; fruit as small as 2–3 cm in diameter may be affected. The disease occasionally develops after the harvest of mature-green fruit. When an affected fruit ripens, the area around the necrosis remains firm and turns from green to yellow, giving the appearance of uneven ripening (see Fig. 236). The affected area also appears woody when cut. Thus, affected fruit are of poor quality.

Fig. 235. External symptoms of graywall. (Courtesy T. A. Zitter)

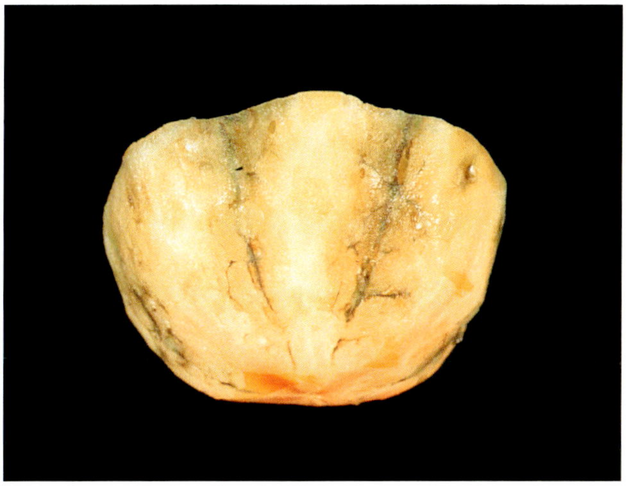

Fig. 236. Internal symptoms of graywall. (Courtesy J. P. Jones)

Etiology and Epidemiology

The blackening of tomato wall tissue has been associated with the presence of active polyphenol oxidase—an enzyme that is not present in healthy tissue. This enzyme is activated by the presence in wall tissue of a relatively large population of bacteria, which are found in greater number in diseased tissue than in healthy tissue.

Tomato plants that develop graywall are also commonly infected with *Tobacco mosaic virus* (TMV), which produces an internal necrosis of fruit. However, graywall will develop on plants free of TMV, and plants resistant to TMV will develop graywall. (See the section on TMV in Part I.)

Increased incidence of graywall has been associated with several environmental conditions. Graywall occurs on plants under low light conditions and in cool weather. Excessive soil moisture and compacted soil also tend to support development of the disease. Finally, the incidence of graywall increases as the ratio of potassium to nitrogen in fertilizers decreases.

Management

Breeders have developed many tomato cultivars that are highly resistant to graywall for culture in both the field and the greenhouse. These cultivars are resistant even under the most favorable environmental conditions for disease. If a susceptible cultivar must be planted, then attention should be paid to the ratio of potassium-to-nitrogen fertility.

Selected References

Conover, R. A. 1949. Vascular browning in Dade County, Florida green-wrap tomatoes. Plant Dis. Rep. 33:283-284.

Hall, C. B., and Dennison, R. A. 1955. Environmental factors influencing vascular browning of tomato fruits. Proc. Am. Soc. Hortic. Sci. 65:353-356.

Jones, J. P., and Alexander, L. J. 1962. Relation of certain environmental factors and tobacco mosaic virus to blotchy ripening of tomatoes. Phytopathology 52:524-528.

Kidson, E. B., and Stanton, D. J. 1953. "Cloud" or vascular browning in tomatoes. I. Conditions affecting the incidence of "cloud." N.Z. J. Sci. Technol. 34:521-530.

Stall, R. E., and Hall, C. B. 1969. Association of bacteria with graywall of tomato. Phytopathology 59:1650-1653.

(Prepared by R. E. Stall)

Glossary

A—acre (43,560 square ft.; 1 acre = 0.40469 hectare)
a.i.—active ingredient
C—Celsius or centigrade
cm—centimeter (1 cm = 0.01 m = 0.3937 in.)
F—Fahrenheit
ft-c—foot-candle
g—gram (1 g = 0.03527 oz)
gal—gallon (1 gal liquid (U.S.) = 3.785 L)
h—hour
ha—hectare (4,046.87 m^2; 1 ha = 2.47 acres)
in.—inch (1 in. = 2.540 cm)
kg—kilogram (1 kg = 2.205 lb)
L—liter (1 L = 1.057 quarts liquid (U.S.))
lb—pound (1 lb = 453.59 g)
m—meter (1 m = 39.37 in.)
mg—milligram (1 mg = 0.001 g)
min—minute
ml—milliliter (1 ml = 0.001 L)
mm—millimeter (1 mm = 0.001 m = 0.03937 in.)
mM—millimolar (concentration of 1 millimole/L)
µg—microgram (1 µg = 10^{-6} g)
µm—micrometer (1 µg = 10^{-6} m)
nm—nanometer (1 nm = 10^{-9} m)
oz—ounce (1 oz = 28.35 g); fluid ounce (1 fl oz (U.S.) = 29.57 ml)
ppm—parts per million
sec—second
T—metric ton (1 t = 1.102 short tons = 2,025 lbs.)

abaxial—directed away from the stem of a plant; pertaining to the lower surface of a leaf (*see* adaxial)
abiotic—pertaining to the absence of life, as diseases not caused by living organisms
abscise—to separate from a plant, as leaves, flowers, and fruits do when they fall
abscission—the shedding of leaves or other plant parts as the result of physical weakness in a specialized layer of cells (the abscission layer), which develops at the base of the structure
acervulus (pl. acervuli)—an erumpent, cushionlike fruiting body bearing conidiophores, conidia, and sometimes setae
acid—having a pH less than 7
acropetal—upward from the base to the apex of a shoot of a plant (*see* basipetal); in fungi, producing spores in succession in the direction of the apex, so that the apical spore is the youngest
acute—with reference to symptoms, developing suddenly (*see* chronic)
adaxial—directed toward the stem of a plant; pertaining to the upper surface of a leaf (*see* abaxial)
adventitious—arising from other than the usual place, as roots from a stem rather than branches of a root
aeciospore—a dikaryotic spore produced in the aecium of a rust fungus; in heteroecious rust fungi, the spore stage that infects the alternate host
aecium (pl. aecia; adj. aecial)—the fruiting body of a rust fungus in which the first dikaryotic spores (aeciospores) are produced
aerial—occurring in air
aerobic—living only in the presence of oxygen
agar—a jellylike material derived from algae and used to solidify liquid culture media; a medium containing agar
alate—winged (*see* apterous)
alkaline—having basic (nonacidic) properties; having a pH greater than 7
alkaloid—any of various nitrogen-containing ring compounds produced by plants and having physiological effects on animals
allele—any of one or more alternative forms of a gene
allelopathy (adj. allelopathic)—the ability of one species to inhibit or prevent the growth of another species through the production of one or more toxic substances
alternate host—one of two kinds of plants on which a parasitic fungus (e.g., a rust fungus) must develop to complete its life cycle
alternative host—a plant other than the main host that can be colonized by a parasite but that is not required for completion of the developmental cycle of the parasite
anaerobic—living in the absence of oxygen
anamorph (adj. anamorphic)—the imperfect state or asexual form in the life cycle of certain fungi, producing asexual spores (e.g., conidia) or no spores (*see* holomorph; teleomorph)
anastomosis (pl. anastomoses)—the fusion of branches of the same or different structures (e.g., hyphae) to make a network
annual—a plant that completes its life cycle and dies within one year (*see* biennial; perennial)
antagonist—an organism or substance that counteracts or limits the action of another organism or substance
anterior—situated toward the front or head (*see* posterior)
anther—the pollen-bearing portion of a flower
antheridium (pl. antheridia)—the male sexual organ (male gametangium) of certain fungi
anthesis—the period of the opening of a flower, during which pollination can occur
anthocyanin—a colored, water-soluble flavenoid pigment in cell sap
anthracnose—disease caused by acervuli-forming fungi (order Melanconiales); characterized by sunken lesions and necrosis
antibiotic—any of various chemical compounds produced by microorganisms and killing or inhibiting the growth of other living organisms
antibody—a protein formed in the blood of warm-blooded animals in response to the presence of an antigen
antigen—any foreign chemical (normally, a protein) that induces antibody formation in warm-blooded animals
antiserum (pl. antisera)—blood serum containing an antibody
apex (pl. apices; adj. apical)—the tip of a root or shoot; contains the apical meristem
aphid—any of numerous species of small, sucking insects of the family Aphididae (order Homoptera) that produce honeydew and injure plants when present in large populations
aplerotic—with reference to oospores, not filling the oogonium
apothecium (pl. apothecia)—open, cuplike or saucerlike, ascus-bearing fruiting body (ascocarp) of ascomycetous fungi; often supported on a stalk
appressed—closely flattened down or pressed against a surface (syn. adpressed)
appressorium (pl. appressoria)—a swollen, flattened portion of a fungal filament that adheres to the surface of a higher plant, providing anchorage for invasion by the fungus
apterous—wingless (*see* alate)
arbuscule (adj. arbuscular)—a branched haustorial structure of certain endomycorrhizal fungi; formed within living cells of the root
arthropod—any member of the phylum Arthropoda, which consists of animals with articulated bodies and limbs, including insects, arachnids, and crustaceans
ascocarp—a sexual fruiting body (e.g., apothecium, ascostroma, cleistothecium, perithecium, pseudothecium) of ascomycetous fungi, producing asci and ascospores (syn. ascoma)

ascogenous—pertaining to ascus-producing hyphae
ascogonium (pl. ascogonia)—a specialized cell that gives rise to hyphae that produce asci
ascoma (pl. ascomata)—*see* ascocarp
ascomycetes (adj. ascomycetous)—a group of fungi in the phylum Ascomycota (kingdom Fungi) characterized by the production of sexual spores (ascopores) within an ascus
ascospore—a sexual spore borne in an ascus
ascostroma (pl. ascostromata)—a fruiting body of ascomycetous fungi containing bitunicate (double-walled) asci in locules (cavities); usually dark and containing multiple locules but sometimes containing only a single locule (*see* pseudothecium)
ascus (pl. asci)—a saclike structure containing ascospores (typically eight) and usually borne in a fruiting body
aseptate—lacking cross-walls (septa); nonseptate; coenocytic
asexual—vegetative; lacking sex organs, gametes, or sexual spores (as in the imperfect or anamorphic stage of a fungus)
asexual reproduction—any type of reproduction not involving the union of gametes and meiosis
asexual stage—the imperfect (anamorphic) stage of certain fungi
attenuate—to narrow; to weaken; to decrease in virulence or pathogenicity
aucuba—a bright-yellow, mosaic leaf variegation of genetic or viral origin
autoecious—with reference to rust fungi, producing all spore forms on one species of host plant (*see* heteroecious)
autotroph—any of various organisms (e.g., plants capable of photosynthesis) that synthesize their nutritive substances from inorganic molecules
avirulent—nonpathogenic; unable to cause disease (*see* virulent)
axenic—culture in the absence of living bacteria or other organisms; pure culture
axil—the angle formed by a leaf petiole and the stem
axillary—pertaining to or placed within an axil

bacilliform—shaped like a short rod with rounded ends
backcross—to cross (mate) an offspring with one of its parents
bactericide—a chemical or physical agent that kills bacteria
bacterium (pl. bacteria)—a prokaryotic, microscopic, single-celled organism having a cell wall and increasing by binary fission
basidiocarp—the sexual fruiting body of fungi in the group basidiomycetes (syn. basidioma)
basidiomycetes (adj. basidiomycetous)—a group of fungi in the phylum Basidiomycota (kingdom Fungi) characterized by the formation of external basidiospores on basidia
basidiospore—a haploid (1*n*) sexual spore produced on a basidium
basidium (pl. basidia; adj. basidial)—a structure on which basidiospores are produced externally
basipetal—situated down from the apex toward the base of a shoot of a plant; developing in the direction of the base, so that the apical part is oldest (*see* acropetal)
bicellular—two-celled
biennial—a plant that produces seed and dies at the end of its second year of growth (*see* annual; perennial)
binary fission—a type of asexual reproduction in which two cells (usually of similar size and shape) are formed by the growth and division of one cell
binucleate—having two nuclei per cell (*see* multinucleate; uninucleate)
bioassay—any test (assay) using a living organism
biocide—a compound toxic to all forms of life
biological control—exploitation of the natural competition, parasitism, or antagonism of organisms for the management of pests and pathogens (syn. biocontrol)
biotic—pertaining to living organisms; with reference to disease, caused by a living organism
biotroph—an organism that can live and multiply only on another living organism (syn. obligate parasite) (*see* necrotroph)
biotype—a subdivision of a species, subspecies, or race based on some identifiable physiological trait, such as a particular virulence pattern
bitunicate—having two walls
blight—the sudden, severe, and extensive spotting, discoloration, wilting, or destruction of leaves, flowers, stems, or entire plants
blotch—an irregularly shaped, usually superficial spot or blot
botryose—shaped like a bunch of grapes
bract—a reduced leaf associated with a flower or inflorescence; a modified leaf from the axil of which a flower arises
breeding line—a plant strain used in a plant breeding program and usually containing one or more desirable agronomic or breeding characteristics
broadcast application—application of fertilizer by spreading or scattering it on the soil surface
buffer—a substance capable in solution of keeping hydrogen-ion concentration constant and thereby avoiding rapid changes in acidity or basicity of a solution
bursa—an extension or flap of cuticle at the side of the male nematode sex organ; used for orienting the body during mating

calcareous—rich in calcium carbonate (lime)
callus—specialized tissue that forms over a wound or cut in a plant (e.g., cork cambium may form, and the cells produced will gradually seal the wound)
cambium—a lateral meristematic layer of stems and roots, giving rise to secondary xylem, secondary phloem, and parenchyma; responsible for secondary growth
canker—a plant disease characterized (in woody plants) by the death of cambium tissue and malformation or loss of bark and (in nonwoody plants) by the formation of sharply delineated, dry, necrotic, localized lesions on the stem; a lesion caused by such a disease, particularly in woody plants
canopy—the expanded, leafy top of a plant or plants
capsid—the protective layer of protein surrounding the nucleic acid core of a virus; the protein molecules that make up this layer (syn. coat protein)
carbohydrate—any of various chemical compounds composed of carbon, hydrogen, and oxygen (e.g., sugars, starches, cellulose)
carpel—the ovule-bearing structure of a flower in angiosperms
catenulate—with reference to the spores of certain fungi, formed in a chain
causal agent—the organism or agent that produces a given disease
cellulose—a carbohydrate comprising the primary cell wall substance
certification scheme—a governmentally supervised procedure of seed propagation to ensure high quality, varietal purity, and freedom from disease
chelate—relating to or having a ring structure that usually contains a metal ion held by coordination bonds
chimera, chimaera—a plant or organ consisting of two or more genetically different tissues
chlamydospore—a thick-walled or double-walled, asexual resting spore formed from hyphal cells (terminal or intercalary) or by transformation of conidial cells; functions as an overwintering stage
chlorophyll (adj. chlorophyllous)—any of a group of green pigments found in chloroplasts and critical to photosynthesis
chloroplast—a disklike structure containing chlorophyll in which photosynthesis occurs in green plants
chlorosis (adj. chlorotic)—the failure of chlorophyll development, caused by disease or a nutritional disturbance; the fading of green plant parts to light green, yellow, or white
chromosome—the structure that contains the genes of an organism (in eukaryotes, chromosomes are in the nucleus and can be visualized with an optical microscope as threads or rods during meiosis and mitosis; in bacteria, the chromosome is usually a single circle of DNA and cannot be visualized with an optical microscope)
chronic—with reference to symptoms, slow-developing, persistent, or recurring (*see* acute)
circulative—with reference to viruses, passing through the gut and circulating in the body of an insect vector before being transmitted to a host
circulative transmission—virus transmission in which the virus must accumulate within or pass through the lymphatic system of an insect vector before it can be transmitted to a plant
clamp connection—a bridge- or buckle-like hyphal protrusion in basidiomycetous fungi, formed at cell division and connecting the newly divided cells
clavate—club shaped (syn. claviform)
cleistothecium (pl. cleistothecia)—a spherical ascocarp that is closed at maturity
coalesce—to grow together into one body, as enlarging lesions may join together to form a single spot

coenocytic—having multiple nuclei embedded in cytoplasm without cross-walls; nonseptate
colonize—with reference to plant pathogens, to infect and ramify through plant tissue
colony—a microorganism growing in mass, especially as a pure culture
conidiogenous—producing and bearing conidia
conidiophore—a simple or branched hypha on which conidia are produced
conidium (pl. conidia)—an asexual, nonmotile fungal spore that develops externally or is liberated from the cell that formed it
cortex (adj. cortical)—the region of parenchyma tissue between the epidermis and phloem in stems and roots; the region beneath the rind of a sclerotium
cotyledon—the primary embryonic leaf within the seed, in which nutrients for the new plant are stored (monocots have one cotyledon; dicots have two)
crown—the junction of the root and stem of a plant, usually at the soil line; in grafted woody plants, the rootstock portion of the plant near the soil surface
cultivar (abbr. **cv.**)—a cultivated variety of a plant species, resulting from deliberate genetic manipulation and having recognizable characteristics, such as color; shape of flowers, fruits, and seeds; height; and form (syn. variety)
cultural practices—the methods by which plants are grown (e.g., application of nutrients, irrigation, cultivation, and so forth)
culture—the growth and propagation of microorganisms on nutrient media; the growth and propagation of living plants
cupulate—cuplike or cup shaped
cuticle (adj. cuticular)—the noncellular outer layer of an insect or nematode; the water-repellent, waxy layer of epidermal cells of plant parts, such as leaves, stems, and fruits
cv. (cultivar)—a cultivated variety of a plant species, resulting from deliberate genetic manipulation and having recognizable characteristics such as color; shape of flowers, fruits, and seeds; height; and form (syn. variety)
cyst—in fungi, a resting structure in a protective membrane or shell-like enclosure; in nematodes, the egg-laden carcass of a female nematode; in bacteria, a specialized cell enclosed in a thick wall; often dormant and resistant to environmental conditions
cytoplasm—the living protoplasm in a cell, not including the nucleus

damping-off—the death of a seedling before or shortly after emergence caused by decomposition of the root or lower stem (it is common to distinguish between preemergence and postemergence damping-off)
decortication—the loss of cortex caused by rot
defoliation—the loss of leaves from a plant, whether normal or premature
dehiscent—opening by breaking into parts
demicyclic—pertaining to the life cycle of rust fungi (e.g., many species of *Gymnosporangium*) that lack the urediniospore (repeating) stage (*see* macrocyclic; microcyclic)
desiccate—to dry out
diagnostic (n. diagnosis)—pertaining to a distinguishing characteristic important for the identification of a disease or disorder
dichotomous—branching, often successively, into two more or less equal arms
dicot—a dicotyledon, or plant having two cotyledons (*see* monocot)
dieback (v. die back)—the progressive death of shoots, leaves, or roots, beginning at the tips
differential host—a plant host that develops disease symptoms that distinguish between strains or races of a plant pathogen (syn. differential cultivar)
digitate—having lobes radiating from a common center
dikaryon (adj. dikaryotic)—an organism having two sexually compatible haploid nuclei per cell that divide simultaneously in the reproductive phase called dikaryophase
dilution end point—the stage of a serial dilution of cells or preparations at which growth or infection from a standard sample of the suspension no longer occurs
dimorphic—having two distinct shapes or forms
diploid—having two complete sets of chromosomes ($2n$ chromosomes) (*see* haploid; polyploid)
disease—the abnormal functioning of an organism
disease cycle—the succession of events and interactions among a host, a parasite, and the environment, from initial infection through pathogenesis to overseasoning, until another infection occurs
disinfest—to kill pathogens that have not yet initiated disease or other contaminating microorganisms that occur in or on inanimate objects (e.g., soil or tools) or on the surface of plant parts (e.g., seed)
dispersal—the spread of infectious material (inoculum) from diseased to healthy plants (syn. dissemination)
dissemination—*see* dispersal
distal—situated away from the point of attachment or origin or away from the main body (*see* proximal)
DNA—deoxyribonucleic acid; the double-stranded, helical molecule that contains genetic code information; consists of repeating units, or nucleotides, each of which is composed of deoxyribose (a sugar), a phosphate group, and a purine (adenine or guanine) or a pyrimidine (thymine or cytosine) base
dolipore septum—a type of cross-wall in fungi in the group basidiomycetes characterized by distinctive swellings and membranes in association with the septal pore
dominant—pertaining to a phenotypic trait that is expressed in hybrid progeny of diploid organisms even if contributed by only one of the parents (*see* recessive)
dormancy (adj. dormant)—a condition of suspended growth and reduced metabolism of an organism; generally induced by internal factors or environmental conditions as a mechanism of survival
dormant—resting; living in a state of reduced physiological activity
dorsal—situated toward the back or top of the body (ant. ventral)
dwarfing—underdevelopment of a plant or plant organs caused by disease, inadequate nutrition, or unfavorable environmental conditions

echinulate—having small spines projecting from the cell wall
ecology—the study of the interactions between individual organisms, groups of organisms, and organisms and their environments
ectoparasite—a parasite that feeds from the exterior of its host (*see* endoparasite)
ectotrophic—pertaining to fungal development primarily over the root surface
edema, **oedema**—blisters produced on leaves or other plant parts under conditions of high moisture and restricted transpiration (syn. intumescence)
effuse (adj.)—stretched out, especially with reference to a filmlike growth
egg mass—a group of eggs held together by a gelatinous matrix
ELISA—enzyme-linked immunosorbent assay; a serological test in which the sensitivity of the reaction is increased by attaching an enzyme that produces a colored product in response to one of the reactants
ellipsoidal—elliptical in the plane section
embryo—an organism in the early stages of development, such as a young plant in the seed or a nematode before hatching from the egg
emergence—the growth of a seedling shoot through the surface of the soil
enation—an abnormal outgrowth from the surface of a stem or leaf
encapsidate—to enclose viral nucleic acid in a protein coat
encyst—to form a cyst or protective covering
endemic—native to a particular place; pertaining to a low and steady level of natural disease occurrence
endocarp—the inner layer of a fruit wall (*see* pericarp)
endoconidium (pl. endoconidia)—a conidium produced inside a hypha or conidiophore
endodermis—the layer of cells between the vascular tissue and the cortex of roots
endogenous—arising from inside (*see* exogenous)
endoparasite—a parasite that lives and feeds inside its host (*see* ectoparasite)
endophyte—a plant developing inside another organism; any of the various endoparasitic fungi associated with grass species
endoplasm (adj. endoplasmic)— the inner granular, relatively fluid part of the cytoplasm of a cell
endosperm—the nutritive tissue formed within the embryo sac of seed plants
enzyme—a protein that catalyzes a specific biochemical reaction
enzyme-linked immunosorbent assay (ELISA)—a serological test in which the sensitivity of the reaction is increased by attaching an

enzyme that produces a colored product in response to one of the reactants
epicotyl—the portion of the stem of a plant embryo or seedling above the node where the cotyledons are attached (*see* hypocotyl)
epidemic—an increase of disease in a population; a general and serious outbreak of disease (*see* epiphytotic)
epidemiology (adj. epidemiological)—the study of factors influencing the initiation, development, and spread of infectious disease; the study of disease in populations
epidermis (adj. epidermal)—the surface layer of cells of leaves and other plant parts
epinasty—the abnormal downward curling of a leaf, leaf part, or stem
epiphyllous—located on the upper surface of a leaf
epiphyte—an organism growing on a plant surface but not as a parasite
epiphytotic—pertaining to the widespread and destructive outbreak of a plant disease
eradicant—a chemical used to eliminate a pathogen from a host or environment
eradication—the control of plant disease by eliminating the pathogen after it has become established or by eliminating the plants that carry the pathogen
erumpent—bursting or erupting through the surface of a substrate
ethylene—a plant hormone influencing vegetative growth, fruit ripening, abscission of plant parts, and senescence of flowers
etiolation—the elongation of stems caused by low levels of light intensity
etiology—the study of the causes of diseases
eukaryote (adj. eukaryotic)—any of various organisms whose cells contain a membrane-bound nucleus and other organelles, including all higher plants, animals, fungi, and protists (*see* prokaryote)
exogenous—originating from the outside (*see* endogenous)
extracellular—not occurring or located within a cell
exudate—a liquid excreted by or discharged from diseased tissues, roots, leaves, and fungi

f. sp. (pl. ff. spp.)—*forma specialis* (pl. *formae speciales*); a taxonomic group within a pathogenic species, defined by its host range (members of different *formae speciales* infect different groups of plants)
facultative—capable of changing lifestyle (e.g., from saprophytic to parasitic or the reverse)
facultative parasite—an organism that is normally saprophytic but capable of living as a parasite
facultative saprophyte—an organism that is normally parasitic but capable of living as a saprophyte
fallow—pertaining to cultivated land kept free from a crop or weeds during the normal growing season
fasciation—the malformation of shoots or floral organs manifested as enlargement and flattening, as if several parts were fused
fascicle (adj. fasciculate)—a small group, bundle, or cluster
fastidious—with reference to prokaryotic organisms, having special requirements for growth and nutrition
feeder root—a fine root that absorbs water and dissolved nutrients
filament (adj. filamentous)—a thin, flexible, threadlike structure
filamentous—threadlike (syn. filiform)
filiform—long, needlelike (syn. filamentous)
flaccid—wilted; lacking turgor
flagellum (pl. flagella)—a hairlike, whiplike, or tinsel-like appendage of a motile cell, bacterium, or zoospore that provides locomotion
fleck—a minute, discolored spot in green tissue
flexuous—having turns or windings; capable of bending
flocculation—aggregation into a loose, fluffy mass
focus (pl. foci)—a small area of diseased plants within a larger plot or field
foliar—pertaining to leaves
forma specialis (pl. *formae speciales*)—a taxonomic group within a pathogenic species, defined by its host range (members of different *formae speciales* infect different groups of plants); abbr. f. sp. (pl. ff. spp.)
fructification—any of the various spore-bearing organs formed by macro- and microfungi
fruiting body—any of the various complex, spore-bearing structures of fungi
fumigant (v. fumigate)—a gas or volatile substance used to kill or inhibit the growth of microorganisms or other pests
Fungi Imperfecti—a group of fungi lacking a sexual stage; a group comprising the asexual stages of fungi in the groups ascomycetes and basidiomycetes (syn. deuteromycetes)
fungicide (adj. fungicidal)—a chemical or physical agent that kills or inhibits the growth of fungi
fungistat (adj. fungistatic)—a compound that inhibits fungal growth or sporulation but does not kill fungi
fusiform—spindle shaped; tapering at each end

gall—an abnormal swelling or localized outgrowth, often roughly spherical, produced by a plant as a result of attack by a fungus, bacterium, nematode, insect, or other organism (syn. tumor)
gametangium (pl. gametangia)—a cell containing gametes or nuclei that act as gametes
gamete—sex cell
gelatinous—resembling gelatin or jelly
gene—a unit located on a chromosome and controlling the transmission of a heritable characteristic
genetic—pertaining to heredity or heritable characteristics
genetically modified organism (GMO)—an organism possessing a gene from another species; having been the subject of genetic engineering
geniculate—bent, like a knee
genome—the complete genetic information of an organism or virus
genotype—the genetic constitution of an individual or group; a class or group of individuals sharing a specific genetic makeup (*see* phenotype)
genus (pl. genera)—a taxonomic category that consists of species that are closely related structurally or phylogenetically (the genus or generic name is the first name in a Latin binomial)
geotropic—plant growth directed toward the force of gravity (e.g., roots)
germ tube—a hypha resulting from an outgrowth of the spore wall and cytoplasm after germination
germinate (n. germination)—to begin growth, as of a seed, spore, sclerotium, or other reproductive body
germplasm—the bearer of heredity material (often loosely applied to cultivars and breeding lines)
giant cell—an enlarged, multinucleate cell formed in a root by repeated nuclear division without cell division; induced by secretions of certain sedentary plant-parasitic nematodes
girdle—to circle and cut through a stem or the bark and outer few rings of wood, disrupting the phloem and xylem
globose—nearly spherical
glycoprotein—a conjugated protein in which the nonprotein group is carbohydrate
graft transmission—the transmission of a pathogen from one host plant to another through fusion of living tissue from the diseased host with living tissue of a healthy host
gram-negative bacteria—bacteria that are stained red or pink after treatment with Gram's stain
gram-positive bacteria—bacteria that are stained violet or purple after treatment with Gram's stain
gravid—containing an egg or eggs; capable of depositing eggs
guttation—the exudation of watery, sticky liquid from hydathodes, especially along leaf margins

haploid—having a single complete set of chromosomes (*see* diploid; polyploid)
hardiness—the ability to withstand stress
haulm—stem or stalks, collectively
haustorium (pl. haustoria)—a specialized branch of a parasite formed inside a host cell to absorb nutrients
herbaceous—pertaining to primary, soft, nonwoody tissue of a plant or plant part; having the characteristics of an herb
herbicide—any of various chemicals that kill plants or inhibit their growth (e.g., a weed or grass killer)
hermaphrodite (adj. hermaphroditic)—having both male and female reproductive organs in a single individual
heteroecious—with reference to rust fungi, passing different stages of the life cycle on alternate hosts (*see* autoecious)
heterokaryon (adj. heterokaryotic)—a cell with genetically different nuclei
heterothallism (adj. heterothallic)—a fungal life cycle in which sexual reproduction can occur only in the presence of genetically different mycelia (*see* homothallism)

heterozygous—having the same form (allele) of a gene on different chromosomes (*see* homozygous)
histopathology—study of the pathology of cells and tissues; the microscopic changes characteristic of disease
holomorph—a fungus having sexual and asexual forms in its life cycle, considered in all its forms (*see* anamorph; teleomorph)
homothallism (adj. homothallic)—a fungal life cycle in which sexual reproduction occurs in a single, self-fertile thallus (*see* heterothallism)
homozygous—having the same form (allele) of a gene on homologous chromosomes (*see* heterozygous)
honeydew—a sugary ooze or exudate often secreted by aphids; a characteristic symptom of ergot
host plant—a living plant attacked by or harboring a parasite or pathogen, from which the invader obtains part or all of its nourishment
host range—the range of plants on which an organism (particularly a parasite) feeds
hull—the outer coat of a seed
hyaline—transparent or nearly so; translucent; colorless
hybrid (v. hybridize)—the offspring of two individuals of different genotypes
hydathode—an epidermal leaf structure specialized for the secretion or exudation of water; the opening at the terminus of a leaf vein
hydrolysis (v. hydrolyze)—a chemical reaction in which a compound decomposes via a reaction with water; involves the splitting of a bond and the addition of hydrogen and oxygen
hymenium—the continuous, spore-bearing layer of a fungal fruiting body
hyperplasia (adj. hyperplastic)—an abnormal increase in the number of cells, often resulting in the formation of galls or tumors
hypersensitive—extremely or excessively sensitive, often with reference to an extreme reaction to a pathogen
hypertrophy (adj. hypertrophic)—an abnormal increase in the size of cells in a tissue or organ, often resulting in the formation of galls or tumors
hypha (pl. hyphae; adj. hyphal)—a tubular filament of a fungal thallus or mycelium; the basic structural unit of a fungus
hypocotyl—the portion of the stem below the cotyledons and above the root (*see* epicotyl)

icosahedral (n. icosahedron)—having 20 faces, like polyhedral virus particles (*see* isometric)
immune—not capable of being infected by a given pathogen
immunogenic—able to induce the production of antibodies
imperfect state—the anamorph, or asexual form in the life cycle of certain fungi, in which asexual spores (e.g., conidia) or no spores are produced (*see* perfect state)
in vitro—in glass, on an artificial medium, or in an artificial environment; not within a living host
in vivo—within a living organism
inclusion body—a structure developed within a plant cell as a result of infection by a virus; often useful in identifying the virus
indicator plant—a plant in which specific or distinctive symptoms develop in reaction to a pathogen or certain environmental conditions; used to detect or identify the pathogen or determine the effects of the environmental conditions
infection—the process in which an organism enters, invades, or penetrates and establishes a parasitic relationship with a host plant
infection court—a site in or on a host plant where infection can occur
infection cushion—an organized mass of hyphae formed on the surface of a plant and producing numerous infective hyphae
infection focus—the initial site of infection, generally with reference to a population of plants
infection peg—the specialized, narrow hyphal strand located on the underside of an appressorium and penetrating host cells (syn. penetration peg)
infectious—with reference to disease, capable of spreading from plant to plant
infective—able to attack a host and cause infection; with reference to vectors, carrying or containing a pathogen and able to transfer it to a host plant
infest (n. infestation)—to attack as a pest (used especially for insects and nematodes); to contaminate (as with microorganisms); to be present in large numbers
inflorescence—a flower or flower cluster
initial inoculum—inoculum (usually from an overwintering source) that initiates disease in the field, as opposed to inoculum that spreads disease during the growing season (syn. primary inoculum)
injury—damage caused by transitory interaction with an agent such as an insect, chemical, or unfavorable environmental condition
inoculate (n. inoculation)—to place inoculum in an infection court; to insert a pathogen into healthy tissue
inoculum (pl. inocula)—a pathogen or its parts capable of causing infection when transferred to a favorable location
inoculum density—a measure of the number of propagules of a pathogenic organism per unit of area or volume
intercalary—inserted within (e.g., located along a hypha, as opposed to located at the end of a hypha)
intercellular—between or among cells
intercostal—between veins (of a leaf) (syn. interveinal)
intercrop—to grow two or more crops simultaneously on the same area of land
internode (adj. internodal)—the portion of a stem between two successive nodes
interveinal—between veins (of a leaf)
intracellular—through or within cells
intumescence—blisters produced on leaves or other plant parts under conditions of high moisture and restricted transpiration (syn. edema)
irradiation—exposure to radiant energy
isolate—(n.) a culture or subpopulation of a microorganism separated from its parent population and maintained under controlled conditions; (v.) to remove from soil or host material and grow in pure culture
isometric—with reference to virus particles, having an icosahedral structure and thus appearing approximately round

juvenile—an immature form that appears similar to the adult stage but is usually smaller and is not sexually mature, as in nematodes or insect species that undergo gradual metamorphosis

knot—a gall or localized abnormal swelling

labile—constantly undergoing or likely to undergo change; unstable
lamina—the expanded part of a leaf (*see* petiole)
land race—a locally developed strain of a plant
larva (pl. larvae)—the immature wormlike or caterpillar stage in the life cycle of an insect that undergoes complete metamorphosis
latent—present but not manifested or visible, as a symptomless infection
latent infection—an infection that does not have visible symptoms
latent period—the time elapsed between infection and the appearance of symptoms or production of new inoculum (sometimes synonymous with incubation period); the time elapsed after the acquisition of a pathogen by a vector but before the pathogen can be transmitted
lateral bud—a bud that forms on a stem at the axil of a leaf
leaching—the loss of water-soluble plant nutrients from the soil, resulting from rain or irrigation
leaf spot—a lesion typically restricted in development after reaching a characteristic size
leafhopper—any of various species of mobile insects with sucking mouthparts in the order Hemiptera
leaflet—one of the separate blades or divisions of a compound leaf
legume—a simple, dry, dehiscent fruit developing from a simple pistil and splitting at maturity along two seams; any plant in the family Fabaceae (formerly Leguminosae)
lenticel—a natural opening in the surface of a stem, tuber, fruit, or root, permitting gas exchange
lenticular—lens shaped (convex on both faces)
lesion—a localized diseased area or wound
life cycle—the cyclical stages in the growth and development of an organism
lignin—a complex organic substance or group of substances that impregnate the cell walls of xylem vessels and certain other plant cells; constituting wood
local lesion—a small, restricted lesion; often the characteristic reaction of a differential cultivar to a specific pathogen, especially in response to mechanical inoculation with a virus

locule (adj. locular)—a small cavity or compartment within an organ or part of an organism
lumen (pl. lumina)—the central cavity of a cell or other structure

macerate—to cause disintegration of tissues by separation of cells; to soften by soaking
macroconidium (pl. macroconidia)—the larger of two kinds of conidia formed by certain fungi (*see* microconidium)
macrocyclic—pertaining to the life cycle of rust fungi that typically exhibit all five spore stages (pycniospores, aeciospores, urediniospores, teliospores, and basidiospores) in the course of their development (*see* demicyclic; microcyclic)
manual transmission—the spread or introduction of inoculum to an infection court by hand
marl—a type of soil rich in lime formed in the bottom of a lake or swamp
mating types—compatible strains (usually designated + and – or A and B), both of which are necessary for sexual reproduction in heterothallic fungi
mechanical injury—injury of a plant part by abrasion, mutilation, or wounding
mechanical transmission—the spread or introduction of inoculum to an infection court by human manipulation, accompanied by physical disruption of host tissues (wounding)
medium (pl. **media**)—a mixture of organic or inorganic chemical compounds and water providing the nutrients needed for the growth of a microorganism in vitro; with reference to higher plants, a mixture of fertilizers and other components in which plants are grown
meiosis—nuclear division in which the number of chromosomes per nucleus is halved (i.e., the diploid state is converted to the haploid state) (*see* mitosis)
melanin (adj. melanoid)—a brown-black pigment
meristem (adj. meristematic)—plant tissue characterized by frequent cell division, producing cells that become differentiated into specialized tissues
meristem culture—an aseptic culture of a plant or plant part from a portion of the meristem
mesophyll—the central, internal, nonvascular tissue of leaves, consisting of the palisade and spongy mesophyll
messenger RNA (abbr. mRNA)—a form of RNA that carries information to direct the synthesis of protein
metabasidium—the part of the basidium in which meiosis occurs
metabolite—any chemical participating in metabolism; a nutrient
microclimate—weather conditions on a small scale, as at the surface of a plant or within a field
microconidium (pl. microconidia)—the smaller of two kinds of conidia formed by certain fungi (*see* macroconidium)
microcyclic—pertaining to the life cycle of rust fungi that produce only teliospores and basidiospores (*see* demicyclic; macrocyclic)
microflora—the combination of all of the microorganisms in a particular environment
microorganism—an organism small enough that it can be seen only with the aid of a microscope (syn. microbe)
microsclerotium (pl. microsclerotia)—a microscopic, dense aggregate of darkly pigmented, thick-walled hyphal cells
middle lamella—the layer between the walls of adjacent plant cells; consists largely of pectic substances
midrib—the central, thickened vein of a leaf
migratory—moving from place to place on a plant or from plant to plant when feeding (*see* sedentary)
mildew—a thin coating of mycelial growth and spores on the surfaces of infected plant parts
mitochondrion (pl. mitochondria)—a cellular organelle outside the nucleus; functions in respiration
mitosis—nuclear division in which the number of chromosomes per nucleus remains the same (*see* meiosis)
MLO—mycoplasmalike organism (*see* phytoplasma)
mold—any microfungus with conspicuous, profuse, or woolly superficial growth (mycelium or spore masses) on various substrates; commonly grow on damp or decaying matter and on the surfaces of plant tissues, especially with reference to economically important saprobes
mollicute—any of a group of prokaryotic organisms bounded by a flexuous membrane and lacking a cell wall (*see* phytoplasma; spiroplasma)
molt—to shed a cuticle or body encasement during a phase of growth
monocot—a monocotyledon, or a plant having only one cotyledon, such as species of grasses, including grain crops and corn (*see* dicot)
monoculture—the cultivation of plants of the same species in close proximity and with few or no other plant species present
monoecious—having male and female reproductive organs on a single individual
monogenic—determined by a single gene (*see* polygenic)
morphology (adj. morphological)—the study of the form of organisms; the form and structure of organisms
mosaic—a disease symptom characterized by nonuniform coloration, with intermingled normal, light-green and yellowish patches; usually caused by a virus (often used interchangeably with mottle)
motile—capable of self-propulsion by means of flagella, cilia, or amoeboid movement
mottle—a disease symptom characterized by light and dark areas in an irregular pattern; usually caused by a virus (often used interchangeably with mosaic)
mucilaginous—viscous; slimy
mulch—a layer of material, such as organic matter or plastic film, applied to the surface of the soil for purposes such as retention of water and inhibition of weeds
multinucleate—having more than one nucleus per cell (*see* binucleate; uninucleate)
multiseptate—having many septa (cross-walls)
muriform—having transverse and longitudinal septa
mutation—an abrupt heritable or genetic change in a gene or individual as a result of an alteration in a gene or chromosome or an increase in chromosome number
mycelium (pl. mycelia; adj. mycelial)—the mass of hyphae constituting the body (thallus) of a fungus
mycoparasite—a fungus that attacks another fungus
mycoplasmalike organism (abbr. MLO)—*see* phytoplasma
mycorrhiza (pl. mycorrhizae; adj. mycorrhizal)—a symbiotic association between a nonpathogenic or weakly pathogenic fungus and the roots of a plant
mycotoxin—any poisonous compound produced by a fungus

necrosis (adj. necrotic)—the death of cells or tissue; usually accompanied by blackening or browning
necrotroph—a parasite that typically kills and obtains its energy from dead host cells (*see* biotroph)
needle cast—a disease of conifers that causes the needles to fall
nematicide—an agent that kills nematodes; usually a chemical
nematode—any of various unsegmented roundworms (animals) that is parasitic on plants or animals or free living in soil or water
node (adj. nodal)—the enlarged portion of a shoot at which leaves or buds arise
nodule—a small knot or irregular, rounded lump; with reference to leguminous plants, a structure located on the root and containing nitrogen-fixing bacteria
noninfectious disease—a disease caused by an abiotic agent and not capable of being transmitted from plant to plant
nonpersistent transmission—virus transmission in which a virus is acquired and transmitted by a vector after a short feeding time and retained by the vector for only a short time (syn. stylet-borne transmission)
nonseptate—lacking cross-walls; coenocytic
nymph—the juvenile stage in the life cycle of an insect with incomplete metamorphosis; superficially resembles the adult

obclavate—shaped like an upside-down club
obligate parasite—an organism that can grow only as a parasite in association with its host plant and cannot be grown in artificial culture media (syn. biotroph)
obovate—egg shaped with the wide end outward
obovoid—egg shaped with the narrow end outward
obtuse—rounded or blunt; pertaining to an angle greater than 90°
oedema—*see* edema
oogonium (pl. oogonia)—a female gametangium of fungi in the group oomycetes containing one or more gametes
oomycetes (adj. oomycetous)—a group of funguslike organisms in the phylum Oomycota (kingdom Chromista); typically has nonseptate mycelium, asexual sporangia and zoospores, and sexual oospores

oospore—the thick-walled, sexually derived resting spore of organisms in the group oomycetes
ooze—a mass of bacterial cells mixed with host fluids
organelle—any of various membrane-bound structures contained within cells and having a specialized function (e.g., mitochondria and chloroplasts)
ostiole (adj. ostiolate)—a pore; the opening in the papilla or neck of a perithecium, pseudothecium, or pycnidium through which spores are released
ovary—the female reproductive structure of an organism; in plants, the enlarged basal portion of the pistil, containing the ovules and developing into the fruit
overseason—to survive or persist from one planting season to the next
oversummer—to survive or persist through the summer
overwinter—to survive or persist through the winter
oviposit—to deposit or lay eggs with an ovipositor
ovule—an enclosed structure that after fertilization becomes a seed; an egg contained within an ovary
ozone—a highly reactive form of oxygen (O3) injurious to plants

palisade parenchyma—the tissue located beneath the upper epidermis of leaves; composed of elongate, tubular cells arranged upright in the manner of posts in a palisade fortification
papilla (pl. papillae; adj. papillate)—a small, blunt projection
paraphysis (pl. paraphyses)—a hairlike cell within a fungal fruiting structure
parasexual—pertaining to the recombination of genetic characters without a sexual process
parasite (adj. parasitic)—an organism that lives in intimate association with another organism, on which it depends for its nutrition; not necessarily a pathogen
parenchyma (adj. parenchymatous)—the soft tissue of living plant cells with undifferentiated, thin, cellulose walls
parthenogenesis (adj. parthenogenetic)—reproduction by the development of an unfertilized egg
pasteurization—the process by which a material (usually a liquid) is treated with heat to eliminate selected harmful microorganisms
pathogen (adj. pathogenic)—a disease-producing organism or agent
pathogenesis—the initiation and development of disease
pathogenicity—the ability to cause disease
pathology—the study of diseases
pathotype—a subdivision of a pathogenic species characterized by its pattern of virulence or avirulence to differential host varieties
pathovar (abbr. pv.)—a subdivision of a plant-pathogenic bacterial species, defined by its host range (a pathovar of a bacterial species is analogous to a *forma specialis* of a fungal species)
pectin (adj. pectic)—a water-soluble substance that binds cells in plant tissues
pectolytic—capable of dissolving pectic substances
pedicel—a small, slender stalk; a stalk bearing an individual flower, inflorescence, or spore
peduncle—the stalk or main stem of an inflorescence; part of an inflorescence or fructification
penetration—the initial invasion of a host by a pathogen
penetration peg—the specialized, narrow hyphal strand located on the underside of an appressorium and penetrating host cells (syn. infection peg)
perennial—a plant that survives for several to many years (*see* annual; biennial)
perfect—sexual; capable of sexual reproduction
perfect state—the teleomorph, or sexual form in the life cycle of certain fungi (*see* imperfect state)
pericarp—the outer layer of a seed or fruit (*see* endocarp)
peridium (pl. peridia; adj. peridial)—the covering of the spore-bearing organ in many fungi
perithecium (pl. perithecia)—the flask-shaped or subglobose, thin-walled fruiting body (ascocarp) of an ascomycetous fungus, containing asci and ascospores and having a pore (ostiole) at its apex, through which ascospores are expelled or released
peritrichous—having hairs or flagella distributed over the whole surface
persistent transmission—virus transmission in which a virus is acquired and transmitted by a vector after a relatively long feeding time and remains transmissible for a prolonged period while in association with the vector (syn. circulative transmission)
pest—any organism that damages plants or plant products
pesticide—any of various chemicals used to control pests
petiole—the stalk portion of a leaf (*see* lamina)
pH—the negative logarithm of the effective hydrogen ion concentration; a measure of acidity (pH 7 is neutral; values less than pH 7 are acidic, and values greater than pH 7 are alkaline)
phenolase—an enzyme that oxidizes phenolic compounds
phenotype—the composite of observable physical qualities of an organism produced by the interaction of its genotype with the environment
phialide—the end cell of a conidiophore with one or more open ends, through which a basipetal succession of conidia develops
phialospore—a conidium produced on a phialide
phloem—the food-conducting, food-storing tissue in the vascular system of roots, stems, and leaves
photochemical oxidant—any of various highly reactive compounds formed by the action of sunlight on less toxic precursors
photosynthate—a chemical product of photosynthesis
photosynthesis—the manufacture of carbohydrates from carbon dioxide and water in the presence of chlorophylls; uses light energy and releases oxygen
phycomycete—an obsolete term for any member of a group of fungi that lacks cross-walls (septa) in its mycelium
phyllody—a disorder in which floral organs are transformed into leaf-like structures
phyllotaxy—the arrangement of leaves on a stem in relation to one another
physiological race—a group of individuals belonging to the same species and differing from other members of the species in their behavior or other characteristics but not in morphology; a physiological form
phytoalexin—any of various substances produced by higher plants in response to chemical, physical, or biological stimuli and inhibiting the growth of certain microorganisms
phytopathology—the study of plant diseases (syn. plant pathology)
phytoplasma—any of various plant-parasitic pleomorphic mollicutes (prokaryotes lacking cell walls) found in phloem tissue and not capable of growth on artificial nutrient media (previously called mycoplasmalike organisms (MLOs))
phytotoxic—harmful to plants; usually with reference to chemicals
pigment—a colored compound, such as chlorophyll, in the cells of plants or fungi
pinnate—featherlike; having parts arranged along two sides of an axis
pistil—the ovule-bearing organ of a plant, consisting of the ovary and its appendages (e.g., the style and stigma)
pith—parenchymatous tissue occupying the center of a stem
plant pathology—the study of plant diseases (syn. phytopathology)
plasmodium (pl. plasmodia)—a naked multinucleate mass of protoplasm moving and feeding in amoeboid fashion
plastid—any of major organelles found in the cells of plants and algae that, in many cases, serve specific metabolic and physiological purposes
pleomorphic—able to assume various shapes and perhaps sizes; having a life cycle characterized by a succession of distinctly different forms
plumule—a rudimentary shoot of the plant embryo
polar—located at one end (pole) of the cell
pollen—male sex cells produced by the anthers of flowering plants or the cones of seed plants
pollination—the transfer of pollen from anther to stigma or from a staminate cone to an ovulate cone
polygenic—pertaining to or governed by many genes (*see* monogenic)
polyploid—having three or more complete sets of chromosomes (*see* haploid; diploid)
positive-sense RNA—RNA that can serve directly as messenger RNA
posterior—situated toward the back or tail (*see* anterior)
predispose (n. predisposition)—to make prone to infection and disease
primary inoculum—inoculum (usually from an overwintering source) that initiates disease in the field, as opposed to inoculum that spreads disease during the season (syn. initial inoculum)
primary leaf—the first true leaf that emerges from a plant following the cotyledons

primary root—a root that develops directly from the radicle of an embryo rather than from a crown or node
primary symptom—a symptom produced soon after infection from a virus (see secondary symptom)
primordium (adj. primordial)—the rudimentary or initiating portion from which a plant part is formed
prokaryote (adj. prokaryotic)—any of various organisms lacking internal membrane-bound organelles and a distinct nucleus, such as bacteria and mollicutes (*see* eukaryote)
promycelium (pl. promycelia)—in rust and smut fungi, a germ tube issuing from a teliospore and bearing basidiospores
propagative virus—a virus that multiplies within its arthropod vector
propagule—any part of an organism capable of independent growth
protectant—an agent (usually a chemical) applied to a plant surface in advance of exposure to a pathogen for the purpose of preventing infection
protein—any of numerous nitrogen-containing organic compounds containing combinations of amino acids
protoplasm—the living contents of a cell
protoplast—the living contents of a cell, exclusive of the cell wall
proximal—situated toward or near the point of attachment or main body (*see* distal)
pseudothecium (pl. pseudothecia)—a perithecium-like fruiting body containing asci and ascospores dispersed rather than in an organized hymenium; an ascostroma with a single locule (cavity) containing bitunicate asci
pupa (pl. pupae; v. pupate)—the quiescent stage between the larval and adult stages of certain insects
pustule—a small, blisterlike elevation of the plant epidermis that forms as spores emerge
pv.—pathovar; a subdivision of a plant-pathogenic bacterial species defined by its host range (a pathovar of a bacterial species is analogous to a *forma specialis* of a fungal species)
pycnidiospore—a spore (conidium) produced in a pycnidium
pycnidium (pl. pycnidia)—an asexual, globose or flask-shaped fruiting body of certain imperfect fungi, producing conidia
pycniospore—a haploid, sexually derived spore formed in the pycnium of a rust fungus (syn. spermatium)
pycnium (pl. pycnia)—the globose or flask-shaped haploid fruiting body of rust fungi, bearing receptive hyphae and pycniospores (syn. spermagonium)
pyriform—pear shaped

quarantine—an enforced isolation imposed to prevent the spread of disease; pertaining to plant disease, control of the transport of plants or plant parts to prevent the spread of pests or pathogens
quiescent—dormant; inactive

race—a subgroup or biotype within a species or variety that is distinguished from other races by virulence, symptom expression, or host range but not by morphology
rachis—the elongated main axis of an inflorescence
radicle—the part of the plant embryo that develops into the primary root
receptacle—the structure of a flower that bears the reproductive organs
receptive hypha—in rust fungi, the part of a pycnium (spermagonium) that receives the nucleus of a pycniospore (spermatium)
recessive—pertaining to a phenotypic trait that is expressed in the progeny of diploid organisms only if contributed by both parents (*see* dominant)
reducing sugars—sugars with free carbonyl groups (e.g., fructose); formed from the hydrolysis of complex sugars
reniform—kidney shaped
resinosis—the exudation of resin
resistant (n. resistance)—possessing properties that prevent or impede disease development (*see* susceptible)
respiration—a series of chemical reactions that make energy available through oxidation of carbohydrates and fat
resting spore—a spore (often thick walled) that can remain alive in a dormant state for some time, germinating later and capable of initiating infection
reticulate—having netlike markings
rhizome—a mostly horizontal, jointed, fleshy, often elongated, usually underground stem
rhizomorph—a macroscopic ropelike strand of compacted tissue formed by certain fungi
rhizosphere—the microenvironment in the soil immediately around a plant's roots
ribonucleic acid (RNA)—any of several nucleic acids composed of repeating units of ribose (a sugar), a phosphate group, and a purine (adenine or guanine) or a pyrimidine (uracil or cytosine) base, transcribed from DNA, and involved in translation to proteins
ribosome—a subcellular protoplasmic particle made up of one or more RNA molecules and several proteins; involved in protein synthesis
ring spot—a disease symptom characterized by yellowish or necrotic rings enclosing green tissue, as in some plant diseases caused by viruses
RNA—ribonucleic acid; any of several nucleic acids composed of repeating units of ribose (a sugar), a phosphate group, and a purine (adenine or guanine) or a pyrimidine (uracil or cytosine) base, transcribed from DNA and involved in translation to proteins
rogue—to remove and destroy individual plants that are diseased, infested by insects, or otherwise undesirable
root cap—a group of cells that protects the growing tip of a root
root hair—a threadlike, single-celled outgrowth from a root epidermal cell
rosette—a disease symptom characterized by short, bunchy growth habit caused by a shortening of internodes with no comparable reduction in leaf size
rot—softening, discoloration, and often disintegration of plant tissue as a result of fungal or bacterial infection
rotation—the practice of growing different kinds of crops in succession in the same field
rugose—wrinkled; roughened
runner—a slender, horizontal stem that grows close to the soil surface (syn. stolon)
runner plants—new plants produced asexually on a runner (stolon)
russet—a brownish, roughened area on the skin of fruit, resulting from cork formation
rust—any of several diseases caused by specialized fungi in the group basidiomycetes, some of the spores of which are a rusty color

sanitation—the destruction or removal of infected or infested plants or plant parts; the decontamination of tools, equipment, containers, work space, hands, and so on
saprobe—*see* saprophyte
saprophyte (adj. saprophytic)—an organism that obtains nourishment from nonliving organic matter (syn. saprobe)
satellite virus—a virus that accompanies another virus and depends on it for its multiplication
scab—a roughened, crustlike diseased area on the surface of a plant organ
scald—a necrotic condition in which tissue is usually bleached and has the appearance of having been exposed to high temperature
sclerenchyma (adj. sclerenchymatous)—tissue made up of thick-walled plant cells
sclerotium (pl. sclerotia)—a vegetative, resting body of a fungus composed of a compact mass of hyphae with or without included host tissue; usually has a darkened rind
scorch—a symptom that suggests the action of flame or fire on the affected tissue, often at the margins of leaves
scutellum (adj. scutellar)—the cotyledon of a grass embryo
secondary infection—an infection resulting from the spread of infectious material from a primary or secondary infection without an intervening inactive period
secondary inoculum—inoculum produced by an infection established earlier in the same growing season
secondary organism—an organism that multiplies in already diseased tissue but is not the primary pathogen
secondary root—a branch of a primary root
secondary rot—rot caused by a secondary organism
secondary symptom—a symptom of infection by a virus that emerges after the initial symptoms (*see* primary symptom)
sedentary—remaining in a fixed location (*see* migratory)
seed—a ripened ovule consisting of an embryo and stored food enclosed by a seed coat
seed treatment—the application of a biological agent, chemical substance, or physical treatment to seed to protect the seed or resulting plant from pathogens or to stimulate germination or plant growth

seedborne—carried on or in seed
selective medium—a culture medium containing substances that inhibit or prevent the growth of certain species of microorganisms
senesce (adj. senescent; n. senescence)—to decline, as with maturation, age, or disease stress
sepal—one of the modified leaves composing a calyx
septate—having septa (cross-walls)
septum (pl. septa; adj. septate)—a dividing wall; in fungi, a cross-wall
serology (adj. serological)—a method using the specificity of the antigen–antibody reaction for the detection and identification of antigenic substances and the organisms that carry them
sessile—with reference to leaves, leaflets, flowers, florets, fruits, ascocarps, basidiocarps, and so on, lacking a stalk, petiole, pedicel, stipe, or stem; with reference to nematodes, permanently attached and not capable of moving about
seta (pl. setae)—a bristle- or hairlike structure, usually deep yellow or brown and thick walled
sexual spore—a spore produced during the sexual cycle of a fungus
sheath—the lower part of a grass leaf, which clasps the culm; a membranous cover
shot hole—a symptom in which small lesions on a leaf fall out, giving the leaf the appearance of being hit by buckshot
sieve element—a tube-shaped living cell in the phloem, functioning in the transport of dissolved organic substances in the plant (syn. sieve tube element)
sign—an indication of disease from direct observation of a pathogen or its parts on a host plant (*see* symptom)
sinuous—having curves, bends, or turns
slime molds—saprophytic organisms in the class Myxomycetes forming vegetative amoeboid plasmodia and spores
smut—any of several diseases caused by fungi in the group basidiomycetes that typically release masses of black, dusty teliospores at maturity
soft rot—a mushy disintegration of fleshy fruit or vegetables as a result of bacterial or fungal infection
soil drench—the application of a solution or suspension of a chemical to the soil, especially a pesticide to control soilborne pathogens
soilborne—carried on or beneath the soil surface
solarization—a disease management practice in which soil is covered with polyethylene sheeting to increase heating by sunlight and thus control soilborne plant pathogens
sooty mold—black, nonparasitic, superficial fungal growth; often develops on honeydew produced by aphids and other phloem-feeding insects
sorus (pl. sori)—a compact fruiting structure, especially with reference to spore masses of rust and smut fungi
sp. (pl. spp.)—species (the singular and plural abbreviations are used only after genus names; the singular form is used to refer to an undetermined species, and the plural form is used to refer to two or more species without naming them individually)
spermagonium (pl. spermagonia)—a structure in which male reproductive cells are produced; in rust fungi, the pycnium, a globose or flask-shaped haploid fruiting body composed of receptive hyphae and spermatia (pycniospores)
spermatium (pl. spermatia)—a male sex cell; a nonmotile male gamete; a haploid male gamete; in rust fungi, a pycniospore
spicule—the copulatory organ of a male nematode
spikelet—the spikelike appendage comprised of one or more reduced flowers and associated bracts; a unit of inflorescence in grasses; a small spike
spiroplasma—any of various spiral-shaped, plant-pathogenic mollicutes (prokaryotes lacking cell walls)
sporangiophore—the sporangium-bearing body of a fungus
sporangiospore—a nonmotile, asexual spore borne in a sporangium
sporangium (pl. sporangia)—a saclike fungal structure, the entire contents of which are converted into an indefinite number of asexual spores
spore—the reproductive structure of fungi and some other organisms, containing one or more cells; a bacterial cell modified to survive an adverse environment
sporidium (pl. sporidia)—the basidiospore of rust fungi, smut fungi, and other members of the group basidiomycetes
sporocarp—a spore-bearing fruiting body
sporodochium (pl. sporodochia)—a superficial, cushion-shaped, asexual fruiting body consisting of a cluster of conidiophores
sporulate—to produce spores
sporulating—producing and (often) liberating spores
spot—a small necrotic area occurring as a disease symptom on a leaf, flower, or stem
stamen (adj. staminal)—the male structure of a flower, composed of a pollen-bearing anther and a filament, or stalk
staminate flower—a male flower
stele—the central cylinder of vascular tissue, especially with reference to roots
stem pitting—a viral disease symptom characterized by depressions on the stem
sterigma (pl. sterigmata)—a small, usually pointed projection that supports a spore
sterile—unable to reproduce sexually; free of living microorganisms
sterile fungus—a fungus that is not known to produce any kind of spores
sterilization (adj. sterilized)—the total destruction of living organisms by various means, including heat, chemicals, and irradiation
stigma—the portion of a flower that receives pollen and on which pollen germinates
stipe—a stalk
stippling—a pattern of small dots or speckles on leaves where chlorophyll is absent
stipule—a small, leaflike appendage at the base of a leaf petiole; usually occurs as one of a pair
stolon—a slender, horizontal stem that grows close to the soil surface (syn. runner); in fungi, a hypha that grows horizontally along a surface
stoma, stomate (pl. stomata; adj. stomatal)—a structure composed of two guard cells and the opening between them in the epidermis of a leaf or stem; functions in gas exchange
strain—a distinct form of an organism within a species or virus, differing from other forms of the species biologically, physically, or chemically
striate (n. striations)—to mark with delicate lines, grooves, or ridges
stroma (pl. stromata)—a compact mass of mycelium (with or without host tissue) that supports fruiting bodies or in which fruiting bodies are embedded
stunting—a reduction in the height of plants resulting from a progressive reduction in the lengths of successive internodes or a decrease in their number
style—the slender part of many pistils, located between the stigma and the ovary and through which the pollen tube grows
stylet—the stiff, slender, hollow feeding organ of plant-parasitic nematodes or sap-sucking insects, such as aphids and leafhoppers
stylet-borne transmission—virus transmission in which a virus is acquired and transmitted by a vector after a short feeding time and is retained by the vector for only a short time (syn. nonpersistent transmission)
subepidermal—located or occurring just below the epidermis
subgenomic RNA—a piece of viral RNA, shorter than the entire genome of the virus, found in cells infected by the virus and sometimes encapsidated
substrate—the substance on which an organism lives or from which it obtains nutrients; a chemical substance acted upon, often by an enzyme
sunscald—the injury of plant tissues burned or scorched by direct sunlight
suscept—a susceptible plant
susceptible (n. susceptibility)—prone to develop disease when infected by a pathogen (*see* resistant)
symbiosis (adj. symbiotic)—a mutually beneficial association of two different kinds of organisms
symbiont—either of the organisms in a symbiotic association
symptom—an indication of disease by a reaction of the host (e.g., canker, leaf spot, wilt) (*see* sign)
symptomatology—the study of disease symptoms
symptomless carrier—a plant that is infected with a pathogen (usually a virus) but has no obvious symptoms
syncytium (pl. syncytia)—a multinucleate structure in root tissue formed by the dissolution of common cell walls induced by secretions of certain sedentary plant-parasitic nematodes (e.g., cyst nematodes)

synergism (adj. synergistic)—an interaction in which the total effect of the interacting factors is greater than the additive effects of the factors acting individually

synnema (pl. synnemata)—compact or fused, generally upright conidiophores, with branches and spores that form a headlike cluster (syn. coremium)

systemic—pertaining to a disease in which the pathogen (or a single infection) spreads generally throughout a plant; pertaining to chemicals that spread internally through a plant

taproot—a primary root that grows vertically downward and from which smaller lateral roots branch

taxonomy (adj. taxonomic)—the science of naming and classifying organisms

teleomorph—the perfect state, or sexual form in the life cycle of certain fungi (*see* anamorph; holomorph)

teliospore—a thick-walled resting spore produced by some fungi (notably rust and smut fungi) from which a basidium is produced

telium (pl. telia)—the fruiting body (sorus) that produces teliospores of a rust fungus

testa (pl. testae)—seed coat

thallus—the vegetative body of a fungus

thermal inactivation point—the lowest temperature at which heating for a limited period (usually 10 min) is sufficient to cause a virus to lose its infectivity or an enzyme to lose its activity

tiller—a lateral shoot, culm, or stalk arising from a crown bud; common in grasses

tissue—a group of cells, usually of similar structure, that perform the same or related functions

tissue culture—an in vitro method of propagating healthy cells from plant tissues

titer—a measure of the concentration of a substance in a solution

tolerance (adj. tolerant)—the ability of a plant to endure an infectious or noninfectious disease, adverse conditions, or chemical injury without serious damage or yield loss; with reference to pesticides, the amount of chemical residue legally permitted on an agricultural product entering commercial channels, usually measured in parts per million (ppm)

tomentose—covered with a dense mat of hair

toxicity—the capacity of a substance to interfere with the vital processes of an organism

toxin—any poisonous substance of biological origin

transgenic—possessing a gene from another species; having been the subject of genetic engineering

translocation—the movement of water, nutrients, chemicals, or food materials within a plant

translucent—clear enough to allow the transmission of light

transmit (n. transmission)—to spread or transfer, as in spreading an infectious pathogen from plant to plant or from one generation of plants to another

transovarial passage—the passage of a virus through the eggs or offspring of its vector and then to the next generation of host plants

transpiration—the loss of water by evaporation from leaf surfaces and through stomata

trichome—a plant epidermal hair, of which several types exist

triturate—to grind, as with a mortar and pestle

truncate—to end abruptly, as though an end has been cut off

tuber—an underground stem adapted for storage; typically produced at the end of a stolon

tumor—an abnormal swelling or localized outgrowth, often roughly spherical, produced by a plant as a result of attack by a fungus, bacterium, nematode, insect, or other organism (syn. gall)

turgid—swollen or inflated; plump or swollen as a result of internal water pressure

turgidity—the state of being rigid or swollen as a result of internal water pressure

tylosis (pl. tyloses)—a balloonlike extrusion of parenchyma cells into the lumina of contiguous vessels, partially or completely blocking them

ultrastructure—the submicroscopic structure of a macromolecule, cell, or tissue

unicellular—one celled (*see* multicellular)

uniflagellate—having one flagellum

uninucleate—having one nucleus per cell (*see* binucleate; multinucleate)

urediniospore, **urediospore**, **uredospore**—the asexual, dikaryotic, often rust-colored spore of a rust fungus, produced in a uredinium; the repeating stage of a heteroecious rust fungus, in which it is capable of infecting the host plant on which it was produced

uredinium (pl. uredinia), uredium (pl. uredia)—the fruiting body (sorus) in which urediniospores of a rust fungus are produced

vacuole—a generally spherical organelle within a plant cell, bound by a membrane and containing dissolved materials, such as metabolic precursors, storage materials, and waste products

variegation—a pattern of two or more colors in a plant part, as in a green and white leaf

variety (adj. varietal)—a plant type within a species, resulting from deliberate genetic manipulation and having recognizable characteristics, such as color; shapes of flowers, fruit, and seeds; height; and form (syn. cultivar)

vascular—pertaining to fluid-conducting tissues (xylem and phloem) in plants

vascular bundle—a strand of conductive tissue; usually composed of xylem and phloem (in leaves, small bundles are called *veins*)

vascular cylinder—the cylinder of vascular tissue in stems and roots (syn. stele)

vascular wilt disease—a xylem disease that disrupts the normal uptake of water and minerals, resulting in wilting and yellowing of foliage

vector—a living organism (e.g., insect, mite, bird, higher animal, nematode, parasitic plant, human) that is able to carry and transmit a pathogen; in genetic engineering, a cloning vehicle, or self-replicating DNA molecule, such as a plasmid or virus, used to introduce a fragment of foreign DNA into a host cell

vegetative—pertaining to the somatic or asexual parts of a plant, which are not involved in sexual reproduction

vegetative propagation—asexual reproduction; in plants, the use of cuttings, bulbs, tubers, or other vegetative parts to grow new plants

vein—a small vascular bundle in a leaf

vein banding—a symptom of virus disease in which the regions along veins are darker green than the tissue between veins

vein clearing—disappearance of green color in or around leaf veins

vermiform—worm shaped

vertical resistance—resistance thought to be conferred by a single gene or a few genes and completely protecting a host but only against specific races or strains (genetic variants) of a pathogen

vesicle—the thin sac in which zoospores are differentiated and from which they are released; the bulbous head terminating the conidiophores of *Aspergillus* spp.; a structure formed by an endomycorrhizal fungus within living cells of the root

vessel—the water-conducting structure of xylem tissue, with pit openings in end walls

viable (n. viability)— alive; capable of germination (e.g., of seeds, fungus spores, sclerotia, etc.); capable of growth

virescence—a state or condition in which normally white or colored tissues (e.g., flower petals) become green

virion—a complete virus particle

viroid—any of various small, unencapsidated (naked), circular, single-stranded RNA molecules capable of causing disease in plants

viroplasm—an inclusion in a virus-infected cell, where viral components are synthesized and virus particles are assembled

virulence—the degree or measure of pathogenicity; the relative capacity to cause disease

virulent—pathogenic; having the capacity to cause disease (*see* avirulent)

viruliferous—virus laden, usually with reference to insects or nematodes as vectors

virus—any of numerous submicroscopic, intracellular, obligate parasites consisting of a core of infectious nucleic acid (either RNA or DNA) usually surrounded by a protein coat

volunteer—a self-set plant; seeded by chance

water soaking—a disease symptom in which plant tissues or lesions appear wet, dark, and usually sunken and translucent

whorl—a circular arrangement of like parts

wild type—the phenotype characteristic of most individuals of a species under natural conditions

wilt—the drooping of leaves and stems caused by a lack of water (inadequate water supply or excessive transpiration); a vascular disease in which normal water uptake is interrupted

witches'-broom—a disease symptom in which many weak shoots arise at or close to the same point, producing abnormal massed, brushlike growth

xylem—the water- and mineral-conducting, food-storing, supporting tissue of a plant

yellows—a plant disease characterized by chlorosis and stunting

zonate—pertaining to the targetlike development of a tree canker, characterized by successive perennial rings of callus; pertaining to any symptom appearing in concentric rings

zoosporangium—a sporangium (spore case) bearing zoospores

zoospore—a fungal spore with flagella; capable of locomotion in water

zygomycetes (adj. zygomycetous)—a group of fungi in the phylum Zygomycota (kingdom Fungi) that produce resistant, spherical sexual spores following fusion of hyphal gametangia

zygospore—the sexual resting spore formed from the union of gametangia of fungi in the group zygomycetes

Index